AF149814

UNITEXT for Physics

Series Editors
Michele Cini, University of Rome Tor Vergata, Roma, Italy
Stefano Forte, University of Milan, Milan, Italy
Guido Montagna, University of Pavia, Pavia, Italy
Oreste Nicrosini, University of Pavia, Pavia, Italy
Luca Peliti, University of Napoli, Naples, Italy
Alberto Rotondi, Pavia, Italy
Paolo Biscari, Politecnico di Milano, Milan, Italy
Nicola Manini, University of Milan, Milan, Italy
Morten Hjorth-Jensen, Department of Physics and Astronomy, University of Oslo, Oslo, Norway
Alessandro De Angelis, Physics and Astronomy, INFN Sezione di Padova, Padova, Italy

UNITEXT for Physics series publishes textbooks in physics and astronomy, characterized by a didactic style and comprehensiveness. The books are addressed to upper-undergraduate and graduate students, but also to scientists and researchers as important resources for their education, knowledge, and teaching.

Giampaolo Co'

Concepts in Quantum Many-Body Physics

Springer

Giampaolo Co'
Dip. Matematica e Fisica "Ennio De Giorgi"
Università del Salento and INFN sez.
Lecce, Italy

ISSN 2198-7882 ISSN 2198-7890 (electronic)
UNITEXT for Physics
ISBN 978-3-032-08919-9 ISBN 978-3-032-08920-5 (eBook)
https://doi.org/10.1007/978-3-032-08920-5

© The Editor(s) (if applicable) and The Author(s), under exclusive license to Springer Nature Switzerland AG 2026

This work is subject to copyright. All rights are solely and exclusively licensed by the Publisher, whether the whole or part of the material is concerned, specifically the rights of translation, reprinting, reuse of illustrations, recitation, broadcasting, reproduction on microfilms or in any other physical way, and transmission or information storage and retrieval, electronic adaptation, computer software, or by similar or dissimilar methodology now known or hereafter developed.
The use of general descriptive names, registered names, trademarks, service marks, etc. in this publication does not imply, even in the absence of a specific statement, that such names are exempt from the relevant protective laws and regulations and therefore free for general use.
The publisher, the authors and the editors are safe to assume that the advice and information in this book are believed to be true and accurate at the date of publication. Neither the publisher nor the authors or the editors give a warranty, expressed or implied, with respect to the material contained herein or for any errors or omissions that may have been made. The publisher remains neutral with regard to jurisdictional claims in published maps and institutional affiliations.

This Springer imprint is published by the registered company Springer Nature Switzerland AG
The registered company address is: Gewerbestrasse 11, 6330 Cham, Switzerland

If disposing of this product, please recycle the paper.

Preface

This book systematically organises the notes from the Theoretical Physics of Matter course that I taught for approximately 15 years at the University of Salento. While the course name may seem somewhat generic, the content I presented over the years, and what this book encompasses, can be more accurately described as Quantum Many-Body Physics.

In this book, I aim to offer a comprehensive overview of the theories developed to describe quantum systems composed of interacting particles. Each of these theories has its own advantages and disadvantages when compared to the others. I believe it is important to consider all of them, as doing so sheds light on various aspects of the many-body problem that may remain obscured if one restricts their focus to just one or a selected few of these theories.

In the book, I have organised the various theories into three categories: theories inspired by field theory, theories based on statistical mechanics, and effective theories. Each category emphasises a particular aspect of the quantum many-body problem, which is then explored using specific theoretical frameworks.

In my presentation, I focused more on clarifying the physics underlying the complex mathematical framework rather than strictly adhering to mathematical rigor. I apologise in advance for any gaps in this regard, which can be addressed by referring to the bibliography.

I did not pay much attention on the applications of various theories on realistic cases. This is because the results of these applications are constantly evolving and are best explored through updated review articles. In contrast, the formulation of the theories is well-established and can be effectively presented in a manual format.

The intended audience for these notes consists of graduate students enrolled in physics courses. In other words, this book is designed for students who possess a solid understanding of basic non-relativistic quantum mechanics and its straightforward applications. Specifically, I assume that the reader is familiar with the solutions to the Schrödinger equation for a particle in a central potential, the quantization of angular momenta and the rules for summing them, the treatment of identical particles, their classification into fermions and bosons, and the associated quantum statistics.

Few words on the notation adopted. The operators are identified with a hat to distinguish them from the functions. This rules is not adopted when the form of the operator is made explicit, for example when it is written as product or sum of derivatives. We use a relatively small number of acronyms whose meanings are listed in a Table at the beginning. The symbols which maintain the same meaning all through the notes are listed in a Table.

Finally, I would like to express my gratitude to my students, who inspired me to write these notes. As they studied the material to succeed, they unknowingly helped me identify errors, poorly written sections, and inconsistencies. Without their valuable input, this book would not have been possible.

Lecce, Italy
September 2025

Giampaolo Co'

Contents

Part II Theories Inspired to Quantum Field Theories

Acronyms

AFDMC	Auxilary Field Diffusion Monte Carlo (Sect. 4.5)
CBF	Correlated Basis Function (Chap. 11)
CCM	Coupled Cluster Method (Chap. 13)
DFT	Density Functional Theory (Sect. 15)
FG	Fermi Gas (Sect. 2.3)
FHNC	Fermi Hypernetted Chain (Sect. 11.3)
GFMC	Green Function Monte Carlo (Sect. 4.4)
HF	Hartree-Fock (Sects. 10.4 and 15.2)
HNC	Hypernetted Chain (Sect. 11.2)
IPM	Independent Particle Model (Chap. 2)
KS	Khon-Sham (Sect. 15.3.2)
MF	Mean-Field (Chap. 2)
OBDM	One-body density matrix (Sect. 11.3)
ONR	Occupation number representation (Chap. 5)
QBA	Quasi Boson Approximation (Sect. 16.3)
QCD	Quantum Chromodynamics
QED	Quantum Electrodynamics
RFHNC	Renormalised Fermi Hypernetted Chain (Sect. 11.4)
RPA	Random Phase Approximation (Sect. 10.6 and Chap. 16)
SOC	Single Operator Chain (Sect. 11.3.1)
TBDF	Two-body distribution function (Sect. 11.2)
TDA	Tamm Dankoff Approximation (Sect. 16.2)
UCOM	Unitary Correlation Operator Method (Chap. 12)
VMC	Variational Monte Carlo (Sect. 4.3)

Symbols

$\hat{a},\ \hat{a}^{+}$	Creation and destruction operators (Chap. 5)
A	Number of particles
E	Energy of the interacting particles system
$\mathscr{E}$	Energy of the non-interacting particles system
G	Green's function (Chaps. 8 and 9)
$\mathscr{G}$	Effective interaction from Brueckner's theory (Chap. 8)
$\hat{h}$	Single-particle hamiltonian (Chap. 2)
$\hat{H}$	Total hamiltonian (Chap. 2)
$\hat{H}_0$	One-body hamiltonian, of mean-field (Chap. 2)
$\hat{H}_1$	Perturbative term, interaction term, of the hamiltonian $\hat{H}$ (Chap. 2)
$\hat{\mathbb{I}}$	The identity operator
$\hat{\mathbb{N}}$	Normal ordered product operator (Chap. 5)
$\hat{T}$	Kinetic energy operator of the hamiltonian (Chap. 2)
$\hat{\mathbb{T}}$	Time ordered product operator (Chap. 5)
$\hat{V}$	Interaction Potential (Chap. 2)
$\mathcal{V}$	Volume of the system (Chap. 2)
Y_{lm}	Spherical harmonics (Chap. 2)
$\mathcal{Y}_{ljm}$	Spin spherical harmonics (Chap. 2)
Φ	Eigenstate of $\hat{H}_0$, mean-field state, Slater determinant for fermions (Chap. 2)
ϕ_i	Single-particle wave function eigenstate of $\hat{h}$ (Chap. 2)
Ψ	Eigenstate of $\hat{H}$ (Chap. 2)
$\hat{\psi},\ \hat{\psi}^{+}$	Field operators (Chap. 5)
ρ	Number density of the particles (Chap. 2)
ρ_ϵ	Density of states (Chap. 2)
Σ	Self-energy (Chap. 10)
Ω	Polar angular coordinates (Chap. 2)
Z	Atomic number

Chapter 1
Introducing the Many-Body Problem

The modern description of the universe given by physicists reduces every natural phenomenon to the interaction of six leptons and six quarks mediated by four fundamental interactions: gravitation, electromagnetic, strong and weak nuclear forces. This picture, so concise and powerful, is the result of studies inspired to the most extreme reductionism: the understanding of a complex system goes through the identification of its components and of their interactions. Decomposing a complex system in its parts is a much simpler task than that of reconstructing, or even building from scratch, the system starting from its components [1].

This last job is the goal of the many-body theories. The necessary preconditions to carry out this task are:
(a) definition of the theoretical framework,
(b) choice of the fundamental degrees of freedom, or in other words the basic components of the system,
(c) definition of their interaction.
These items are not necessary independent of each other.

The theoretical framework can be, for example, that of the Classical Physics, of the non-relativistic Quantum Mechanics or of the Quantum Field Theory. Evidently, the choice is strictly related to the kind of system under investigation. The description of a planetary system in terms of Quantum Field Theory is not at all convenient. On the other hand, it would be very difficult to use Classical Physics for the description of a barion in terms of quarks and antiquarks.

In the nomenclature currently used in physics, many-body theory indicates a theory developed in the framework of the non-relativistic Quantum Mechanics. Eventual relativistic effects are treated as a correction, or a perturbation, of the main non-relativistic theory.

The validity of the non-relativistic Quantum Mechanics is limited to those microscopic phenomena where the energies into play are much smaller than the rest masses of the components of the system. These are atoms, molecules, quantum liquids, metals, crystals and also atomic nuclei, which we shall call simply nuclei henceforth.

© The Author(s), under exclusive license to Springer Nature Switzerland AG 2026
G. Co', *Concepts in Quantum Many-Body Physics*, UNITEXT for Physics,
https://doi.org/10.1007/978-3-032-08920-5_1

From the pragmatical point of view, this means that the many-body problem consists in solving the Schrödinger equation for a system of identical, and interacting, particles.

The second precondition requires the definition of the fundamental degrees of freedom, which are, in other words, the particles composing the system. Also in this case, the choice is related to an economy principle in the description of the problem. As it has been already pointed out, at present, the fundamental bricks of the matter are leptons, quarks, the bosons mediating the four fundamental interactions and the Higgs boson. Even though it is correct from the fundamental point of view, the use of these physical entities to describe every microscopic many-body system is not convenient. A more useful description of the system is obtained by identifying the largest components of which it is possible to ignore the internal structure without losing relevant information about the system as a whole. This choice depends on the energy range under investigation compared to the energy required to excite the chosen degree of freedom. For example, the dynamics of the nucleus at few MeV is well studied by considering the nucleons, protons and neutrons, as basic degrees of freedom. The description of the liquid helium, at few meV, is carried out by using the helium atoms as fundamental particles.

The third precondition to be defined, the interaction between the basic degrees of freedom, is, clearly, related to the second precondition. The choice of the degrees of freedom defines what is the interaction to be used, which is not directly one of the four fundamental forces quoted above. The relation between particles and their interaction is the topic of the Chap. 3.

After having chosen theoretical framework, fundamental degrees of freedom and their interaction, the many-body problem is well defined and it is only matter of solving the differential, or better to say, the integro-differential, equations which describe the system. In the case of our interest it is matter of solving the many-body Schrödinger equation. Formally, it is a purely technical problem, but, in reality, the problem can be solved without making approximation only for a limited number of cases. For more general treatments it is necessary to formulate approximations which simplify the problem. The development of these approximations, their validity, the possibility of controlling, and to improve, them at will are the topics of the many-body theories. This task is accomplished by investigating deeply the problem. This means understanding the relations between the various theoretical and physical quantities in order to identify those which are relevant and, if it possible, to separate them from what is negligible. A problem that is formally only technical leads to studying the physical characteristics of the many-body system in terms of its components.

In the study of the many-body systems of different nature, it comes out the remarkable fact that the interactions between the particles show some common features, which are analogous for the interaction between nucleons, and between atoms or molecules. For this reason, the approximated techniques aimed to solve the many-body Schrödinger equation are, in fact, independent of the system under study. These techniques are universal, in the sense that they can be applied without using other assumptions concerning the system. The only differences are the characteristics of the particles and the interactions which are external inputs of the theories. There is an important discriminating fact in the structure of the particles and this is related to their bosonic or fermionic structure. Apart from this important feature, the theories developed have validity ranges which span on 6 orders of magnitude for what concerns the distances, from the nano to the femto meter, and about 12 orders of magnitudes for what the energy is concerned, from the meV up to the GeV.

There is not a unique way of tackling the many-body problem. In this book, the variety of approaches are divided in three main groups. The first group is that of the theories inspired to the Quantum Field Theory. These theories are the most commonly used and they define a specific language. The second group is that of the theories inspired to the Statistical Mechanics. Finally a third group is that composed by the so-called Effective Theories. These three groups of theories focus their attentions on different aspects of the many-body problem. As usual, studying a problem from different points of view allows a better understanding of the various aspects pointed out with different emphasis by the each specific perspective.

Reference

1. P.W. Anderson, More is different. Science **177**, 393 (1972)

Part I
Basic Information

Chapter 2
Mean-Field Models

2.1 General Properties

The mean-field (MF) model serves as the foundation for all many-body theories. It is also the simplest approach to the problem, as it makes an approximation that reduces the complex many-body problem into many one-body problems.

In the framework of the non-relativistic Quantum Mechanics, a very general expression of the hamiltonian describing the many-body system is

$$\hat{H} = \sum_{i=1}^{A} \left(-\frac{\hbar^2}{2m_i} \nabla_i^2 + \hat{V}_0(i) \right) + \frac{1}{2} \sum_{i,j=1}^{A} \hat{V}(i, j) + \cdots \,, \tag{2.1}$$

where A is the number of particles, each of them with mass m_i.

In expression (2.1), the term containing the Laplace operator ∇_i^2 represents the kinetic energy. The term $\hat{V}_0(i)$ is a generic potential acting on each particle, while $\hat{V}(i, j)$ describes the interaction between two particles. The dots indicates the presence of more complex interaction terms, which will be discussed in Chap. 3. These additional terms are not considered at this stage. The use of potential terms implies an instantaneous interaction between the particles, a concept valid only in non-relativistic framework.

Let us consider, for example, the case of the electrons of an atomic system. In this case, the potential term $\hat{V}_0(i)$ of the hamiltonian (2.1) is the electrostatic potential generated by the nucleus

$$\hat{V}_0(i) = -\frac{e^2}{4\pi\epsilon_0} \frac{Z}{r_i} \,, \tag{2.2}$$

where e and ϵ_0 are respectively the unitary charge and the vacuum permittivity, r_i the distance between the position of the electron and that of the nucleus which is at the center of the coordinate system, and Z represents the number of the protons of

© The Author(s), under exclusive license to Springer Nature Switzerland AG 2026
G. Co', *Concepts in Quantum Many-Body Physics*, UNITEXT for Physics,
https://doi.org/10.1007/978-3-032-08920-5_2

the nucleus, i.e. the atomic number which, obviously, is the same as the number of electrons. The interaction potential between two electrons is

$$\hat{V}(i,j) = \frac{e^2}{4\pi\epsilon_0}\frac{1}{r_{ij}} , \tag{2.3}$$

where $r_{ij} = |\mathbf{r}_i - \mathbf{r}_j|$ is the distance between the two electrons. In this, atomic, case we have $A = Z$ in the expression (2.1).

In the case of an atomic nucleus, the basic degrees of freedom are the nucleons and in the hamitonian (2.1) we have $\hat{V}_0(i) = 0$ and $\hat{V}(i,j)$ represents the strong interaction between two nucleons.

By adding and subtracting to the expression (2.1) an average potential $\hat{U}(i)$, which acts on a single particle at the time, we obtain:

$$\hat{H} = \underbrace{\sum_i^A \left(-\frac{\hbar^2}{2m_i}\nabla_i^2 + \hat{V}_0(i) + \hat{U}(i) \right)}_{\hat{H}_0} + \underbrace{\frac{1}{2}\sum_{i,j}^A \hat{V}(i,j) - \sum_i^A \hat{U}(i)}_{\hat{H}_1} . \tag{2.4}$$

The term $\hat{H}_0$ is a sum of terms that operate on a single particle at the time, specifically on the i-th particle. We define each term of this sum as single-particle hamiltonian $\hat{h}(i)$,

$$\hat{H}_0 = \sum_i \hat{h}(i) = \sum_i^A \left(-\frac{\hbar^2}{2m_i}\nabla_i^2 + \hat{V}_0(i) + \hat{U}(i) \right) . \tag{2.5}$$

The fundamental approximation of the MF model involves neglecting the term $\hat{H}_1$, known as the residual interaction, in the expression (2.4). By doing so, the many-body problem is transformed into a sum of one-body problems. This model is also referred to as the Independent Particle Model (IPM), as the particles described by $\hat{H}_0$ do not interact with one another.

The fact that the hamiltonian $\hat{H}_0$ is a sum of independent terms implies that its eigenstates can be built as a product of the eigenstates of $\hat{h}(i)$

$$\hat{h}(i)|\phi_i\rangle = \epsilon_i|\phi_i\rangle , \tag{2.6}$$

therefore

$$\hat{H}_0|\Phi\rangle = \left(\sum_i \hat{h}(i) \right)|\Phi\rangle = \mathcal{E}|\Phi\rangle , \tag{2.7}$$

where

$$|\Phi\rangle = |\phi_1\rangle|\phi_2\rangle\cdots|\phi_A\rangle . \tag{2.8}$$

For fermions, the antisymmetry of the global wave function under the exchange of two particles implies that the wave function $|\Phi\rangle$ has to be described as sum of antisymmetrized products of one-particle wave functions. This solution is known in the literature as Slater determinant [1]

$$|\Phi\rangle = \frac{1}{\sqrt{A!}} \det\{|\phi_i\rangle\} \ . \tag{2.9}$$

The MF potential $\hat{U}$ included in $\hat{H}_0$ is typically chosen based on phenomenological considerations. For instance, in many-electron atoms, $\hat{U}(i)$ represents a potential that accounts for the screening effect generated by the inner electrons on the motion of the outer ones. In the case of atomic nuclei, the most commonly used forms of $\hat{U}(i)$ are the harmonic oscillator potential and the Woods-Saxon potential

$$\hat{U}(r) = \frac{-U_0}{1 + \exp(\frac{r-R}{a})} \ , \tag{2.10}$$

where U_0, R and a are real and positive constants whose values are selected with a procedure involving a comparison with empirical data.

In the following, we present the solution of the MF problem for some specific expressions of $\hat{U}$. We consider separately systems with spherical and translational symmetries.

2.2 Spherical Symmetry

In the MF model, the solution to the many-body problem involves solving the one-body Schrödinger equation (2.6) for each individual particle.

Atoms and nuclei are systems that can be conveniently described using a set of spherical coordinates, along with a MF potential that is symmetric with respect to rotations of the coordinate system. The spherical symmetry of the potential implies that $\hat{U}(\mathbf{r}) = \hat{U}(r)$, where $r = |\mathbf{r}|$. It is convenient to search for solutions to the one-body Schrödinger equation of the form

$$\phi(\mathbf{r}) = \sum_{n,l,\mu,\sigma} R_{nl}(r) Y_{l\mu}(\Omega) \chi_\sigma \ , \tag{2.11}$$

where n, l, μ, σ are the quantum numbers identifying the various terms of the wave function: n is the principal quantum number, l is the quantum number indicating the orbital angular momentum, μ its projection on the quantization axis and σ the spin projection on this axis. We indicated with $Y_{l\mu}$ the spherical harmonics and with $\Omega \equiv (\theta, \phi)$ the angular part of the polar spherical coordinates. The Pauli spinor of the fermion, electron or nucleon, is indicated as

$$\chi_{1/2} = \begin{pmatrix} 1 \\ 0 \end{pmatrix} \; ; \; \chi_{-1/2} = \begin{pmatrix} 0 \\ 1 \end{pmatrix} \; . \tag{2.12}$$

The techniques used to solve the one-body Schrödinger equation with a spherically symmetric potential are well established; see, for example, [2]. The Laplace operator is expressed in spherical polar coordinates, allowing us to separate the terms related to r, the distance from the origin, from those associated with the angular coordinates Ω. The eigenstates of the angular part are the spherical harmonics $Y_{l,\mu}(\Omega)$, with eigenvalues given by $l(l+1)\hbar^2$. By substituting these eigenstates and their corresponding eigenvalues into the Schrödinger equation, we obtain an expression that, from the operator perspective, depends solely on r,

$$\left[\frac{\hat{p}_r^2}{2m} + \frac{l(l+1)\,\hbar^2}{2mr^2} + \hat{U}(r)\right]\left[R_{nl}(r)Y_{l\mu}(\Omega)\chi_\sigma\right] = \epsilon_{nl}\left[R_{nl}(r)Y_{l\mu}(\Omega)\chi_\sigma\right] \; , \tag{2.13}$$

where the explicit expression of the $\hat{p}_r^2$ operator is

$$\hat{p}_r^2 R_{nl}(r) = -\hbar^2 \frac{1}{r^2}\frac{d}{dr}\left(r^2 \frac{d}{dr} R_{nl}(r)\right) = -\hbar^2\left(\frac{d^2}{dr^2} R_{nl}(r) + \frac{2}{r}\frac{d}{dr} R_{nl}(r)\right) \; . \tag{2.14}$$

By using a potential $\hat{U}$ depending only on r, and not on the particle's spin, we obtain the expression

$$\frac{d^2}{dr^2} R_{nl}(r) + \frac{2}{r}\frac{d}{dr} R_{nl}(r) + \left[\frac{2m}{\hbar^2}(\epsilon_{nl} - U(r)) - \frac{l(l+1)}{r^2}\right] R_{nl}(r) = 0 \; . \tag{2.15}$$

This technique, which expands the wave function in spherical harmonics, is applicable to any potential that depends solely on r. The result of this approach is the reduction of a three-dimensional differential equation to a one-dimensional differential equation. Once we have obtained the expression (2.15), we can then specify the particular dependence of the potential on r. Below, we will discuss cases of potentials that are relevant for the many-body systems to be considered in the upcoming chapters.

2.2.1 Constant Potential

The constant potential is characteristic of systems with translational symmetry, such as the infinite Fermi gas or a free particle. In these cases, the value of the constant can be set to zero. The approach to this problem using the spherical symmetry technique described above is quite useful, as the eigenstates can serve as a basis for expanding more complex situations.

Let us consider a spherical system whose dimensions are defined by R. The potential is

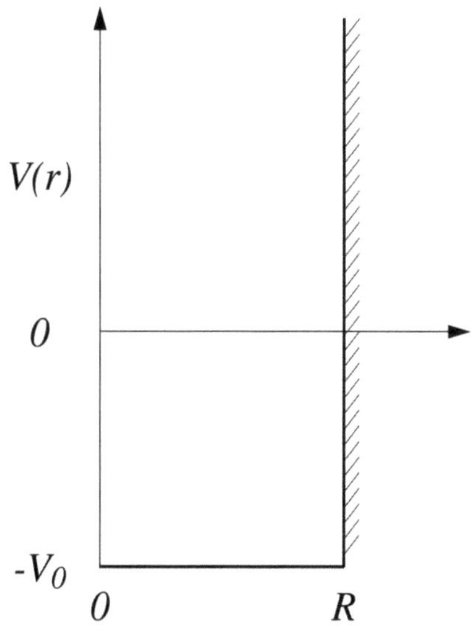

Fig. 2.1 Infinite well potential

$$\hat{U}(r) = -V_0 \text{ for } r \leq R \text{ and } \hat{U}(r) = \infty \text{ for } r > R \ , \tag{2.16}$$

as it is shown in Fig. 2.1. The Eq. (2.15) is defined in the range $0 \leq r \leq R$. By dividing Eq. (2.15) by k^2 defined as

$$k^2 = \frac{2m}{\hbar^2}(\epsilon_{nl} + V_0) \ , \tag{2.17}$$

we obtain the following equation which depends on the dimensionless variable $\rho = kr$

$$\frac{d^2}{d\rho^2}R_{nl}(\rho) + \frac{2}{\rho}\frac{d}{d\rho}R_{nl}(\rho) + \left[1 - \frac{l(l+1)}{\rho^2}\right]R_{nl}(\rho) = 0 \ . \tag{2.18}$$

This differential equation is well known in the literature [3]. Two types of independent solutions are the spherical Bessel functions $j_l(\rho)$ and the spherical Neumann functions $n_l(\rho)$. The former are regular at the origin, while the latter are irregular. Due to the physical significance of the wave function, only the spherical Bessel functions must be considered. The explicit expressions of the first two spherical Bessel functions are

$$j_0(\rho) = \frac{\sin\rho}{\rho} \ ; \ j_1(\rho) = \frac{\sin\rho}{\rho^2} - \frac{\cos\rho}{\rho} \ , \tag{2.19}$$

and, for $l > 0$, the following recurrence relation is valid

$$(2l+1)\,j_l(\rho) = \rho\,[j_{l+1}(\rho) + j_{l-1}(\rho)] \ . \tag{2.20}$$

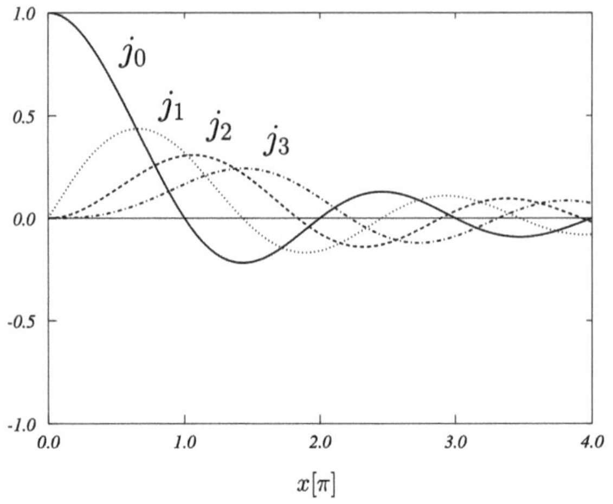

Fig. 2.2 The first four spherical Bessel functions $j_l(x)$

Table 2.1 First zeros of the spherical Bessel functions, see Eq. (2.21)

$X_{1,0}$	$X_{1,1}$	$X_{1,2}$	$X_{2,0}$	$X_{1,3}$	$X_{2,1}$	$X_{2,2}$	$X_{3,0}$
3.1416	4.4934	5.7634	6.2832	6.9879	7.7252	9.0995	9.4248

From a physical perspective, since the potential approaches infinity for $r \geq R$, it is necessary for j_l to be zero at the point $r = R$ and, obviously, for $r > R$. This means $j_l(kR) = j_l(X_{nl}) = 0$, and, because of the definition of k,

$$\frac{2m}{\hbar^2}(\epsilon_{nl} + V_0)R^2 = X_{nl}^2 \ ; \ \epsilon_{nl} = \frac{\hbar^2}{2m}\frac{X_{nl}^2}{R^2} - V_0 \ . \tag{2.21}$$

The zeros of the wave function depend on the principal quantum number n and on the orbital one l. For example, for j_0 the zeros are integer multiple of π. Equation (2.21) clearly indicates that, in this case, all the eigenvalues are discrete (Table 2.1).

2.2.2 *Three-Dimensional Harmonic Oscillator*

In this case, the expression of the potential is

$$\hat{U}(r) = \frac{1}{2}m\omega^2 r^2 \ . \tag{2.22}$$

Since $r^2 = x^2 + y^2 + z^2$ it is possible to solve the problem by rewriting the Schrödinger equation in cartesian coordinates. In this way, the differential equation becomes separable in the three coordinates, allowing the eigenvalues to be constructed as the product of the eigenfunctions of a one-dimensional harmonic oscillator obtained for each of the three coordinates. The overall eigenvalue is then given by the sum of the individual eigenvalues in the three directions (Fig. 2.2).

Clearly, the harmonic oscillator potential is of central type, therefore it is possible to use the expansion in spherical harmonics. In this case the radial part of the wave function is proportional to the Laguerre polynomials [2, 3]. The energy eigenvalues can be expressed by using the quantum numbers identifying the single particle basis in spherical coordinates (2.11) as

$$\epsilon_{nl} = \hbar\omega\left(N + \frac{3}{2}\right) = \hbar\omega\left(N_x + N_y + N_z + \frac{3}{2}\right) = \hbar\omega\left(2(n-1) + l + \frac{3}{2}\right) . \tag{2.23}$$

The first two terms express the eigenvalues in terms of the quantum numbers derived from the solution obtained in cartesian coordinates, while the last expression utilises the quantum numbers from the solution in spherical coordinates. In this latter case, $n - 1$ indicates the number of nodes in the wave function (with the minimum value of n being 1), and l is the quantum number associated with the orbital angular momentum. The energy value depends on N, a number that can be varied by changing both n and l. States with the same energy, even if characterised by different quantum numbers, are referred to as degenerate. A degeneracy is termed casual or accidental when it arises from a specific choice of potential rather than from the global symmetry properties of the problem. This latter type of degeneracy is related to the μ quantum number of the spherical harmonics and is present in all problems exhibiting spherical symmetry.

A list of quantum number values that produce the same energy value, as given in Eq. (2.23), is shown in Table 2.2. It is evident that the harmonic oscillator generates sequences of eigenstates that are accidentally degenerate. Notably, all states with the same energy share the same parity, which changes when the energy eigenvalue is increased by one step of $\hbar\omega$.

Table 2.2 Examples of combinations of quantum numbers that produce the same energy values for the harmonic oscillator potential, as described in Eq. (2.23). The various energy levels are identified using traditional spectroscopic notation. The last column indicates the parity of the states, given by $(-1)^l$

$\epsilon_{nl}\,(\hbar\omega)$	N	n	l		n	l		Π
3/2	0	1	0	1s				+1
5/2	1	1	1	1p				−1
7/2	2	2	0	2s	1	2	1d	+1
9/2	3	2	1	2p	1	3	1f	−1

2.2.3 Coulomb Potential

In MF description of an atomic system with many electrons, the interactions between electrons are neglected, and only the interaction of each individual electron with the nucleus, located at the center of the coordinate system, is taken into account. Additionally, the potential generated by the nucleus can be adjusted by incorporating a MF potential that accounts for the screening effect caused by the presence of other electrons. If the electron is very close to the nucleus, the effective potential is the bare Coulomb potential

$$\lim_{r_i \ll R} \left[\hat{V}_0(r_i) + \hat{U}(r_i) \right] = -\frac{e^2}{4\pi\epsilon_0} \frac{Z}{r_i} , \tag{2.24}$$

where R is the atomic radius. At distances far from nucleus the electron is sensitive to a potential screened by the presence of N electrons

$$\lim_{r_i \simeq R} \left[\hat{V}_0(r_i) + \hat{U}(r_i) \right] = -\frac{e^2}{4\pi\epsilon_0} \frac{Z-(N-1)}{r_i} . \tag{2.25}$$

The bare Coulomb potential (2.24) generates eigenvalues of the energy which are independent of l and of μ [2],

$$\epsilon_n = -\frac{1}{2} mc^2 \frac{(Z\alpha)^2}{n^2} , \tag{2.26}$$

where α is the fine structure constant

$$\alpha = \frac{e^2}{4\pi\epsilon_0} \frac{1}{\hbar c} \simeq \frac{1}{137} , \tag{2.27}$$

and n is the principal quantum number. For a given value of n the angular momentum quantum number l can assume the values $0, 1, \ldots, n-1$.

In Table 2.3, levels with accidental degeneracy are presented. Each level has a degeneracy associated with the values of the third component of angular momentum, characterised by the quantum number μ, which can take on $2l+1$ values. Additionally, since the potential does not depend on the direction of the electron spin, the

Table 2.3 Level scheme generated by the Coulomb potential. The energies are identified by the quantum number n, therefore there is an accidental degeneracy for different values of l

	n	l		n	l		n	l	
ϵ_1	1	0	$1s$						
ϵ_2	2	0	$2s$	2	1	$2p$			
ϵ_3	3	0	$3s$	3	1	$3p$	3	2	$3d$

overall degeneracy is $2(2l+1)$. Based on this understanding, and taking into account the Pauli exclusion principle, the periodic table of elements is constructed.

2.2.4 Spin-Orbit Potential

At the end of this section dedicated to the solution of the one-body Schrödinger equation with central potentials, we will discuss the treatment of the spin-orbit term in the potential. It is now well established that relativity is the source of the effects arising from the interaction between orbital angular momentum and the half-integer spin of fermions. While this concept has long been accepted in atomic physics, it was not quantitatively understood in nuclear physics until the early 1980s [4].

For our purposes, the idea is to describe the spin-orbit effects in the framework of the non-relativistic Quantum Mechanics. We add to one of the spherical potentials presented so far a new term proportional to the spin-orbit coupling

$$\hat{V}(r) = \hat{U}_c(r) - \frac{2a}{\hbar^2}\hat{\mathbf{l}} \cdot \hat{\sigma} \ , \tag{2.28}$$

where a is a real and positive constant. The scalar product between the orbital angular momentum operator $\hat{\mathbf{l}}$ and the spin operator $\hat{\mathbf{s}} = \hat{\sigma}/2$ is due the fact that the hamiltonian is a scalar operator, and this is the simplest type of coupling between the two vector operators describing the orbital angular momentum and the spin.

The presence of a spin-dependent term in the hamiltonian requires a change in the procedure used to solve the Schrödinger equation. We used so far an expansion of the wave functions (2.11) in terms of the spherical harmonics, eigenstates of the $\hat{\mathbf{l}}^2$ operator. The Pauli spinors χ, eigenstates of $\hat{\mathbf{s}}^2$, were factorised. So far, the presence of these latter terms was irrelevant since the potentials adopted did not contain spin dependent terms. The present situation is more complicated due to the presence of a spin-dependent term in the hamiltonian.

Let us considers the total angular momentum of the fermion obtained as a sum of the orbital angular momentum and of the spin $\hat{\mathbf{j}} = \hat{\mathbf{l}} + \hat{\mathbf{s}}$. This definition implies

$$\hat{\mathbf{j}}^2 = (\hat{\mathbf{l}} + \hat{\mathbf{s}})^2 = \hat{\mathbf{l}}^2 + \hat{\mathbf{s}}^2 + 2\,\hat{\mathbf{l}} \cdot \hat{\mathbf{s}} \ , \tag{2.29}$$

from which one obtains an expression for $\hat{\mathbf{l}} \cdot \hat{\mathbf{s}}$ depending on the squares of the three angular momenta under consideration

$$\hat{\mathbf{l}} \cdot \hat{\mathbf{s}} = \frac{1}{2}\left(\hat{\mathbf{j}}^2 - \hat{\mathbf{l}}^2 - \hat{\mathbf{s}}^2\right) \ . \tag{2.30}$$

At this point, it is convenient to consider expressions of the eigenfunction of the hamiltonian of the form

$$\phi_{nljm}(\mathbf{r}) = R_{nlj}(r) \sum_{\mu\sigma} \langle l\, \mu\, \frac{1}{2}\, \sigma | j\, m \rangle Y_{l\mu}(\Omega) \chi_\sigma = R_{nlj}(r) \mathcal{Y}_{ljm}(\Omega) \ , \quad (2.31)$$

where spherical harmonics and Pauli spinors are connected by the Clebsch-Gordan coefficients and form the so-called the spin spherical harmonics [5] which are eigenstates of the following operators

$$\hat{\mathbf{j}}^2 \mathcal{Y}_{ljm}(\Omega) = j(j+1)\, \hbar^2\, \mathcal{Y}_{ljm}(\Omega) \ \ ; \ \ \hat{j}_z \mathcal{Y}_{ljm}(\Omega) = m\, \hbar\, \mathcal{Y}_{ljm}(\Omega) \ , \quad (2.32)$$

$$\hat{\mathbf{l}}^2 \mathcal{Y}_{ljm}(\Omega) = l(l+1)\, \hbar^2\, \mathcal{Y}_{ljm}(\Omega) \ \ ; \ \ \hat{\mathbf{s}}^2 \mathcal{Y}_{ljm}(\Omega) = \frac{3}{4}\, \hbar^2\, \mathcal{Y}_{ljm}(\Omega) \ , \quad (2.33)$$

The spin-orbit term inserted in the hamiltonian gives the result

$$\begin{aligned} \hat{\mathbf{l}} \cdot \hat{\mathbf{s}}\, \mathcal{Y}_{ljm}(\Omega) &= \frac{1}{2} \left[\hat{\mathbf{j}}^2 - \hat{\mathbf{l}}^2 - \hat{\mathbf{s}}^2 \right] \mathcal{Y}_{ljm}(\Omega) \\ &= \frac{1}{2} \left[j(j+1) - l(l+1) - \frac{3}{4} \right] \hbar^2 \mathcal{Y}_{ljm}(\Omega) \ , \end{aligned} \quad (2.34)$$

and, since $j = l \pm 1/2$, we obtain

$$\begin{aligned} &\text{for } j = l + \frac{1}{2} \quad \left[\left(l + \frac{1}{2} \right) \left(l + \frac{3}{2} \right) - l^2 - l - \frac{3}{4} \right] = l, \\ &\text{for } j = l - \frac{1}{2} \quad \left[\left(l - \frac{1}{2} \right) \left(l + \frac{1}{2} \right) - l^2 - l - \frac{3}{4} \right] = -(l+1) \ . \end{aligned} \quad (2.35)$$

This means that ϵ^c_{nl}, the energy obtained by considering only the term $\hat{U}_c$ in Eq. (2.28), is modified as

$$\begin{aligned} \epsilon_{nlj} &= \epsilon^c_{nl} + a(l+1) \ \text{ for } j = l - \frac{1}{2}, \\ \epsilon_{nlj} &= \epsilon^c_{nl} - al \ \text{ for } j = l + \frac{1}{2}. \end{aligned} \quad (2.36)$$

In Eq. (2.28) we defined the constants such as for $\alpha > 0$ the energy with $l + 1/2$ is smaller than ϵ^c_{nl}, and the contrary for $l - 1/2$. This is what happens in nuclear physics [6]. In atomic physics the effect of the spin-orbit term is inverted [7] (Fig. 2.3).

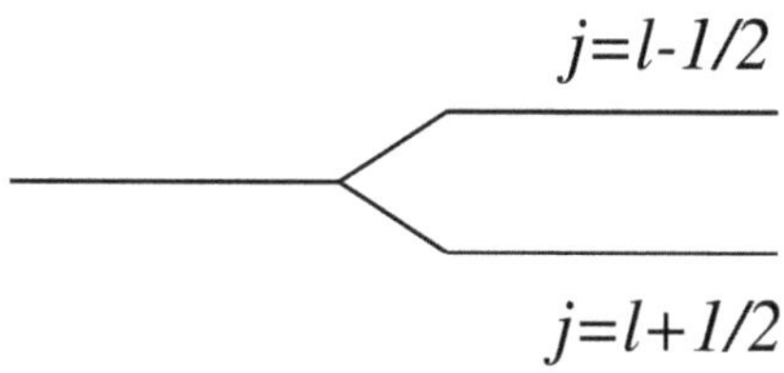

Fig. 2.3 Splitting of the single particle level because of the action of the spin-orbit term in the potential

2.3 Translational Symmetry

Baryons, nuclei, atoms, and molecules are many-body systems that can be effectively described using rotational symmetry. In these systems, even if they are deformed, it is possible to identify a point around which the matter distribution is centered, which can be considered the center of the MF.

In condensed matter systems, translational symmetry predominates. A basic structure of the system is periodically repeated in three cartesian directions, making it impossible to identify a central point. The fundamental MF model for these types of systems assumes that the potential $\hat{U}$ is constant. This fermionic system is commonly referred to as a Fermi gas. It serves as a toy model that is homogeneous, occupies an infinite volume, and consists of an infinite number of non-interacting fermions. Since the energy scale is arbitrary, it is possible to set $\hat{U} = 0$ without loss of generality. In this case, the one-body Schrödinger equation is

$$-\frac{\hbar^2}{2m_i}\nabla_i^2\phi_i(\mathbf{r}) = \epsilon_i\phi_i(\mathbf{r}) \ , \tag{2.37}$$

with

$$\phi_{k_i}(\mathbf{r}) = \frac{1}{\sqrt{\mathcal{V}}}e^{i(\mathbf{k}_i\cdot\mathbf{r})}\chi_\sigma\chi_\tau \ , \tag{2.38}$$

where $\mathcal{V}$ is the volume, and χ are the Pauli spinors related to the spin of the fermion and, eventually, to its isospin. The third components of spin and isopin are indicated as σ e τ, respectively. The physical quantities of interest are those that are independent of $\mathcal{V}$.

The symmetry of the problem induces to consider the system contained in a cubic box of side $\mathrm{L} = \mathcal{V}^{1/3}$ with periodic boundary conditions

$$\phi_i(x+\mathrm{L}, y, z) = \phi_i(x, y+\mathrm{L}, z) = \phi_i(x, y, z+\mathrm{L}) = \phi_i(x, y, z) \ . \tag{2.39}$$

Since

$$\phi_i(\mathbf{r}=0) = \frac{1}{\sqrt{\mathcal{V}}}\chi_\sigma\chi_\tau \ , \tag{2.40}$$

in order to satisfy the periodic conditions we have to impose

$$e^{ik_x \mathrm{L}} = e^{ik_y \mathrm{L}} = e^{ik_z \mathrm{L}} = 1 \ , \tag{2.41}$$

and this implies

$$k_x = \frac{2\pi}{\mathrm{L}} n_x \ ; \ k_y = \frac{2\pi}{\mathrm{L}} n_y \ ; \ k_z = \frac{2\pi}{\mathrm{L}} n_z \ , \tag{2.42}$$

where n_x, n_y, n_z are integer numbers. By defining $\mathbf{n} \equiv (n_x, n_y, n_z)$, we can write

$$d^3 \mathbf{n} = \frac{\mathrm{L}^3}{(2\pi)^3} d^3 \mathbf{k} = \frac{\mathrm{L}^3}{(2\pi\hbar)^3} d^3 \mathbf{p} \ . \tag{2.43}$$

The evaluation of the density of states begins by considering the expression for the energy of the system within an infinitesimally small energy interval, specifically between the values ϵ and $\epsilon + d\epsilon$. In non-relativistic mechanics, the momentum is related to the wave number by the relation $\mathbf{p} = m\mathbf{v} = \hbar\mathbf{k}$, therefore, $\epsilon = m\mathbf{v}^2/2 = \mathbf{p}^2/2m = \hbar^2\mathbf{k}^2/2m$. The energy differential is

$$d\epsilon = d\left(\frac{1}{2} m\mathbf{v}^2\right) = \frac{1}{2} m\, 2\, v dv = v\, m\, dv = v\, dp \ , \tag{2.44}$$

with $v = |\mathbf{v}|$ and $p = |\mathbf{p}|$. The density of states is

$$\begin{aligned} \rho_\epsilon &= \frac{d^3\mathbf{n}}{d\epsilon} = \frac{\mathrm{L}^3}{(2\pi\hbar)^3} d^3\mathbf{p} \frac{1}{v\, dp} \\ &= \frac{\mathcal{V}}{(2\pi\hbar)^3} p^2 d\Omega_p dp \frac{1}{v\, dp} = \frac{\mathcal{V}}{(2\pi\hbar)^3} \frac{p^2}{v} d\Omega_p \ , \end{aligned} \tag{2.45}$$

where Ω_p indicates the spherical coordinates identifying the direction of $\mathbf{p}$. This is the expression of the density of states commonly adopted in the calculations of the scattering cross sections [8].

Let us calculate the number density and the kinetic energy per particle of the Fermi gas. The normalisation of the single-particle wave functions (2.38) implies that each integral involving these wave functions must be multiplied by the factor $\mathcal{V}/(2\pi)^3$. In effect,

$$\begin{aligned} \langle \phi_a | \phi_b \rangle &= \frac{\mathcal{V}}{(2\pi)^3} \int d^3 r \frac{1}{\sqrt{\mathcal{V}}} e^{-i(\mathbf{k}_a \cdot \mathbf{r})} \frac{1}{\sqrt{\mathcal{V}}} e^{i(\mathbf{k}_b \cdot \mathbf{r})} \\ &= \frac{\mathcal{V}}{(2\pi)^3} \frac{(2\pi)^3 \delta(\mathbf{k}_a - \mathbf{k}_b)}{\mathcal{V}} = 1 \ . \end{aligned} \tag{2.46}$$

The particle density, or number density, in the Fermi gas can be evaluated as

$$\rho(\mathbf{r}) = \sum_a^A |\phi_a(\mathbf{r})|^2 \ , \tag{2.47}$$

where A is the number of fermions composing the system. This expression is valid only within the MF model. It can be derived by calculating the mean value of the density operator between two Slater determinants. A more intuitive understanding of its validity can be achieved by considering that the interpretation of $|\phi_a(\mathbf{r})|^2$ is the probability of finding a particle, characterised by the a quantum numbers, in an infinitesimal volume around $\mathbf{r}$. In a situation where the individual particles do not interact, the probabilities for each particle are independent of one another. Therefore, the overall probability of finding a particle, regardless of its quantum numbers, in the infinitesimal volume around $\mathbf{r}$ is simply the sum of the individual probabilities.

Each single-particle state is characterised by its energy and spin, and possibly also by isospin components, with each state occupied by a single fermion. At zero temperature, when the system is in its ground state, all single-particle states below a certain energy value, known as the Fermi energy ϵ_{F}, are filled, while all states with energy greater than ϵ_{F} remain unoccupied. In our case, each state is associated with a wave number that is directly related to its energy. The above statements can also be expressed in terms of a maximum momentum value, referred to as the Fermi momentum $\mathbf{p}_{\mathrm{F}}$, and the corresponding wave number $\mathbf{k}_{\mathrm{F}}$. The relation between Fermi energy and wave number is

$$\epsilon_{\mathrm{F}} = \frac{\hbar^2}{2m}\mathbf{k}_{\mathrm{F}}^2 \ . \tag{2.48}$$

As consequence of these considerations the particle density is calculated as

$$\begin{aligned} \rho(\mathbf{r}) &= \sum_{a \le k_{\mathrm{F}}} |\phi_a(\mathbf{r})|^2 = \frac{\mathcal{V}}{(2\pi)^3} \mathcal{D} \int_0^{k_{\mathrm{F}}} d^3k \frac{1}{\sqrt{\mathcal{V}}} e^{-i(\mathbf{k}\cdot\mathbf{r})} \frac{1}{\sqrt{\mathcal{V}}} e^{i(\mathbf{k}\cdot\mathbf{r})} \\ &= \frac{\mathcal{D}}{(2\pi)^3} \int_0^{k_{\mathrm{F}}} d^3k \ . \end{aligned} \tag{2.49}$$

In this expression $\mathcal{D}$ indicates the degeneracy factor related to the characteristics of the fermions forming the system. In the case of an electron gas, the wave function of each particle is characterised by its momentum and by the orientation of the spin. In the case of nuclear matter there is an additional quantum number related to the third component of the isospin distinguishing protons from neutrons. Therefore, for electrons we have

$$\mathcal{D} = \sum_{\sigma=\pm 1/2} \chi_\sigma^\dagger \chi_\sigma = 2 \ , \tag{2.50}$$

and for the nucleons

$$\mathcal{D} = \sum_{\sigma=\pm 1/2} \chi_\sigma^\dagger \chi_\sigma \sum_{\tau=\pm 1/2} \chi_\tau^\dagger \chi_\tau = 4 \ , \tag{2.51}$$

The integral of Eq. (2.49) is given by

$$\rho(\mathbf{r}) = \frac{\mathcal{D}}{(2\pi)^3}\int_0^{k_F} d^3k = \frac{\mathcal{D}}{(2\pi)^3}\int_0^{k_F} k^2 dk \int d\Omega_k$$
$$= \frac{\mathcal{D}}{(2\pi)^3}\frac{4}{3}\pi k_F^3 = \frac{\mathcal{D}}{2\pi^2}\frac{k_F^3}{3} \,, \tag{2.52}$$

with

$$\int d\Omega_k \equiv \int_{-1}^{1} d(\cos\theta)\int_0^{2\pi} d\phi = 4\pi \,. \tag{2.53}$$

The density is independent of **r**, as expected, since we have assumed that the system is homogeneous. In other words, by choosing the plane waves (2.37) as the single-particle basis, the homogeneity of the system is inherently implied. More interestingly, the particle density depends on the cube of the Fermi momentum.

Let us calculate the mean kinetic energy of a single particle

$$\langle\phi_{a'}|\frac{-\hbar^2}{2m}\nabla^2|\phi_a\rangle = \frac{\mathcal{V}}{(2\pi)^3}\int d^3r\frac{1}{\sqrt{\mathcal{V}}}e^{-i(\mathbf{k}_{a'}\cdot\mathbf{r})}\left(\frac{-\hbar^2}{2m}\nabla^2\right)\frac{1}{\sqrt{\mathcal{V}}}e^{i(\mathbf{k}_a\cdot\mathbf{r})}$$
$$= \frac{\hbar^2 k_a^2}{2m} \,. \tag{2.54}$$

This expression confirms the validity of the non-relativistic relation between energy and momentum $\epsilon_a = p_a^2/2m$. The total kinetic energy of the system is

$$\mathcal{K} = \sum_{a\le k_F}\langle\phi_a|\frac{-\hbar^2}{2m}\nabla^2|\phi_a\rangle = \mathcal{D}\frac{\mathcal{V}}{(2\pi)^3}\int_0^{k_F} d^3k\,\frac{\hbar^2k^2}{2m}$$
$$= \mathcal{D}\frac{\mathcal{V}}{(2\pi)^3}\frac{\hbar^2 4\pi}{2m}\int_0^{k_F} k^2\,k^2\,dk = \mathcal{D}\frac{\mathcal{V}}{(2\pi)^3}\frac{\hbar^2 4\pi}{2m}\frac{k_F^5}{5}$$
$$= \mathcal{D}\frac{\mathcal{V}}{(2\pi)^3}\frac{\hbar^2 4\pi}{2m}\frac{k_F^2}{5}\left(\frac{2\pi^2 3\rho}{\mathcal{D}}\right) = \frac{3}{5}\mathcal{V}\rho\frac{\hbar^2k_F^2}{2m} = \frac{3}{5}A\epsilon_F \,, \tag{2.55}$$

therefore, the kinetic energy per particle is

$$\frac{\mathcal{K}}{A} = \frac{3}{5}\epsilon_F \,. \tag{2.56}$$

Let us calculate the system pressure by considering that the total energy of a Fermi gas is given by the kinetic energy, the potential $\hat{U}$ is a constant renormalising the total value,

$$E = \mathcal{K} = \frac{3}{5}A\epsilon_F = \frac{3}{5}A\frac{\hbar^2k_F^2}{2m} = \frac{3}{5}A\frac{\hbar^2}{2m}\left(\frac{2\pi^2 3}{\mathcal{D}}\frac{A}{\mathcal{V}}\right)^{2/3} = \mathcal{S}\,\mathcal{V}^{-2/3} \,, \tag{2.57}$$

where we have used the expression (2.52) of the density and we called $\mathcal{S}$ the part multiplying the volume $\mathcal{V}$. The pressure can be expressed as

$$P = -\left(\frac{\partial E}{\partial \mathcal{V}}\right)_A = -\mathcal{S}\left(-\frac{2}{3}\right)\mathcal{V}^{-2/3}\mathcal{V}^{-1} = \frac{2}{3}\frac{E}{\mathcal{V}} = \frac{2}{3}\frac{1}{\mathcal{V}}\frac{3}{5}A\epsilon_F = \frac{2}{5}\epsilon_F\rho. \quad (2.58)$$

The compressibility K is the inverse of the compression modulus B, which is expressed in the context of the Fermi gas as follows:

$$B = \frac{1}{-}K = -\mathcal{V}\frac{\partial P}{\partial \mathcal{V}} = -\mathcal{V}\frac{2}{3}\mathcal{S}\,\mathcal{V}^{-5/3}\left(\frac{-5}{3}\right)\mathcal{V}^{-1} = \frac{10}{9}\frac{E}{\mathcal{V}} = \frac{2}{3}\epsilon_F\rho. \quad (2.59)$$

Table 2.4 presents the values of k_F, ϵ_F and B calculated for the electron gases of various crystals. These results were obtained by substituting empirical values of electron densities into Eqs. (2.52), (2.48), and (2.59), respectively. The comparison with measured values of the compression modulus clearly highlights the limitations of the MF model.

The MF models provide a description of many-body systems in which each particle moves independently of the others. This independence is particularly evident in translationally invariant systems, where the total Hamiltonian is the sum of Hamiltonians that describe free particle motion, as shown in Eq. (2.37). However, this simple model incorporates a crucial physical element that limits free particle motion: the Pauli exclusion principle. This principle is sufficient to explain certain phenomena that manifest at the macroscopic level. For instance, the stability of a white dwarf star can be understood using the Fermi gas model [11]. Below, we will discuss another example: the specific heat in metals.

The classical, non-quantum, description of a gas of A point-like, non-interacting particles predicts a specific heat value of $3Ak_B/2$ [12], where k_B is the Boltzmann

Table 2.4 The electron densities, taken from Refs. [9, 10], are empirical data used to obtain the values of k_F, Eq. (2.52), ϵ_F, Eq. (2.48), and B, Eq. (2.59). This latter quantity is compared with measured values B_{exp}

	ρ [10^{22} cm^{-3}]	k_F [10^8 cm^{-1}]	ϵ_F [eV]	B [dyne/cm^2]	B_{exp} [dyne/cm^2]
Li	4.70	1.11	4.75	23.84	11.5
Na	2.65	0.92	3.24	9.17	6.42
K	1.40	0.75	2.12	3.17	2.81
Rb	1.15	0.70	1.86	2.28	1.92
Cs	0.91	0.65	1.59	1.54	1.43
Cu	8.45	1.36	7.02	63.37	134.3
Ag	5.85	1.20	5.50	34.34	99.9
Al	18.06	1.75	11.65	224.74	76.0

constant. However, the measured values of the electronic contribution to the specific heat at room temperature are usually hundreds of times smaller. This discrepancy can be explained by the Pauli exclusion principle.

In the classical description, all electrons in the system contribute to the specific heat, even at low temperatures, when the excitation energies are very small compared to the total ground state energy. In the quantum case, at low temperatures, only those electrons near the Fermi surface can change their states, moving from a state below the Fermi surface (referred to as a hole state) to occupy a state above it (known as a particle state). For relatively small excitation energies, electrons with energies much lower than the Fermi energy cannot change their states, as they would attempt to occupy states that are already filled by other electrons.

Let us derive here below the expression of the specific heat of a gas of free electrons, i.e. for a Fermi gas. By using the expression (2.52) of the particle density, we obtain

$$k_{\mathrm{F}}{}^3 = \frac{2\pi^2}{\mathcal{D}} 3\,\rho \,, \tag{2.60}$$

$$\epsilon_{\mathrm{F}} = \frac{\hbar^2}{2\,m} k_{\mathrm{F}}{}^2 = \left(\frac{2\pi^2}{\mathcal{D}} 3\,\rho\right)^{2/3} \,, \tag{2.61}$$

$$A = \mathcal{V}\rho = \frac{\mathcal{V}\mathcal{D}}{2\pi^2 3}\left(\frac{2\,m}{\hbar^2}\epsilon_{\mathrm{F}}\right)^{3/2} \,, \tag{2.62}$$

therefore, the density of states around the Fermi energy can be expressed as

$$\rho_\epsilon(\epsilon_{\mathrm{F}}) \equiv \left(\frac{dA}{d\epsilon}\right)_{\epsilon=\epsilon_{\mathrm{F}}} = \frac{\mathcal{V}\mathcal{D}}{4\pi^2}\left(\frac{2\,m}{\hbar^2}\right)^{3/2} \epsilon_{\mathrm{F}}^{1/2}. \tag{2.63}$$

The increase of internal energy E of an electron gas when the value of the temperature changes from 0 to T is given by

$$\Delta E \equiv E(T) - E(0) = \int_0^\infty d\epsilon\, \epsilon\, \rho_\epsilon(\epsilon) f(\epsilon, T) - \int_0^\infty d\epsilon\, \epsilon\, \rho_\epsilon(\epsilon) f(\epsilon, 0) \,. \tag{2.64}$$

In the above equation $f(\epsilon, T)$ is the Fermi-Dirac energy distribution

$$f(\epsilon, T) \equiv \frac{1}{\exp\left[(\epsilon - \mu)/\mathrm{k_B}T\right] + 1} \,, \tag{2.65}$$

where T is the absolute temperature of the system and μ is the chemical potential

$$\mu = \frac{\partial E}{\partial A} \,, \tag{2.66}$$

whose value is ϵ_{F} at $T = 0$. The Fermi-Dirac distribution in the limit $T \to 0$ is $\Theta(\epsilon_{\mathrm{F}} - \epsilon)$, where $\Theta(x)$ is the step function equal to 1 for $x > 0$ and to 0 for $x < 0$. The Eq. (2.64) becomes

$$\Delta E \equiv E(T) - E(0) = \int_0^\infty d\epsilon\, \epsilon\, \rho_\epsilon(\epsilon) f(\epsilon, T) - \int_0^{\epsilon_F} d\epsilon\, \epsilon\, \rho_\epsilon(\epsilon) . \tag{2.67}$$

By using the Fermi-Dirac distribution the number of particles can be expressed as

$$A = \int_0^\infty d\epsilon \frac{dA}{d\epsilon} = \int_0^\infty d\epsilon\, \rho_\epsilon(\epsilon) f(\epsilon, T) . \tag{2.68}$$

By multiplying this expression by ϵ_F we obtain

$$\epsilon_F A = \left(\int_0^{\epsilon_F} + \int_{\epsilon_F}^\infty \right) d\epsilon\, \epsilon_F \rho_\epsilon(\epsilon) f(\epsilon, T) = \int_0^{\epsilon_F} d\epsilon\, \epsilon_F \rho_\epsilon(\epsilon) . \tag{2.69}$$

where the last equality has been obtained by considering that the number of particle does not change when the temperature changes from 0 to T, and that the Fermi distribution (2.65) becomes a step function around ϵ_F when $T \to 0$. By adding and subtracting $\epsilon_F A$ in the expression (2.67) of ΔE we obtain

$$\begin{aligned} \Delta E &= \int_0^\infty d\epsilon\, \epsilon\, \rho_\epsilon(\epsilon) f(\epsilon, T) - \left(\int_0^{\epsilon_F} + \int_{\epsilon_F}^\infty \right) d\epsilon\, \epsilon_F \rho_\epsilon(\epsilon) f(\epsilon, T) \\ &\quad - \int_0^{\epsilon_F} d\epsilon\, \epsilon\, \rho_\epsilon(\epsilon) + \int_0^{\epsilon_F} d\epsilon\, \epsilon_F \rho_\epsilon(\epsilon) \\ &= \int_{\epsilon_F}^\infty d\epsilon\, (\epsilon - \epsilon_F)\, \rho_\epsilon(\epsilon) f(\epsilon, T) \\ &\quad + \int_0^{\epsilon_F} d\epsilon\, (\epsilon_F - \epsilon) \rho_\epsilon(\epsilon) \left[1 - f(\epsilon, T)\right] . \end{aligned} \tag{2.70}$$

The term $\rho_\epsilon(\epsilon) f(\epsilon, T) d\epsilon$ represents the number of electrons passing from levels with energy ϵ to levels of energy $\epsilon + d\epsilon$. In the second integral the term $1 - f(\epsilon, T)$ is the probability that an electron is removed from a level with energy ϵ.

The specific heat for an electron gas is given by

$$C_{el} \equiv \frac{dE}{dT} = \int_0^\infty d\epsilon\, (\epsilon - \epsilon_F) \rho_\epsilon(\epsilon) \frac{df(\epsilon, T)}{dT} , \tag{2.71}$$

where we have assumed that only term dependent on the temperature T is f. For temperatures much smaller than the Fermi temperature, i.e. $k_B T \ll \epsilon_F$, it is plausible to consider the density of states almost constant, therefore

$$C_{el} \simeq \rho_\epsilon(\epsilon_F) \int_0^\infty d\epsilon\, (\epsilon - \epsilon_F) \frac{df(\epsilon, T)}{dT} . \tag{2.72}$$

This approximation is well verified if we consider that the typical values of the Fermi temperatures are of the order of $\sim 5 \times 10^4$ K. The evaluation of the integral is presented in the box and it gives the result

$$C_{\text{el}} \simeq \mathrm{k}_{\mathrm{B}}^2\, T \rho_\epsilon(\epsilon_{\mathrm{F}}) \frac{\pi^2}{3} \ . \tag{2.73}$$

In the following we use new variables defined as $\tau = \mathrm{k_B} T$ e $x = (\epsilon - \mu)/(\mathrm{k_B} T)$.

$$\frac{1}{\mathrm{k_B}} \frac{df(\epsilon, T)}{dT} = \frac{df}{d\tau} = \frac{\exp(\frac{\epsilon-\mu}{\mathrm{k_B}T}) \frac{\epsilon-\mu}{(\mathrm{k_B}T)^2}}{\left[\exp\left(\frac{\epsilon-\mu}{\mathrm{k_B}T}\right) + 1\right]^2} = \frac{1}{[e^x + 1]^2} e^x \frac{x}{\tau} \ . \tag{2.74}$$

For the definition of x one has $dx = d\epsilon/\tau$, therefore

$$\begin{aligned} C_{\text{el}} &= \mathrm{k_B} \rho_\epsilon(\epsilon_{\mathrm{F}}) \int_{-\epsilon_{\mathrm{F}}/\tau}^{\infty} (dx\, \tau)(\tau x) \frac{1}{[e^x + 1]^2} e^x \frac{x}{\tau} \\ &\simeq \mathrm{k_B} \tau \rho_\epsilon(\epsilon_{\mathrm{F}}) \int_{-\infty}^{\infty} dx\, x^2 \frac{e^x}{[e^x + 1]^2} = \mathrm{k_B} \tau \rho_\epsilon(\epsilon_{\mathrm{F}}) \frac{\pi^2}{3} \ . \end{aligned} \tag{2.75}$$

In the last equation the lower integration limit has been extended to $-\infty$ since the term e^x is already very small for $x = -\epsilon_{\mathrm{F}}/\tau$ when we consider room temperatures. It only remains to calculate an integral which is well known in the literature.

From the Eq. (2.63) we obtain

$$\rho_\epsilon(\epsilon_{\mathrm{F}}) = \frac{1}{2} \frac{\mathcal{V}\mathcal{D}}{2\pi^2} \left(\frac{2m}{\hbar^2}\right)^{3/2} \epsilon_{\mathrm{F}}^{1/2} = \frac{1}{2} 3 \frac{1}{3} \frac{\mathcal{V}\mathcal{D}}{2\pi^2} \left(\frac{2m}{\hbar^2} \epsilon_{\mathrm{F}}\right)^{3/2} \frac{1}{\epsilon_{\mathrm{F}}} = \frac{3}{2} \frac{A}{\epsilon_{\mathrm{F}}} \ , \tag{2.76}$$

where we used the expression (2.62) for the electron number. By considering that $\epsilon_{\mathrm{F}} = \mathrm{k_B} T_{\mathrm{F}}$, the expression of the specific heat becomes

$$C_{\text{el}} = \frac{\pi^2}{3} \frac{3}{2} \frac{A}{\mathrm{k_B} T_{\mathrm{F}}} \mathrm{k}_{\mathrm{B}}^2\, T = \frac{\pi^2}{2} A \mathrm{k_B} \frac{T}{T_{\mathrm{F}}} \ . \tag{2.77}$$

This result indicates how, contrary to the prediction of the classical statistical mechanics, only a fraction of electrons, proportional to T/T_{F} is excited at the temperature T. These are the electrons close to the Fermi surface.

The prediction of a linear relationship between temperature and specific heat in metals is one of the most significant consequences of Fermi-Dirac statistics, or, in other words, the Pauli exclusion principle for electrons. This result holds true only if the conduction electrons are the sole contributors to the specific heat. In reality, at temperatures above room temperature, the specific heat is primarily dominated by the contributions from the ions that make up the metal. Below room temperature, this ionic contribution behaves proportionally to T^3, and at even lower temperatures, it becomes smaller than that of the electrons, which, as indicated by Eq. (2.77), is linearly dependent on T.

Table 2.5 Values of the linear coefficient γ of the specific heats of some metals, Eq. (2.78). The values are expressed in units of 10^{-4} cal mole^{-1} K^{-2}. From [9]

Element	Theory	Experiment
Cu	1.2	1.6
Ag	1.5	1.6
Au	1.5	1.6
Fe	1.5	12.0
Mn	1.5	40.0
Bi	4.3	0.2

It is common practice to separate the two contributions as

$$c_v = \gamma T + K T^3 \ . \tag{2.78}$$

where γ and K are two real constants. The value of the γ coefficient is obtained by extrapolating the values of c_v for $T \to 0$.

Table 2.5 compares some values of the γ coefficient calculated by using Eq. (2.77) with values taken from measurements. The alkaline metals (Cu, Ag, Au) are relatively well described by the Fermi gas model. This model makes completely wrong predictions for Fe and Mn giving values much smaller than the measured ones. On the contrary, the prediction for the Bi is remarkably larger than the observed value.

References

1. J.C. Slater, The Theory of complex Spectra. Phys. Rev. **34**, 1293 (1929)
2. A. Messiah, *Quantum Mechanics* (North Holland, Amsterdam, 1961)
3. H.W. Wyld, *Mathematical Methods for Physics* (Benjamin, London, 1976)
4. B.D. Serot, J.D. Walecka, Relativistic nuclear many-body theory. Adv. Nucl. Phys. **16**, 1 (1986)
5. A.R. Edmonds, *Angular Momentum in Quantum Mechanics* (Princeton University Press, Princeton, 1957)
6. P. Ring, P. Schuck, *The Nuclear Many-Body Problem* (Springer, Berlin, 1980)
7. B.H. Bransden, C.J. Joachain, *Physics of Atoms and Molecules* (Wiley, New York, 1983)
8. K. Krane, *Introductory Nuclear Physics* (Wiley, New York, 1988)
9. N.W. Ashcroft, N.D. Mermin, *Solid State Physics* (Saunders, Orlando, 1976)
10. C. Kittel, *Introduction to Solid State Physics* (Wiley, New York, 1986)
11. O. Benhar, *Structure and Dynamics of Compact Stars* (Springer Nature Switzerland, Cham (Switzerland), 1986)
12. C. Kittel, H. Kroemer, *Thermal Physics* (W. H. Freeman, New York, 1980)

Chapter 3
Interactions

3.1 Introduction

In the MF models, the interactions between the particles that make up the many-body system are not taken into account. In other words, in these models, the interaction term of the Hamiltonian in Eq. (2.1) is set to zero. In the previous chapter, we demonstrated that by using this approximation, the total Hamiltonian of the system simplifies to a sum of single-particle Hamiltonians. Many-body theories extend beyond MF models by incorporating the interactions between particles.

We have chosen to describe many-body systems within the framework of non-relativistic quantum mechanics. This choice is valid only in situations where the energy values of the phenomena under study are significantly smaller than the rest mass energies of the particles involved. In the non-relativistic framework, the interactions between particles are effectively described using potentials. The concept of potential implies an instantaneous transmission of interaction, regardless of the distance between the particles. However, this assumption is not generally valid, as there is a limiting speed for signal transmission: the speed of light in a vacuum. Nevertheless, under the energy conditions mentioned earlier, relativistic effects are negligible or can be treated as perturbations

In this chapter, we will explore three different types of many-body systems: electron gas, atomic nuclei, and quantum fluids, which include liquids and strongly interacting gases composed of molecules. These systems are effectively described by non-relativistic quantum mechanics. A key characteristic of these systems is that, for each particle, the interaction energy is comparable to the kinetic energy.

© The Author(s), under exclusive license to Springer Nature Switzerland AG 2026

G. Co', *Concepts in Quantum Many-Body Physics*, UNITEXT for Physics,
https://doi.org/10.1007/978-3-032-08920-5_3

3.2 Electron Gas

A good description of metals can be provided by a model in which the valence electrons of the metal atoms become conduction electrons, allowing them to move almost freely throughout the volume of the metal. This concept is encapsulated in the free electron model, which is analogous to a Fermi gas, as discussed in Sect. 2.3.

A more realistic description of a metal requires considering the mutual interactions between electrons, as well as the interactions between the electrons and the ionised atoms that make up the crystal lattice. This situation is quite complex; however, at temperatures close to absolute zero, the Jellium model, also known as the uniform electron gas, provides a good representation of the properties of this many-body system. In this model, the positive charges create a homogeneous background that is uniformly distributed throughout space. This approach highlights the effects generated by the quantum characteristics of the electrons, such as the Pauli exclusion principle and their interactions, independent of the crystal structure.

In this model, the system hamiltonian is

$$\hat{H} = \sum_{i=1}^{A} \frac{-\hbar^2}{2m_i} \nabla_i^2 + \frac{1}{2} \sum_{i,j=1}^{A} \frac{e^2}{4\pi\epsilon_0} \frac{1}{|\mathbf{r}_i - \mathbf{r}_j|}, \tag{3.1}$$

where e is the elementary charge and ϵ_0 the vacuum permittivity. and $|\mathbf{r}_i - \mathbf{r}_j|$ the distance between two interacting electrons.

3.3 Nuclei

The description of nuclear systems is based on the interactions between nucleons. Unlike elementary particles such as electrons, nucleons are composite entities made up of quarks, antiquarks, and gluons. Currently, the most widely accepted theory that describes the strong nuclear interaction which involves gluons and quarks is Quantum Chromodynamics (QCD). This theory is non-perturbative within the energy range relevant to nuclear phenomena, which spans from a few keV to approximately 150 MeV, the threshold for pion emission. Consequently, the well-established techniques used to solve Quantum Electrodynamics (QED), which is perturbative in nature, cannot be applied here.

The QCD field equations are primarily solved using numerical methods by discretising space and time into a set of points. The goal is to create a denser lattice to better represent a continuum space-time. This area of research is highly active and is yielding increasingly precise results, such as improved predictions of hadron masses [1]. These findings support the notion that QCD is the correct framework for understanding the strong interaction. However, describing nuclei in terms of quarks and gluons remains too complex to be practical at this time. While research is ongoing into nuclear systems using lattice QCD techniques [2], even experts express skepticism about the direct applicability of this method to systems with a large number of nucleons.

For the reasons mentioned above, the fundamental degrees of freedom used to describe nuclear systems are mesons and baryons, primarily protons and neutrons, collectively referred to as nucleons. In this context, the interaction between nucleons can be understood as an interaction between aggregates of quarks and gluons, similar to the van der Waals electromagnetic forces that occur between molecules. While QCD serves as an asymptotic limit, it is rarely employed to generate realistic interactions between nucleons.

3.3.1 Two-Body Forces

The nucleon-nucleon interaction is designed to accurately describe the experimental data associated with two-nucleon systems. This includes the deuteron, which is the only bound state, as well as thousands of elastic scattering data points.

The deuteron has the following characteristics [3].

1. It is the only bound state of two nucleons.
2. It is composed by one proton and one neutron.
3. It has a binding energy of about 2.22 MeV.
4. It does not have excited states.
5. It has total spin $S = 1$.
6. It has a dipole magnetic moment whose value in, nuclear magnetons, is

$$\mu_D = 0.8574 = \mu_p + \mu_n - 0.0222 \ ,$$

   where μ_p and μ_n are the dipole magnetic moments of the proton and of the neutron, respectively.
7. It has an electric quadrupole moment of Q=2.82 mb.

In selecting the two-nucleon interaction, the primary source of information is derived from scattering data. We only consider elastic scattering processes, where there is no exchange of energy between the nucleons that could alter their internal structure. Elastic processes are the only ones allowed below the pion production threshold, which is approximately 150 MeV. This threshold represents the maximum relative kinetic energy of the data used to determine the nucleon-nucleon interaction.

Experimental data have been collected from various laboratories. Significant efforts have been made to ensure that this dataset is homogeneous and coherent, and this work is ongoing. Currently, the database used to select the two-nucleon interactions consists of approximately 3,000 data points from proton-proton scattering and about 4,700 data points from proton-neutron scattering.

The elastic cross sections are decomposed using a partial wave expansion. Each partial wave depends on the relative angular momentum of the interacting nucleon pair. From these expanded cross sections, it is possible to extract the phase shifts. The relationship between the total cross section σ and the phase shifts is given by [4]

$$\sigma = \frac{4\pi}{k^2} \sum_{L=0}^{\infty} (2L+1) \sin^2 \delta_L \ ,$$

where k is the wave number of the interacting particle, L is the relative angular momentum of the two-interacting nucleons and δ_L is the related phase shift. Currently, nuclear potentials that can describe these data with a χ^2 per datum of approximately 1 are referred to as *realistic*.

From the analysis of the two-nucleon data, it is possible to determine some general characteristics of the nucleon-nucleon potential.

- *Attraction.*
  Nuclei are bound systems of nucleons, which means that the nucleon-nucleon potential includes an attractive component that is capable of holding the system together.
- *Short-range.* There are several observations suggesting that the two-nucleon interaction remains significant up to distances on the order of 2 fm. Beyond this range, the interaction becomes negligible. This is a key distinction compared to gravitational and electromagnetic interactions, which have infinite interaction ranges.
- *Dependence on spin and isospin.*
  The only bound state of two nucleons is the deuteron, which consists of one proton and one neutron with a total spin of 1. This observation suggests that the nuclear force depends on the orientation of the spins. If the force were independent of spin, we would expect to find deuterons with spin 0 in a ratio of 1/3 compared to those with spin 1. Similarly, if the interaction were independent of isospin, we would also observe bound systems composed of two protons or two neutrons.
- *Non centrality.*
  The deuteron possesses a non-zero electric quadrupole moment, which can only occur if the charge distribution lacks spherical symmetry. When we consider a purely central potential, the ground state of the two-particle system exhibits zero angular momentum, indicating a spherical symmetry within the system. A deformation of this symmetry is possible only if the interaction includes a term that behaves differently based on the relative orientation of the spins of the two nucleons, even when the distance between them remains constant. This is illustrated in Fig. 3.1. The easiest expression of terms of this kind, called tensor terms, is of the type $(\sigma_1 \cdot \mathbf{r})(\sigma_2 \cdot \mathbf{r})$, where σ indicates the nucleon spin and $\mathbf{r}$ is the vector joining the two nucleons.
- *Repulsive core.*
  The analysis of the elastic scattering phase shifts indicates that there is a sign change as the scattering energy increases. This suggests that by enhancing the resolution of the probe, it becomes possible to observe that the potential at short relative distances, specifically those less than 0.5 fm, becomes strongly repulsive.

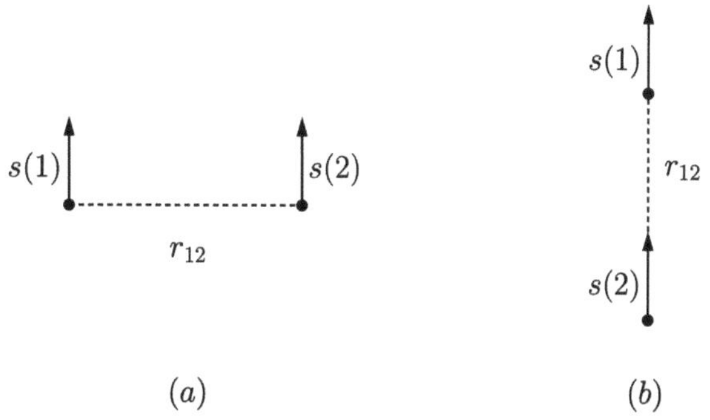

Fig. 3.1 Representation of the tensor term in the two-nucleon interaction. This term is influenced by the relative orientation of the spins of the two nucleons as well as their positions. In both scenarios, the distance between the two nucleons remains constant. However, in case a, the tensor term is zero, while in case b, it attains its maximum value

The modelling of the nucleon-nucleon interaction must consider the following features related to general symmetry properties.

1. *Hermiticity.*
   Globally, the Hamiltonian is hermitian because its eigenvalues, which represent the energies of the system, are observable quantities. Since the kinetic energy term is hermitian, it follows that the potential energy term must also be hermitian.
2. *Scalar expression*
   The hamiltonian is a scalar operator, therefore also the potential term must be a scalar operator.
3. *Invariance for coordinate exchange*
   $\hat{V}(1, 2) = \hat{V}(2, 1)$.
4. *Translational invariance*
   $\hat{V}(\mathbf{r}_1, \mathbf{r}_2) = \hat{V}(|\mathbf{r}_1 - \mathbf{r}_2|) \equiv \hat{V}(r_{12})$.
   The interaction depends only on the distance between the two particles.
5. *Galilean invariance*
   The interaction depends only on the relative momentum between the two particles and it is independent of the motion of the observer.
6. *Invariance under space inversions*
   $\hat{V}(\mathbf{r}, \mathbf{p}) = \hat{V}(-\mathbf{r}, -\mathbf{p})$.
   The strong interaction conserve parity, contrary to the weak interaction.
7. *Time reversal invariance*
   $\hat{V}(\mathbf{p}, \mathbf{S}) = \hat{V}(-\mathbf{p}, -\mathbf{S})$,
   where $\mathbf{S} = \sigma_1 + \sigma_2$ is the total spin of the nuclear pair.
8. *Invariance under space rotations*
   The total angular momentum of the system is conserved.
9. *Invariance under isospin rotations*
   This property indicates that the nucleon-nucleon interaction is the same for proton-proton, neutron-neutron, and proton-neutron scattering, provided that the final states of these processes are comparable. This assertion stems from the fact that nucleons, being fermions, are subject to the Pauli exclusion principle. This principle requires that the overall wave function describing a system of fermions

must be antisymmetric when two of them are exchanged. Since protons and neutrons are different particles, or, in the isospin formalism, possess different third components of isospin, they can access all possible final states during collisions. In contrast, nucleons of the same type can only access specific combinations of partial waves and spin couplings. Modern experiments have revealed small violations of this isospin symmetry, and the updated potential includes terms that account for these effects.

The process of constructing a nucleon-nucleon potential involves assuming a functional form for the potential that includes various parameters. It is essential that this functional form meets the specified properties. The parameter values are selected to best match the experimental data. Below, we present some specific approaches used to construct the two-body nuclear interaction.

Phenomenological potentials. The purpose of constructing these potentials is to utilise them for nuclear structure calculations. There is no intention to explore the underlying physics that determines their characteristics; rather, the focus is solely on fitting the experimental data.

The most straightforward approach involves adopting a potential expression that is a sum of terms dependent on operators that adhere to the symmetries of the Hamiltonian. If we disregard terms that depend on derivatives of the position, the most commonly used expression is:

$$\hat{V}(i,j) = \sum_{p=1,18} v_p(r_{ij})\hat{O}^p_{ij}, \tag{3.2}$$

where $v_p(r_{ij})$ are scalar functions whose dependence on r_{ij}, the distance between the two interacting nucleons, is expressed in terms of analytic functions containing free parameters whose values are chosen to reproduce the experimental data.

In this representation, the so-called central terms are

$$\hat{O}^{p=1,4} = \hat{\mathbb{I}}, \quad \hat{\tau}_i \cdot \hat{\tau}_j, \quad \hat{\sigma}_i \cdot \hat{\sigma}_j, \quad \hat{\sigma}_i \cdot \hat{\sigma}_j \hat{\tau}_i \cdot \hat{\tau}_j \ , \tag{3.3}$$

and the tensor operators are

$$\hat{O}^{p=5,6} = \hat{S}_{ij}, \quad \hat{S}_{ij}\hat{\tau}_i \cdot \hat{\tau}_j \ , \tag{3.4}$$

where the tensor term is defined as

$$\hat{S}_{ij} = 3\frac{\hat{\sigma}_i \cdot \mathbf{r}_{ij}\hat{\sigma}_j \cdot \mathbf{r}_{ij}}{\mathbf{r}^2_{ij}} - \hat{\sigma}_i \cdot \hat{\sigma}_j \ . \tag{3.5}$$

Obviously, the tensor part is represented by the first term of this expression, the second one is added in order to set equal to zero the angular integral of S_{ij}. In the potential also spin-orbit terms are considered

$$\hat{O}^{p=7,8} = \hat{\mathbf{L}} \cdot \hat{\mathbf{S}}, \ \hat{\mathbf{L}} \cdot \hat{\mathbf{S}}\, \hat{\tau}_i \cdot \hat{\tau}_j \ , \tag{3.6}$$

where $\mathbf{L}$ is the relative angular momentum of the nucleon pair. Also quadratic terms of the angular momentum are considered

$$\hat{O}^{p=9,14} = \hat{\mathbf{L}}^2 \, , \ \hat{\mathbf{L}}^2\hat{\tau}_i \cdot \hat{\tau}_j \, , \ \hat{\mathbf{L}}^2\hat{\sigma}_i \cdot \hat{\sigma}_j, \, , \ \hat{\mathbf{L}}^2\hat{\sigma}_i \cdot \hat{\sigma}_j\hat{\tau}_i \cdot \hat{\tau}_j \ . \tag{3.7}$$

The terms violating the rotational isospin invariance are

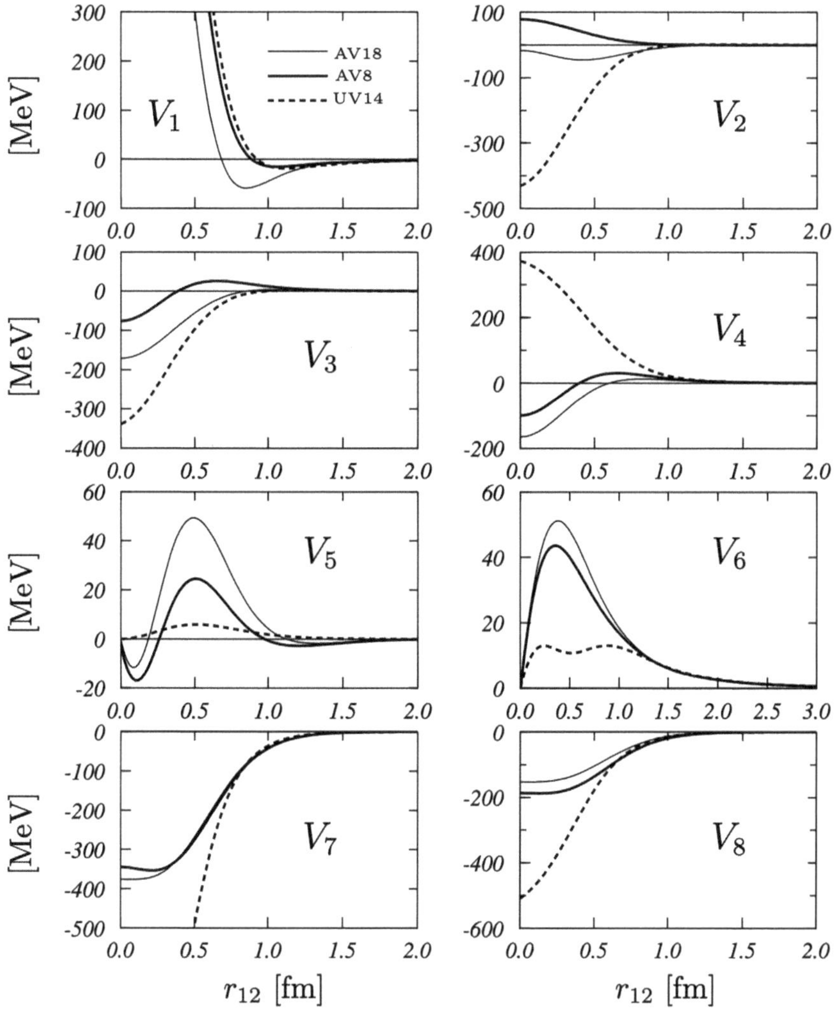

Fig. 3.2 The first 8 terms defined in (3.2) for the Urbana V14 (UV14), Argonne V8* (AV8), and Argonne V18 (AV18) potentials

$$\hat{O}^{p=15,18} = \left[\hat{1},\ \hat{\sigma}_i \cdot \hat{\sigma}_j\ \hat{S}_{ij}\right] \otimes \left[\hat{\tau}_{z,i} + \hat{\tau}_{z,j}\right]. \tag{3.8}$$

This is the typical structure of potentials known as Urbana and Argonne types. The $v_p(r_{ij})$ are designed to encompass three distinct interaction ranges: a short range, up to about 0.5 fm; a medium range, up to about 1 fm; and a long range, up to about 2 fm. Each of these components has a specific functional form, such as a sum of Yukawa functions (as discussed below), and includes parameters whose values are selected to fit the experimental data.

In Fig. 3.2, we present the first eight terms for the Urbana U14 [5], Argonne V18 [6], and Argonne V8* [7] potentials as a function of the two-nucleon distance. Two comments are

Table 3.1 Mesons used in the construction of the nucleon-nucleon potential

Type	J^π	Meson	Mass [MeV]
Pseudoscalar	0^-	π	139.578
		η	548
		η'	958
Vector	1^-	ρ	765
		ω	783
		ϕ	1019
Scalar	0^+	σ	500

warranted. First, although the number of data points to be reproduced is substantial and all the potentials exhibit similar performance during the fitting procedure, with $\chi^2 \sim 1$ per datum, the functions themselves are quite different. This suggests that the data do not impose sufficiently stringent conditions to definitively determine the potential. The data are correlated and can be reproduced through suitable combinations of the various terms. The second observation is that all the parameterisations in the scalar channel V_1 display a strongly repulsive core at short distances, an attractive pocket at intermediate distances, and rapidly approach zero at around 2 fm.

Meson exchange potentials.

A more ambitious approach to constructing the nucleon-nucleon potential builds on the Yukawa concept and shapes the potential in terms of meson exchange. The key idea is that the range of the interaction is determined by the mass of the meson exchanged between the two nucleons. The motion of the meson exchanged is a Yukawa function and it has the expression

$$\hat{V}(r) = \frac{1}{r} \exp\left(-\frac{m}{\hbar c} r\right) ,$$

where m is the mass of the exchanged meson and r the distance between the two interacting nucleons. Evidently, $\hat{V}(r)$ approaches zero more rapidly as the mass of the exchanged meson increases.

The construction of realistic nucleon-nucleon potentials of this type was proposed in the early 1980s [8, 9]. Table 3.1 presents the characteristics of the mesons considered in the development of these potentials. In addition to their mass, each meson is characterised by its spin and intrinsic parity. These properties determine the type of coupling with the nucleon. While the mass defines the interaction ranges, the type of meson-nucleon coupling selects the operator dependence, which in phenomenological potentials is described by the O^p operators.

The goal of this research project is to identify the mesons responsible for the various components of the nucleon-nucleon interaction, both in terms of interaction range and operator dependence. For instance, the pion, or π meson, is the lightest meson and is responsible for the long-range part of the interaction. Furthermore, its pseudoscalar coupling is related, in a non-relativistic reduction, to the tensor-isospin channel $O^6_{ij} = S_{ij}, \tau_i \cdot \tau_j$ of the interaction. Indeed, as shown in Fig. 3.2, the term with the largest interaction range is precisely the one dependent on tensor-isospin, V_6.

This ambitious project must contend with the limitations of our knowledge in meson physics. We have well-defined information regarding the mass, type of nucleon-meson coupling, and the value of the coupling constant only for the π meson, thanks to the extensive pion-nucleon scattering data available. In contrast, the situation for other mesons is less clear. The short lifetimes of these mesons prevent us from conducting scattering experiments. While we do have information on their masses, spin, and parity, which indicates the type of coupling, we lack knowledge of their coupling constants. In constructing the potential, these experimentally unknown coupling constants are treated as free parameters, with their values adjusted to fit the nucleon-nucleon data. The number of free parameters used is nearly the same as that in the phenomenological approach. With this number of free parameters, it is possible to achieve fits to the experimental data with an accuracy comparable to that obtained in the phenomenological approach.

At this point, it is important to emphasise that, in order to accurately reproduce the two-nucleon experimental data, it is necessary to include a hypothetical meson, the σ meson, which has never been experimentally identified. This meson has a scalar coupling type and a mass of precisely 500 MeV. The σ meson accounts for the attractive pocket of the potential at distances of about 1 fm. The necessity of including this hypothetical meson highlights a limitation of this ambitious approach. In reality, the model has been constructed by considering the exchange of a single meson between the two interacting nucleons. However, calculations that also take into account the exchange of two pions suggest that the fictitious σ meson partially simulates the effects of exchanging two π mesons in a relative s wave.

This does not imply that meson exchange potentials lack differences and advantages compared to phenomenological potentials. Their formulations are fully relativistic, making them invariant under Lorentz transformations. This means that by applying charge conjugation operators, it is possible to derive potentials for nucleon-antinucleon and two antinucleon systems. Comparisons with the relatively limited experimental data for these systems further validate the success of this approach.

The use of these potentials in standard nuclear structure calculations is not straightforward. Typically, these calculations require potentials that are non-relativistic and local, meaning they do not include terms related to the derivatives of particle positions. In contrast, meson exchange potentials are relativistic and are usually expressed in momentum space, featuring an explicit dependence on the momentum of each particle; in other words, they are non-local. Before employing these potentials in nuclear structure calculations, a non-relativistic reduction is necessary, along with the elimination of the non-local terms.

Potentials from chiral effective field theories.

The most modern and ambitious method for modeling the two-nucleon potential is derived from the so-called chiral effective field theories [10, 11]. The idea is to construct an effective theory of QCD that is valid at energies on the order of the MeV and respects its symmetries, particularly chiral symmetry. The interaction is formulated by considering pion exchanges at different orders, with each order of expansion incorporating contact terms to ensure compliance with chiral symmetry. Modern potentials achieve a fit to the two-nucleon data with $\chi^2 \simeq 1$ by including terms up to the third perturbative order. Notably, the number of free parameters required to achieve this result is approximately half that used in other approaches. The advantage of chiral potentials over phenomenological and meson-exchange potentials lies not only in the reduced number of parameters but also in the fact that many-body interaction terms, such as three-body interactions (see next section), naturally and consistently emerge from the theory without the need to be added individually.

3.3.2 Three-Body Forces

The fundamental assumption underlying the construction of the nucleon-nucleon potential is that nucleons are considered to have no internal structure. This is clearly an approximation, but it is expected to hold true within the energy range relevant to nuclear physics. The data used to determine the two-nucleon potential primarily come from elastic scattering experiments and observations related to the deuteron.

Since two-nucleon systems are used to construct the interaction, the simplest system in which we can test the predictive power of the potential is the triton. The ^{3}H nucleus consists of one proton and two neutrons. In this system, the electromagnetic interaction is absent, making it an ideal scenario for evaluating the quality of nucleon-nucleon potentials.

The techniques for solving the Schrödinger equations for three- and four-body systems have been standardised in the same manner as those for the two-body case. It is possible to address these problems without relying on approximations. Recently, binding energies of ^{4}He obtained using seven different methods were compared [12]. With the same nucleon-nucleon potential, all results align within the numerical accuracy of the calculations. This clearly demonstrates that the solutions to these few-body problems, when approached without approximation, are well-controlled.

Table 3.2 presents the binding energies of ^{3}H obtained using various nucleon-nucleon potentials, all of which describe the two-nucleon data with the same level of accuracy. The differences in the results arise solely from the choice of potential, as the Schrödinger equation is solved without any approximations. These results confirm the earlier assertion that fitting the two-nucleon data is not sufficiently restrictive for constructing the potential. The main and more general conclusion is that none of the two-body potentials can accurately reproduce the experimental value of the triton binding energy.

We can eliminate the possibility that the discrepancy arises from the challenges associated with solving the three-body problem as various solution methods produce consistent results. Additionally, we can rule out the choice of the two-body potential

Table 3.2 Binding energies of ^{3}H, in MeV, calculated with different two-nucleon potentials and then obtained by inserting a proper three-body force. The experimental value of the ^{3}H binding energy is 8.481 MeV

Potential	2N	2N+3N
CD Bonn	7.953	8.483
Nijm II	7.709	8.477
Nijm I	7.731	8.480
Nijm 93	7.664	8.480
Reid 93	7.648	8.480
AV14	7.683	8.480
AV18	7.567	8.479

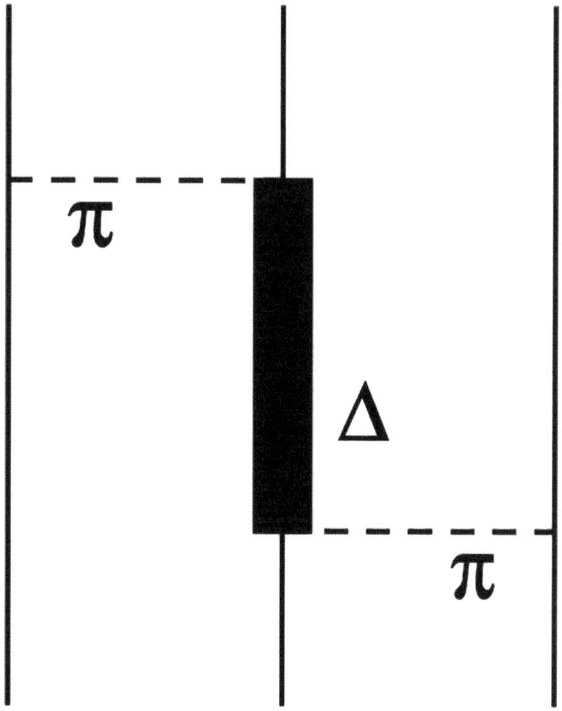

Fig. 3.3 Fujita-Miyazawa three-body interaction

as a contributing factor. As demonstrated in Table 3.2, the deviation from the empirical binding energy value remains significant, regardless of the two-body potential used. Therefore, the source of this discrepancy must be attributed to the fundamental assumptions of the theory, particularly the assumption that nucleons lack internal structure.

In Fig. 3.3, we present an example of a phenomenon not accounted for in the two-nucleon potential. The diagram illustrates a system of three nucleons, represented by the three thin vertical lines. The nucleon in the center interacts with the nucleon to its right by exchanging a π meson, resulting in the formation of a Δ resonance. While in the Δ state, the baryon in the middle interacts with the nucleon to its left and transforms back into a nucleon. Although the initial and final states of this process involve only nucleons, there is a transformation of one nucleon into another baryon. This transformation is possible only because nucleons possess an internal structure that can be altered through interaction. Effects of nucleons changing their structure and interacting with other nucleons while in this altered state, before returning to their original nucleonic form, are also present in two-body interactions. These phenomena are automatically considered during the fitting procedure. However, the process depicted in Fig. 3.3 is not included in the construction of the two-body potential because it involves three particles and must be treated separately.

With the insertion of a three-body term, the hamiltonian becomes

$$\hat{H} = \sum_{i=1}^{A} \frac{-\hbar^2}{2m_i}\nabla_i^2 + \sum_{i<j}^{A} \hat{V}(i,j) + \sum_{i<j<k}^{A} \hat{W}(i,j,k) \ , \tag{3.9}$$

where the first term represents the kinetic energy, the second one is the two-body potential, selected to fit the two-body data, and the third one is the three-body potential. This latter term contains the contribution of the process described in Fig. 3.3, called Fujita-Miyazawa, and, in addition, phenomenological terms whose parameters are selected to reproduce the experimental value of the triton binding energy. As pointed out by the results of Table 3.2, a specific three-body force is associated to each two-body potential.

Since the Hamiltonian that includes only two-body interaction terms is unable to accurately reproduce the simplest three-body system, necessitating the introduction of an *ad hoc* three-body interaction, it is reasonable to question whether this approach should be extended further. For more complex systems, do we need to incorporate four-, five-, or even A-body terms into the Hamiltonian? Fortunately, the Hamiltonian (3.9) is capable of effectively describing nuclear systems that contain more than three nucleons, as we will discuss in Chap. 4.

The significance of many-body forces increases with the likelihood of finding a certain number of particles close enough to interact. It is important to remember that nuclear interactions are short-ranged, meaning that particles must be within a specific range to engage in interactions. In the case of ordinary nuclei, four-body forces are negligible compared to the strength of two- and three-body forces at typical nuclear densities.

3.4 Liquids and Strongly Interacting Gases

The fundamental interaction that forms liquids or strongly interacting gases is the electromagnetic interaction, which is well described by quantum electrodynamics (QED). The QED equations are solved with high accuracy using perturbation techniques. In principle, it is possible to use QED equations to describe the interactions between atoms or molecules by considering the electromagnetic forces among their fundamental components, such as electrons and nuclei. While this approach is theoretically feasible, it is practically very challenging and might be too ambitious if the goal is to accurately capture the properties of systems with many atoms or molecules.

A more pragmatic approach employs a strategy similar to that used to define the nucleon-nucleon interaction. The fundamental interaction used to describe liquids and strongly interacting gases is derived by neglecting the internal structure of what are considered the basic degrees of freedom, in this case, atoms or molecules. The focus is on the interaction between two atoms or two molecules, which differs significantly from the basic Coulomb interaction that generates them.

In condensed matter physics, the forces acting among atoms are classified as (a) valence binding, (b) ionic binding, (c) metallic binding, and (d) van der Waals binding. The forces in categories (a) and (b), which are present in materials such as diamonds and NaCl, are so strong that they lead to the formation of solid structures at room temperature, which we will not discuss here. The metallic bond gives rise to the electron gas, as presented in Sect. 3.1. Our focus will be on van der Waals forces, which are responsible for the fluid state of certain substances at room temperature. These interactions are also significant in quantum liquids, such as superfluid helium, which can exist in both bosonic form (composed of ^{4}He nuclei) and fermionic form (composed of ^{3}He nuclei).

The system we consider consists of well-structured molecules that do not share electrons with other molecules. When the molecules are composed of different types of atoms or atoms with a significant asymmetry in their electron clouds, polarisation of positive and negative charges occurs in space. These molecules possess permanent electric multipoles, such as dipoles and quadrupoles. The van der Waals interaction between molecules can be understood as the interaction between these various electric multipoles.

Even in cases where the centers of mass of positive and negative charges coincide, typically occurring in molecules composed of identical atoms, the interaction can still be described in terms of electric multipoles. This is because the approach of two molecules induces a distortion, or polarisation, of their electron clouds.

The interaction between two atoms or two molecules is negligible at distances much larger than their atomic or molecular dimensions. This is because non-ionised atoms and molecules are neutral. At large distances, the centers of their positive and negative charge distributions appear to overlap. This characteristic classifies these interactions as finite-range interactions, similar to the nucleon-nucleon interaction. At intermediate distances, the interaction can be approximated as that between two electric dipoles, which is primarily attractive. However, at short distances, the electron clouds begin to overlap, leading to repulsion. Figure 3.4 illustrates the potential between two neon atoms. Its behaviour is analogous to that of the scalar part of the nucleon-nucleon potential: it is zero at large distances, attractive at intermediate distances, and exhibits a strong repulsive core at short distances. The approach used to derive this potential involves an analytic expression with free parameters, the values of which are selected to reproduce empirical data describing systems of two atoms or two molecules.

One of the the most used potentials is the that of **Lennard-Jones** [13] which is usually expressed as

$$V(r) = 4\epsilon \left\{ \left(\frac{\sigma}{r}\right)^{12} - \left(\frac{\sigma}{r}\right)^{6} \right\} , \tag{3.10}$$

where r is the distance between the two interacting particles wihch are assumed to be point-like, and ϵ and σ are two free parameters.

The interaction between two molecules is much simpler than that between two nucleons: it is purely scalar, meaning it is independent of the spins of the two molecules and their relative angular momenta. The expression (3.10) is remarkably straightforward, yet it encompasses all the essential physics needed to describe these many-body systems.

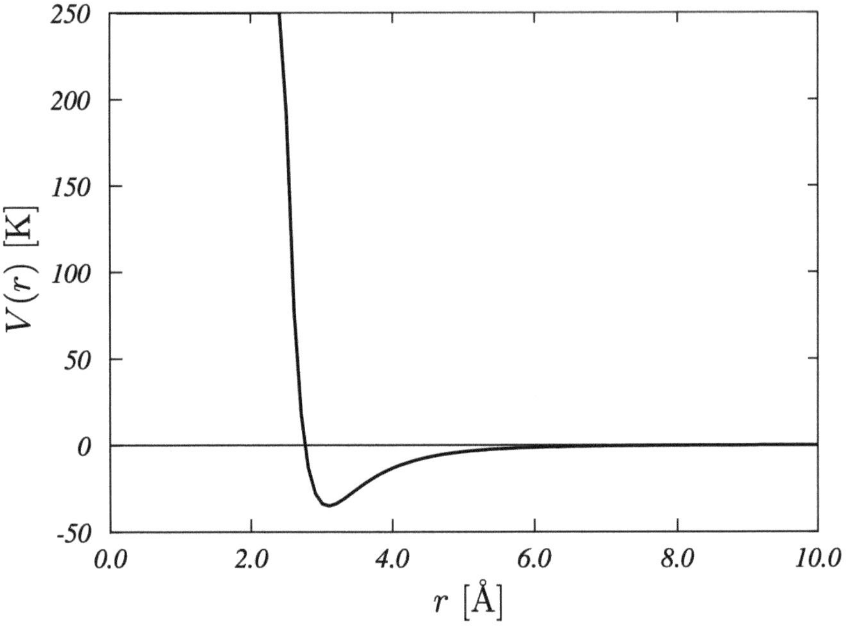

Fig. 3.4 Lennard-Jones potential between two Ne atoms

Table 3.3 Parameters of the Lennard-Jones potential for some atoms and molecules, from [14]

	Ne	Ar	Kr	Xe	N_2	CH
σ (Å)	2.75	3.405	3.60	4.10	3.70	3.82
ϵ (K)	35.6	119.8	171	221	95.1	148.2

In the nuclear case, the interaction chosen, as indicated in Sect. 3.3, is used to describe every nucleus. In contrast, for fluids, it is necessary to specify the interaction parameters for each type of atom or molecule that makes up the system. For example, Table 3.3 presents the values of the parameters selected for several molecules.

So far, we have considered a Hamiltonian that includes only two-body interaction terms. This description of the interaction holds as long as the assumption that the internal structure of the interacting particles can be neglected remains valid. However, we have demonstrated that in the case of nucleons, this assumption is not particularly reliable, necessitating the inclusion of a three-body force in the nuclear Hamiltonian. The assumption that Hamiltonians in molecular physics require only two-body interaction terms is commonly referred to as **pairwise additivity**. The validity of this assumption is extensively analysed in statistical and molecular physics (see, for example, the discussion in Sect. 4.3.b of [15]).

To assess the significance of many-body forces, it is essential to consider two physical quantities. The first one is the excitation energy of the fundamental particle, whose internal structure is neglected. The second quantity is the relative density of the system.

The first quantity must be compared with the energy of the entire system. For instance, the empirical value of the binding energy of nuclear matter at the stability point is approximately 16 MeV per nucleon [16], while the first nucleonic resonance, the Δ, has a mass about 300 MeV greater than that of the nucleon. This indicates that the internal structure of nucleons becomes significant at energies roughly 18 times that of the system. In contrast, the typical energies of superfluid helium are around 0.2 meV [17], while the energy of the first excited state of a helium atom is about 20 eV [18]. This represents a difference of five orders of magnitude. By comparing these values, it becomes clear that the assumption of pairwise additivity is much more valid in liquid helium than in nuclear matter.

The second quantity to consider is the probability that the particles are close enough to interact, which is closely related to the density of the system. The particles must be within the interaction range of the force, which is short-ranged. An estimate of this probability can be obtained by determining how many particles reside within the volume defined by the repulsive short-range core of the interaction. Below, we calculate these numbers for both liquid helium and nuclear matter. The results indicate that, at equilibrium density, there are 0.084 nucleons in the relevant volume, compared to 2.317 helium atoms in the case of liquid helium. From a relative density perspective, liquid helium is significantly denser than nuclear matter. Although the probability of finding three interacting helium atoms is certainly greater than that of finding three interacting nucleons, the energetics suggest that exciting helium atoms in liquid helium is much less probable than exciting a nucleon in nuclear matter. For this reason, three-body forces are rarely used in the description of liquid helium.

Relative densities

In the context of many-body theories, relative densities are crucial. These densities are associated with the number of particles present within the volume defined by the repulsive short-range core of the interaction.

1. *Nuclear matter.*

   In this case the range of the repulsive core is $\sigma \simeq 0.5$ fm [3]. The value of the reference volume is then

   $$\mathcal{V} = \frac{4}{3}\pi\sigma^3 = 0.524\,\text{fm}^3 \ ,$$

   The empirical value of the nuclear matter density at equilibrium is 0.16 nucleons per fm^3. The number of nucleons in the volume of interest is then

   $$N = \rho\mathcal{V} = (0.16)\frac{\text{nucleons}}{\text{fm}^3}(0.524)\,\text{fm}^3 = 0.084\,\text{nucleons} \ .$$

2. *Liquid Helium.*

   The mass density of liquid helium is $1.47 \cdot 10^5$ g m^{-3} [17]. The molecular weight is 4 g mole^{-1}, which is the weight of an Avogadro number of helium atoms. The number density of helium atoms in liquid helium is

$$\rho = \frac{1.47 \cdot 10^5 \mathrm{g}}{\mathrm{m}^3} \frac{6.2 \cdot 10^{23}}{4\mathrm{g}} = \frac{2.263 \cdot 10^{28} \text{ atoms}}{\mathrm{m}^3}.$$

In liquid helium the range of the repulsive core of the interaction i s σ=2.9 Å= 2.9 10^{-10} m. The volume of the sphere of radius σ is $V = 1.024 \cdot 10^{-28}$ m^3. The number of atoms in this volume is $(V\rho)$ = 2.317.

References

1. S. Dürr et al., Ab-initio determination of light hadron masses. Science **322**, 1224 (2008)
2. N. Ishii, S. Aoki, T. Hatsuda, The nuclear force from lattice QCD. Phys. Rev. Lett. **99**, 022001 (2007)
3. K. Krane, *Introductory Nuclear Physics* (Wiley, New York, 1988)
4. A. Messiah, *Quantum Mechanics* (North Holland, Amsterdam, 1961)
5. I. Lagaris, V.R. Pandharipande, Variational calculations of realistic models of nuclear matter. Nucl. Phys. A **359**, 331 (1981)
6. R.B. Wiringa, V.G.J. Stoks, R. Schiavilla, Accurate nucleon-nucleon potential with charge-independence breaking. Phys. Rev. C **51**, 38 (1995)
7. B.S. Pudliner, V.R. Pandharipande, J. Carlson, S.C. Pieper, R.B. Wiringa, Quantum Monte Carlo calculations of nuclei with A $<\sim$ 7. Phys. Rev. C **56**, 1720 (1997)
8. K. Holinde, Two-nucleon forces and nuclear matter. Phys. Rep. **68**, 121 (1981)
9. R. Machleidt, K. Holinde, Ch. Elster, The Bonn Meson-exchange model for the nucleon-nucleon interaction. Phys. Rep. **149**, 1 (1987)
10. S. Weinberg, Nuclear forces from chiral Lagrangians. Phys. Lett. B **251**, 288 (1990)
11. S. Weinberg, Effective chiral Lagrangians for Nucleon-Pion Interactions and Nuclear Forces. Nucl. Phys. B **363**, 3 (1991)
12. H. Kamada et al., Benchmark test calculation of a four-nucleon bound state. Phys. Rev. C **64**, 044001 (2001)
13. J. E. Jones: *On the Determination of molecular Fields -I From the Variation of the Viscosity of a Gas with Temperature*, Proc. R. Soc. London, Ser. A 106 (1924) 441
14. J.A. Pryde, *The liquid State* (Hutchinson, London, 1966)
15. D.L. Goodstein, *States of Matter* (Dover, New York, 1985)
16. O. Benhar, S. Fantoni, *Nuclear Matter Theory* (Taylor and Francis, New York, 2021)
17. J.R. Donnelly, C.F. Barenghi, *The observed Properties of liquid Helium at the saturated Vapor Pressure*, Jour. of Phys. and Chem. Reference Data 27 (1998) 1217
18. B.H. Brandsen, C.J. Joachain, *Physics of Atoms and Molecules* (Prentice Hall, Harlow-London, 2003)

Chapter 4
Solutions Without Approximations

4.1 Introduction

After establishing the theoretical framework of non-relativistic quantum mechanics, defining the degrees of freedom (the particles), and formulating the Hamiltonian (the interaction), the next step in describing the many-body system is to solve the Schrödinger equation. There are techniques available that can solve the Schrödinger equation without making approximations, specifically designed for two, three, and four-body systems. These methods are tailored to handle a specific number of particles, making it challenging to extend them to systems with larger particle counts.

In this chapter, we introduce a technique that solves the many-body Schrödinger equation without any approximations, and it is formulated independently of the number of particles involved. As we will discuss, the only limitations are practical in nature, pertaining to the technological constraints of computational resources.

The advancement of computer capabilities has significantly enhanced the ability to perform complex calculations that require substantial memory and computational time. These technological improvements have facilitated the use of Monte Carlo simulations. The term "Monte Carlo" refers to a method that leverages the computer's ability to generate sequences of random numbers. This feature is utilized in various ways, such as simulating the behaviour of complex systems, including the performance of particle detectors.

The application of Monte Carlo techniques in this context pertains to the ability to perform multidimensional integrals within reasonable execution times. We will first introduce the fundamental concepts behind numerical integration using Monte Carlo methods, followed by implementations related to many-body problems.

© The Author(s), under exclusive license to Springer Nature Switzerland AG 2026

G. Co', *Concepts in Quantum Many-Body Physics*, UNITEXT for Physics,
https://doi.org/10.1007/978-3-032-08920-5_4

4.2 Monte Carlo Numerical Integrations

If we neglect the presence of spin and, eventually, isospin, the many-body wave function $|\Psi\rangle$ is characterised by the $3A$ variables that specify the position of each particle. This implies that the calculation of the energy of the system

$$E = \frac{\langle\Psi|\hat{H}|\Psi\rangle}{\langle\Psi|\Psi\rangle} , \tag{4.1}$$

involves $3A$ dimensional integrals. If we consider a uniform grid of cartesian coordinates and denote the number of points in each dimension of the grid as N, the total number of $3A$-dimensional integrals to be calculated is $N^{3A-1} + 1$. To ensure numerical stability, the number of integration points typically used in standard integration techniques is on the order of 100. Even if we assume a very short execution time for each integral, say 10^{-6} seconds, it becomes clear that using these integration techniques to describe a many-body system presents significant challenges. For a system with $A = 4$, the total computation time would amount to 10^6 seconds, which is approximately four months. For $A = 5$, the required time would extend to about 400 years.

The computation of integrals using Monte Carlo techniques is not ideal for one- or two-dimensional integrals, as it typically requires significantly more computational time than traditional integration methods to achieve similar levels of numerical accuracy. However, Monte Carlo methods are the only techniques that allow for the estimation of multidimensional integrals within a reasonable timeframe.

Below, we present the basic concept of Monte Carlo integration using a one-dimensional example. The extension to higher dimensions is straightforward.

Any definite integral can be expressed as

$$\int_a^b f(u)\,du = (b-a)\int_0^1 f(x)\,dx , \tag{4.2}$$

where we used a new integration variable

$$x = \frac{u-a}{b-a} . \tag{4.3}$$

This indicates that the problem to be addressed involves an integration within the range from 0 to 1, where the computer generates random numbers (Fig. 4.1).

From a numerical perspective, the value of the integral is obtained by summing the values of f calculated for N random values of the integration variable x,

$$I = \int_0^1 f(x)\,dx \simeq \frac{1}{N}\sum_{i=1}^{N} f(x_i) . \tag{4.4}$$

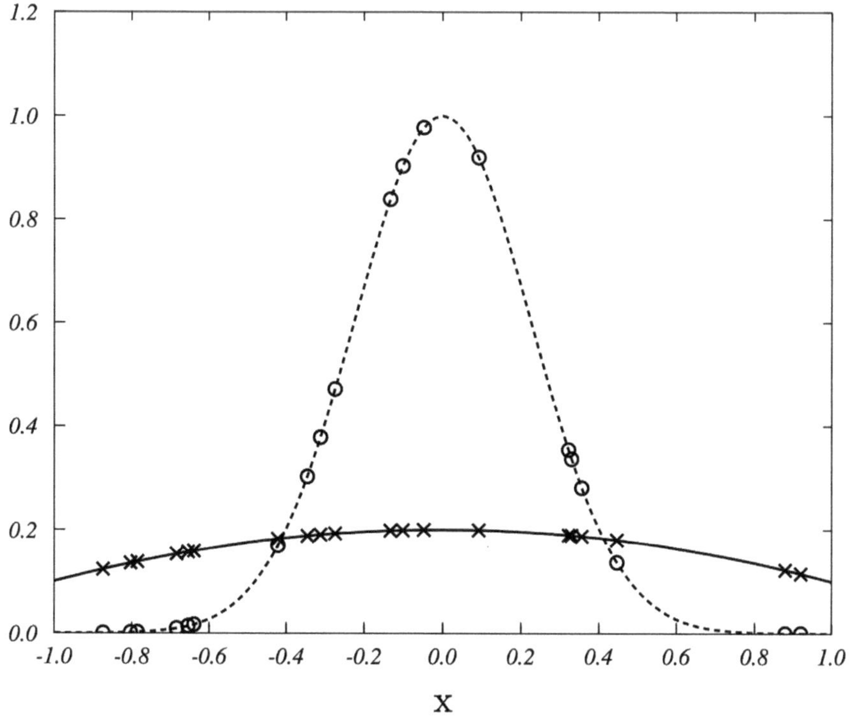

Fig. 4.1 Two functions to be integrated using the Monte Carlo method. The white circles and black crosses represent the points of the two functions, selected by the same random choice of the variable x

The estimate of I improves as the number N increases. The value of N required to obtain an accurate estimate of the integral depends on the behaviour of f within the integration domain. If f is relatively flat, only a few points will be sufficient. In the extreme case where f is constant, a single point is enough.

A helpful strategy for smoothing the behaviour of the function to be integrated is to multiply and divide it by a weight function P. This weight function is defined to be always positive within the integration interval and behaves like a probability density function

$$\int_0^1 P(x)\,dx = 1\,, \tag{4.5}$$

therefore

$$I = \int_0^1 f(x)\,dx = \int_0^1 \frac{f(x)}{P(x)} P(x)\,dx = \int_0^1 F(x)P(x)\,dx\,. \tag{4.6}$$

We define a new variable

$$y(x) = \int_0^x P(x')\,dx' \;;\quad \frac{dy(x)}{dx} = P(x) \;;\quad y(0) = 0 \;;\quad y(1) = 1 \;, \tag{4.7}$$

and then

$$I = \int_0^1 f(x)\,dx = \int_0^1 F(x)P(x)\,dx = \int_0^1 F(x(y))\,dy \simeq \frac{1}{N}\sum_{i=1}^{N} \frac{f(x(y_i))}{P(x(y_i))} \;. \tag{4.8}$$

The optimal choices for P are those that can maintain the ratio f/P nearly constant. A challenging aspect of the procedure is the inversion of the relationship in Eq. (4.7), which may not be straightforward.

At this point, a theorem, called **Theorem of the central limit** [1], describes the distribution of the approximated values of I. The theorem states that, for a function $F(x)$, the probability distribution $\mathcal{P}(I_N)$ of the values I_N, defined as

$$I_N = \frac{1}{N}\sum_{i=1}^{N} F(x_i) \;, \tag{4.9}$$

has a gaussian form

$$\lim_{N\to\infty} \mathcal{P}(I_N) = \frac{1}{\sqrt{2\pi\sigma_N^2}} \exp\left(\frac{(I_N - \langle F\rangle)^2}{2\sigma_N^2}\right) \;. \tag{4.10}$$

In the above expression, we used the quantities defined as

$$\langle F\rangle = \int_0^1 F(x)P(x)dx \;\;;$$
$$\langle F^2\rangle = \int_0^1 F^2(x)P(x)dx \;\;;$$
$$\sigma_N^2 = \frac{1}{N}\left(\langle F^2\rangle - \langle F\rangle^2\right) \;.$$

The theorem is based on the assumption that the values of x_i are independent of one another and have an equal probability of being selected. It indicates that, for sufficiently large N, the value of I follows a Gaussian distribution centered around the expected value. The width of this distribution is denoted by σ_N and scales as $N^{-1/2}$. Therefore,

$$I \simeq \frac{1}{N}\sum_{i=1}^{N} \frac{f(x(y_i))}{P(x(y_i))} \pm \sigma_N . \tag{4.11}$$

This technique can be extended to multidimensional integrals. The Gaussian uncertainty remains independent of the number of dimensions. The goal is to apply this technique to evaluate the energy given in equation (4.1), or, more generally, to calculate the expectation value of a generic operator $\hat{\mathcal{O}}$ between two many-body states.

4.3 Variational Monte Carlo

A first application of Monte Carlo techniques, known as Variational Monte Carlo (VMC), is used to evaluate the ground state energy of a many-body system based on the variational principle.

Searching for the minimum of the energy functional (4.1) within the complete Hilbert space, which is formed by the many-body wave function $|\Psi\rangle$, is equivalent to solving the Schrödinger equation (see Appendix A). In practical applications, the search for the energy minimum is typically conducted by considering only wave functions with a specific functional form. This limitation confines the search to a subspace of the Hilbert space. As a result, the solution to the Schröinger equation is often approximated, leading to an obtained energy value that is always greater than the true eigenvalue of the Hamiltonian $\hat{H}$ obtained without approximations.

In the case of the VMC calculations, the commonly used expression of the trial wave function is

$$|\Psi_T\rangle = F|\Phi\rangle, \tag{4.12}$$

where $|\Phi\rangle$ is a Slater determinant, and F is called correlation function. This many-body correlation function is commonly expressed by using the Jastrow's ansatz [2], i.e. as a product of two-body correlation functions f

$$F = \prod_{i<j}^{A} f(r_{ij}) \ , \tag{4.13}$$

where r_{ij} is the distance between two particles.

In Chap. 3, we highlighted that a common characteristic of interactions in many-body systems is the presence of a strongly repulsive core at short distances between interacting particles. The purpose of the correlation function f is to prevent two particles from coming too close, within the range of the repulsive core. Consequently, f is nearly zero at short relative distances and approaches 1 at distances greater than the interaction range.

The complexity of the nucleon-nucleon interaction, requires, in nuclear physics, the use of correlations depending on various operator terms, in analogy with the expression of the interaction presented in Sect. 3.3.1

$$\hat{F} = \mathcal{S} \prod_{i<j}^{A} \left(\sum_{p} f^p(r_{ij}) \hat{O}_{ij}^p \right) \ , \tag{4.14}$$

where $\mathcal{S}$ is an operator which makes symmetric $\hat{F}$ for the exchange of two fermions, i.e. the two indexes i, j. This is necessary since the Slater determinant Φ is already antisymmetrized for two-fermions exchange, and, in general, the $\hat{O}^p$ operators do not commute with each other. The explicit expressions of the operators are those presented in Sect. 3.3.1.

The wave functions used in VMC calculations are expressed as:

$$|\Psi(\mathbf{r}), S, T\rangle \equiv \sum_{s=1,2^A} \sum_{t=1,2^A} \mathcal{R}_{s,t}(R)\mathcal{X}_s(S)\mathcal{X}_t(T) \ , \tag{4.15}$$

where $\mathcal{R}$ represents the radial component of the wave function, R denotes the set of $3A$ spatial coordinates that describe all the fermions in the system, and $\mathcal{X}_s$ and $\mathcal{X}_t$ correspond to the components of the wave function that account for the spin S and, if applicable, the isospin T terms. This latter part of the wave function encompasses all possible combinations of spin and isospin for the system composed of A fermions. For a nucleus containing Z protons and $A - Z$ neutrons, the total number of these combinations is given by

$$N_{\text{conf}} = 2^A \frac{A!}{Z!(A-Z)!} \ . \tag{4.16}$$

The values of N_{conf} for some nuclei of interest are given in Table 4.1.

The calculation of the energy does not present formal problems and consists in evaluating the expectation value of the hamiltonian

$$E = \langle \hat{H} \rangle = \frac{\langle \Psi_T | \hat{H} | \Psi_T \rangle}{\langle \Psi_T | \Psi_T \rangle} \ . \tag{4.17}$$

The variational principle is utilized by seeking the minimum of this energy functional. The components of the many-body wave function that can vary include the single-

Table 4.1 Number of spin and isospin configurations for some nuclei

Nucleo	Z	N=A-Z	N_{conf}
^{3}H	1	2	24
^{3}He	2	1	24
^{4}He	2	2	96
^{6}He	2	4	960
^{6}Li	3	3	1280
^{8}He	2	6	7168
^{12}C	6	6	3784704
^{16}O	8	8	$8.4 \cdot 10^8$
^{40}Ca	20	20	$1.5 \cdot 10^{23}$
^{48}Ca	20	28	$4.7 \cdot 10^{27}$

particle wave functions ϕ_i that make up the Slater determinant, as well as the scalar functions f related to the correlation.

In the calculation of energy (4.17), or more generally, the expectation value of any operator $\hat{\mathbb{O}}$, the modulus of the wave function $|\Psi|^2$ serves as the weight function $P(x)$ introduced in the previous section. The wave function Ψ of a fermionic system is antisymmetric with respect to the exchange of two particles, meaning that exchanging two particles results in a change of sign for the wave function. Since Ψ is a continuous function of the spatial coordinates, there exist values of $\mathbf{R}$ for which the wave function is zero, and consequently, its square modulus is also zero. Clearly, these points, where $P(x)$ is zero, pose challenges in the calculation of the integral (4.11). This issue, known in the literature as the sign problem [3], is addressed using approximate techniques. This approach involves making certain assumptions in the calculation of fermionic systems that are not present in the bosonic case.

4.4 Green Function Monte Carlo (GFMC)

The theoretical limitations of the Variational Monte Carlo (VMC) calculations stem from the necessity of selecting specific forms for the trial wave function. In the case discussed above, the expression (4.12) presents a trial wave function constructed from a single Slater determinant, which is further multiplied by a correlation function that has its own unique form. In contrast, the Green's Function Monte Carlo (GFMC) technique addresses the solution of the Schrödinger equation without imposing any assumptions about the structure of the many-body wave function. This method is independent of any parameters or assumptions regarding the wave function's form; it relies solely on the Hamiltonian, which describes the interactions involved.

A trial wave function can be expressed as linear combination of eigenstates $|\Psi_n\rangle$ of the hamitonian $\hat{H}$

$$|\Psi_{\mathrm{T}}\rangle = \sum_n D_n |\Psi_n\rangle \ , \tag{4.18}$$

wehre the D_n coefficients are numbers. We construct an operator that, when applied to the trial wave function, isolates the state $|\Psi_n\rangle$ with the smallest energy eigenvalue. This operator depends on an evolution parameter τ, which has the dimensions of imaginary time. We apply this operator to the trial wave function and then take the asymptotic limit as τ approaches infinity. We obtain

$$\lim_{\tau\to\infty} e^{-\frac{\hat{H}-E_0}{\hbar}\tau} |\Psi_{\mathrm{T}}\rangle = \lim_{\tau\to\infty} e^{-\frac{\hat{H}-E_0}{\hbar}\tau} \sum_n D_n |\Psi_n\rangle$$

$$= \lim_{\tau\to\infty} \sum_n e^{-\frac{E_n-E_0}{\hbar}\tau} D_n |\Psi_n\rangle = D_0 |\Psi_0\rangle \ , \tag{4.19}$$

because for $n \neq 0$ $E_n > E_0$, where E_0 is the smallest eigenvalue. If the set of quantum numbers characterising the states $|\Psi_n\rangle$ is the same as those describing the ground state of the system, then E_0 represents the ground state energy.

We can express the evolution of a state in coordinates representation $|R\rangle$ for small increments of the imaginary time as

$$\Psi(R, \tau + d\tau) \equiv \langle R|\Psi(\tau + d\tau)\rangle = \langle R|e^{-(H-E_0)\frac{d\tau}{\hbar}}|\Psi(\tau)\rangle$$
$$= \int dR' \underbrace{\langle R|e^{-(H-E_0)\frac{d\tau}{\hbar}}|R'\rangle}_{\text{propagator}}\langle R'|\Psi(\tau)\rangle \ , \tag{4.20}$$

where we used the completeness relation

$$\int dR'|R'\rangle\langle R'| = 1 \ .$$

We consider the hamiltonian $\hat{H} = \hat{T} + \hat{V}$, and we make the assumption that the potential $\hat{V}$ is diagonal in R, in other words that the potential is local. This means that the operator terms forming the potential do not contain derivatives with respect to R. This is one of the essential requirements of the Monte Carlo calculations.

In order to simplify the calculations it is convenient to use an approximated expression, named Trotter-Suzuki formula, of the evolution operator [4, 5].

$$e^{-(\hat{T}+\hat{V}-E_0)\frac{d\tau}{\hbar}} = e^{-\frac{1}{2}(\hat{V}-E_0)\frac{d\tau}{\hbar}}e^{-\hat{T}\frac{d\tau}{\hbar}}e^{-\frac{1}{2}(\hat{V}-E_0)\frac{d\tau}{\hbar}} + O(d\tau^3) \ , \tag{4.21}$$

which is valid when

$$|d\tau^3(\hat{T}\hat{V} - \hat{V}\hat{T})| \ll 1 \ . \tag{4.22}$$

By using the Trotter-Suzuki formula, and neglecting the terms in $d\tau^3$, we obtain

$$\Psi(R, \tau + d\tau) \simeq$$
$$\int \langle R|e^{-\frac{1}{2}(\hat{V}-E_0)\frac{d\tau}{\hbar}}|R\rangle\langle R|e^{-\hat{T}\frac{d\tau}{\hbar}}|R'\rangle\langle R'|e^{-\frac{1}{2}(\hat{V}-E_0)\frac{d\tau}{\hbar}}|R'\rangle\Psi(R', \tau)dR'$$
$$= \int e^{-\left[\frac{1}{2}(\hat{V}(\mathbf{R})+\hat{V}(\mathbf{R}'))-E_0\right]\frac{d\tau}{\hbar}}\langle R|e^{-\hat{T}\frac{d\tau}{\hbar}}|R'\rangle\Psi(R', \tau)\,dR' \ . \tag{4.23}$$

The calculation of the evolution propagator defined in Eq. (4.20) is presented in the box here below.

We define the propagators

$$G_V(R \leftarrow R') = e^{-\left[\frac{1}{2}(\hat{V}(\mathbf{R})+\hat{V}(\mathbf{R}'))-E_0\right]\frac{d\tau}{\hbar}} \ , \tag{4.24}$$
$$G_0(R \leftarrow R') = \langle R|e^{-\hat{T}\frac{d\tau}{\hbar}}|R'\rangle \ . \tag{4.25}$$

We obtain the expression of the free propagator G_0 by using a Schrödinger-like equation depending on the imaginary time τ,

$$\hbar\frac{\partial}{\partial\tau}\Phi(R,\tau) - \frac{\hbar^2}{2m}\nabla^2\Phi(R,\tau) = 0 \; . \tag{4.26}$$

Let us formally solve this equation by using the Fourier transform of Φ in $n = 3A$ dimensions

$$\tilde{\Phi}(K,\tau) = \int dR\, e^{-i\mathbf{K}\cdot\mathbf{R}}\Phi(R,\tau) \; , \tag{4.27}$$

which satisfies the equation

$$\hbar\frac{\partial}{\partial\tau}\tilde{\Phi}(K,\tau) + \frac{\hbar^2K^2}{2m}\tilde{\Phi}(K,\tau) = 0 \; , \tag{4.28}$$

with

$$\tilde{\Phi}(K,\tau) = \tilde{\Phi}(K,0)\exp\left(-\frac{\hbar^2K^2}{2m}\frac{\tau}{\hbar}\right) \equiv \tilde{\Phi}(K,0)\tilde{g}(K,\tau) \; , \tag{4.29}$$

and the obvious definition of $\tilde{g}$. For the folding theorem of the Fourier transforms [6], Eq. (4.29) can be considered as the result of the folding product between Φ and g expressed in coordinates space, therefore, we can write

$$\Phi(R,\tau) = \int \Phi(R',0)g(R'-R,\tau)dR' \; , \tag{4.30}$$

This expression allows us to obtain g in coordinates space as anti-transform of $\tilde{g}$

$$g(R,\tau) = \frac{1}{(2\pi)^{n/2}}\int dK e^{i\mathbf{K}\cdot\mathbf{R}}e^{-\frac{\hbar^2\tau}{2m}K^2} = \frac{\exp\left(-2mR^2/4\hbar^2\tau\right)}{\left(2\frac{\hbar^2\tau}{2m}\right)^{n/2}} \; , \tag{4.31}$$

therefore

$$\Phi(R,\tau) = \frac{1}{(\hbar^2\tau/m)^{n/2}}\int dR' \exp\left[-\frac{m}{2\hbar^2\tau}(R-R')^2\right]\Phi(R',0) \; , \tag{4.32}$$

The free propagator can be expressed as

$$G_0(R \leftarrow R',\tau) = \left(\frac{\hbar^2\tau}{m}\right)^{-n/2} \exp\left[-\frac{m}{2\hbar^2\tau}(R-R')^2\right] . \tag{4.33}$$

By using the expressions of the free and interacting propagators, the evolution of the wave function given in Eq. (4.23) can be expressed as

$$\Psi(R,\tau+\Delta\tau) \simeq \int dR' \exp\left\{\left[-\frac{1}{2}\left(\hat{V}(R')+\hat{V}(R)\right) - E_0\right]\frac{\Delta\tau}{\hbar}\right\} \times \left(\frac{\hbar^2}{m}\Delta\tau\right)^{-n/2} \exp\left[-\frac{m}{2\hbar^2\Delta\tau}(R-R')^2\right]\Psi(R',\tau) \; . \tag{4.34}$$

Table 4.2 Energies, expressed in K, of liquid He^4 calculated with VMC and GFMC by using different types of trial wave functions. Exp. indicates the experimental value of the energy. Data from [8]

Method	Trial wave function	Energy (K)
VMC	McMillan	–5.72(2)
VMC	PPA	–5.93(1)
VMC	Shadow	–6.24(4)
VMC	McMillan + 3B	–6.65(2)
VMC	OPT + 3B	–6.79(1)
GFMC	McMillan	–7.12(3)
Exp.	–	–7.14

To calculate the expectation value of an operator, it is common practice to adopt a mixed representation that includes both the trial state and the time-evolved state. We denote the set of spatial coordinates at time τ_n as R_n and

$$\mathbf{P}_n = R_n R_{n-1} R_{n-2} \cdots R_0 \ . \tag{4.35}$$

We express the expectation values of the operator $\hat{\mathbb{O}}$ in the mixed representation as

$$\begin{aligned} \langle \hat{\mathbb{O}} \rangle_{\text{mix}} &= \frac{\langle \Psi_{\text{T}} | \hat{\mathbb{O}} | \Psi(\tau) \rangle}{\langle \Psi_{\text{T}} | \Psi(\tau) \rangle} \\ &= \frac{\int d\mathbf{P}_n \Psi_{\text{T}}^{\dagger}(R_n)\, \hat{\mathbb{O}}\, G(R_n, R_{n-1}) \cdots G(R_1, R_0) \Psi_{\text{T}}(R_0)}{\int d\mathbf{P}_n \Psi_{\text{T}}^{\dagger}(R_n) G(R_n, R_{n-1}) \cdots G(R_1, R_0) \Psi_{\text{T}}(R_0)} \ , \end{aligned} \tag{4.36}$$

where the propagator G is given by the product of G_V and G_0. With good approximation the expectation value of $\hat{\mathbb{O}}$ is given by [7]

$$\langle \hat{\mathbb{O}} \rangle \simeq \langle \hat{\mathbb{O}} \rangle_{\text{mix}} + \left[\langle \hat{\mathbb{O}} \rangle_{\text{mix}} - \langle \hat{\mathbb{O}} \rangle_{\text{T}} \right] \ . \tag{4.37}$$

In Tables 4.2 and 4.3, we present examples of Monte Carlo calculations comparing the binding energies of liquid helium systems obtained through VMC and GFMC methods, utilizing different types of trial wave functions. The results indicate a significant sensitivity of the VMC outcomes to the choice of the trial wave function, denoted as Ψ_{T}. In contrast, the GFMC results are independent of this choice and yield lower energy values than those derived from the VMC calculations. This serves as a quantitative confirmation of the limitations associated with the variational principle. Notably, the bosonic liquid, He^4, exhibits greater binding than the fermionic liquid, He^3. Furthermore, the agreement between the GFMC results and the experimental values is remarkable (Fig. 4.2).

Table 4.3 Energies, expressed in K, of liquid He^3 calculated with VMC and GFMC by using different types of trial wave functions. Exp. indicates the experimental value of the energy. Data from [8]

Method	Trial wave function	Energy (K)
VMC	McMillan	–1.08(2)
VMC	2B + 3B	–1.61(3)
VMC	2B + BF	–1.55(4)
VMC	2B + 3B + BF	–2.15(3)
GFMC	2B + 3B + BF	–2.44(4)
Exp.	–	–2.47

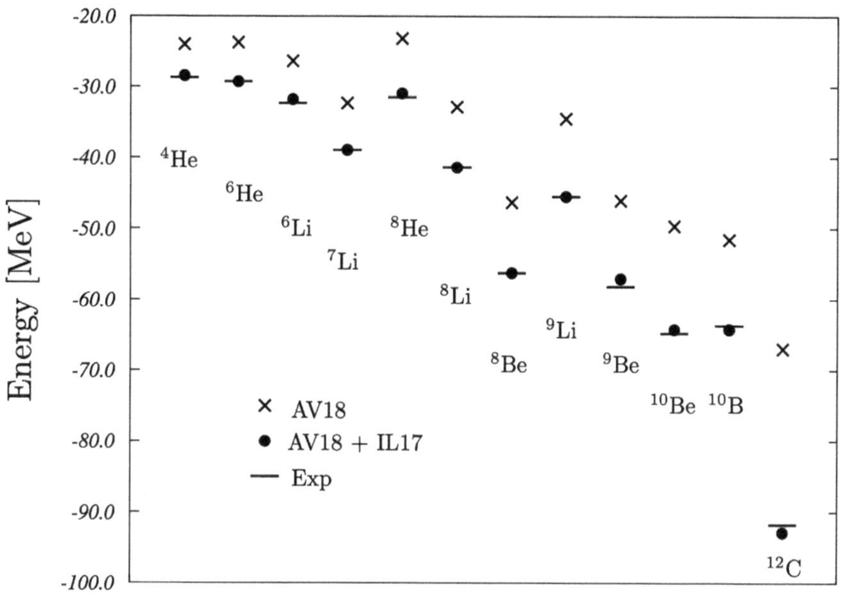

Fig. 4.2 GFMC results for some nuclei [7]. The crosses show the energies obtained by considering only a two-body interaction (Argonne V18). The points include the three-body interaction (IL-17). The lines indicate the experimental values

Similar calculations have been conducted for light nuclei. We summarise the GFMC results for the ground states of several light nuclei as reported in Ref. [7]. The crosses represent results obtained using only a two-body nucleon-nucleon interaction. In contrast, the points illustrate the outcomes when three-body interaction terms are incorporated into the nuclear Hamiltonian. The inclusion of the three-body interaction enhances the binding of the system and leads to an excellent agreement with experimental values.

4.5 Auxiliary Field Diffusion Monte Carlo (AFDMC)

The primary limitation of the GFMC method is the computational effort required, which increases exponentially with the number of particles due to the numerous spin and isospin configurations. In the expression (4.34) for the wave function calculated at time $\tau + \Delta\tau$, the potential term $\hat{V}(\mathbf{R})$ appears in the exponent. This potential includes terms that depend quadratically on the spin and isospin of the interacting particles. The quadratic form arises because two particles are involved in the interaction. These terms contribute to the rapid increase in the number of spin and isospin configurations. The situation becomes even more complex when considering the terms associated with the three-body force.

This problem is addressed using an approximate expression known as the Hubbard-Stratonovich transformation [9]. In this expression, propagators that include quadratic operator terms can be approximated as propagators that contain only linear terms of the operator

$$e^{-dt\frac{\hat{\mathcal{O}}^2}{2}} \simeq \frac{1}{\sqrt{2\pi}} \int dx e^{-\frac{x^2}{2}} e^{-x\,\hat{\mathcal{O}}\,\sqrt{-dt}} \quad , \tag{4.38}$$

where the variables x are called auxiliary fields, and $\hat{\mathcal{O}}$ is a generic operator.

To utilize the expression (4.38), it is helpful to separate the interaction into two components: a scalar part (SI) and another that depends on spin and isospin (SD)

$$\hat{V}(\mathbf{R}) = \hat{V}_{\mathrm{SI}}(\mathbf{R}) + \hat{V}_{\mathrm{SD}}(\mathbf{R}) \quad , \tag{4.39}$$

where the $\hat{V}_{\mathrm{SD}}$ term contains the full operator dependence with $\hat{O}_{ij}^{p>1}$. By using various transformations it is possible to rewrite $V_{\mathrm{SD}}(\mathbf{R})$ as [7]

$$\hat{V}_{\mathrm{SD}}(\mathbf{R}) \simeq \sum_n \lambda_n(\mathbf{R})\hat{O}_n^2, \tag{4.40}$$

where the coefficients λ_n are numerical values, and the summation is taken over all spin and isospin operator channels. By utilizing this expression for $\hat{V}_{\mathrm{SD}}$ and applying the Hubbard-Stratonovich transformation, as shown in Eq. (4.38), we can rewrite Eq. (4.34) as follows:

$$\begin{aligned}\Psi(R, \tau + \Delta\tau) \simeq &\int dR' \left(\frac{\hbar^2}{m}\Delta\tau\right)^{-n/2} \exp\left[-\frac{m}{2\hbar^2\Delta\tau}(R - R')^2\right] \\ &\exp\left\{\left[-\frac{1}{2}\left(\hat{V}_{\mathrm{SI}}(R') + \hat{V}_{\mathrm{SI}}(R)\right) - E_0\right]\frac{\Delta\tau}{\hbar}\right\} \\ &\times \prod_n \frac{1}{\sqrt{2\pi}} \int dx_n e^{-\frac{x_n^2}{2}} e^{-x\,\hat{\mathcal{O}}\,\sqrt{-\lambda_n(\mathbf{R}')\Delta\tau}} \quad \Psi(R', \tau) \ .\end{aligned} \tag{4.41}$$

The key difference between the expressions (4.34) and (4.41) lies in the presence of an additional integral over the x_n variables in the latter expression. The effort required to compute this additional integral is offset by the linear presence of the $\hat{\mathbb{O}}_n$, which significantly reduces the number of spin and isospin configurations that need to be calculated.

The AFDMC method enables calculations of nuclear systems with more than 12 particles. However, challenges arise when dealing with systems composed of different types of particles, particularly in the nuclear context due to the presence of isospin terms. Consequently, systems consisting solely of neutrons are preferred. Results obtained with a finite number of neutrons (66) have been compared with those derived from other microscopic theories that address infinite neutron matter [7]. This comparison indicates a strong agreement between the results produced by different theories for this specific many-body system.

4.6 Conclusions

The results of the GFMC calculations provide a quantitative and positive response to the question regarding the validity of the assumptions made to address the many-body problem, as presented in Chap. 1. The Schrödinger equation effectively describes many-body systems in a non-relativistic regime. The key point is that no approximations should be made in the pursuit of a solution. Even a simple approximation, such as that associated with the variational principle, can undermine the accurate description of the many-body system. Unfortunately, the GFMC method has significant computational limitations, making it necessary to develop approximate methods for solving the many-body Schrödinger equation. Understanding and controlling these approximations is essential for the application of these theories.

References

1. O. Kallenberg, *Foundations of modern Probability* (Springer, Berlin, 1997)
2. R. Jastrow, Many-body Problem with strong Forces. Phys. Rev. **98**, 1479 (1955)
3. G. Pan, Z. Y. Meng: *Sign Problem in Quantum Monte Carlo Simulation*, The Encyclopedia of Condensed Matter Physics, 2nd edition, Volume 1, Pag. 879-893, Editor: T. Chakraborty, Elsevier, Amsterdam (2024)
4. H.F. Trotter, On the Product of Semi-Groups of Operators. Proc. Amer. Math. Soc. **10**, 545 (1959)
5. M. Suzuki, Generalized Trotter's formula and systematic approximants of exponential operators and inner derivations with applications to many body problems. Commun. Math. Phys. **51**, 183 (1976)
6. H.W. Wyld, *Mathematical Methods for Physics* (Benjamin, Reading, 1976)
7. J. Carlson et al., Quantum Monte Carlo methods for nuclear physics. Rev. Mod. Phys. **87**, 1067 (2015)

8. K. E. Schmidt, D. E. Ceperley: *Monte Carlo Techniques for Quantum Fluids, Solids and Droplets*, in The Monte Carlo Method in Condensed Matter Physics, ed. K. Binder, Topics in Applied Physics, Vol 71, Springer-Verlag, (1992)
9. J. Hubbard, Calculation of Partition Functions. Phys. Rev. Lett. **3**, 77 (1959)

Part II
Theories Inspired to Quantum Field Theories

Chapter 5
Occupation Number Representation

5.1 Introduction

The use of creation and annihilation operators has been introduced in Quantum Field Theories where the number of particles is not fixed, allowing for the creation of particle-antiparticle pairs. Our study focuses on phenomena that occur at energies much lower than the rest masses of the particles in the system, which means that the creation of particle-antiparticle pairs is not feasible. However, in the analysis of many-body systems, the formalism developed using creation and annihilation operators proves to be very useful. As we will discuss below, the physical interpretation of the action of these operators differs significantly from that defined in field theory.

We will concentrate solely on the case of fermions. Even though the treatment of bosonic systems is not particularly challenging, it does not simply involve a straightforward modification of the fermionic case.

5.2 Slater Determinants

A system of A identical fermions can be described by the time-independent Schrödinger equation

$$\hat{H}\,\Psi(x_1, x_2 \ldots x_A) = E\,\Psi(x_1, x_2 \ldots x_A) \ , \tag{5.1}$$

where x indicates all the quantum numbers characterising each particle: position ($\mathbf{r}$), spin (σ), isospin (τ) and eventually other quantum numbers such as flavor, color. Since we deal with fermions, the global wave function Ψ must be antisymmetric for the exchange of two particles:

$$\Psi(\ldots x_i, \ldots, x_j \ldots) = -\Psi(\ldots x_j, \ldots, x_i \ldots) \ . \tag{5.2}$$

© The Author(s), under exclusive license to Springer Nature Switzerland AG 2026

G. Co', *Concepts in Quantum Many-Body Physics*, UNITEXT for Physics,
https://doi.org/10.1007/978-3-032-08920-5_5

Each eigenfunction of $\hat{H}$ can be written as linear combination of a complete set of orthonormal functions that form a basis, such as those composed of Slater determinants Φ_i,

$$\Psi = \sum_i C_i \Phi_i \ , \tag{5.3}$$

where the C_i coefficients are numbers. The Slater determinant is built by products of single particle wave functions. As it has been discussed in Chap. 2, these single-particle wave functions, which form a basis, are obtained by solving a single-particle Schrödinger equation:

$$\hat{h}_\nu \phi_\nu(x) = \epsilon_\nu \phi_\nu(x) \ , \tag{5.4}$$

where ν is the set of quantum numbers characterising each single-particle state.

Once the single-particle wave functions ϕ_ν are determined, the Slater determinant for A particles can be constructed as follows

$$\Phi(x_1, \ldots x_A) = \frac{1}{\sqrt{A!}} \sum_P (-)^P P \phi_{\nu_1}(x_1)\phi_{\nu_2}(x_2)\ldots\phi_{\nu_N}(x_A) \ , \tag{5.5}$$

where P indicates the permutations between the indexes of the coordinates. The $\sqrt{A!}$ factor gives the normalisation

$$\langle \Phi \mid \Phi \rangle = 1 \ . \tag{5.6}$$

Equation (5.5) can be written as

$$\Phi(x_1 \ldots x_A) = \frac{1}{\sqrt{A!}} \begin{vmatrix} \phi_{\nu_1}(x_1) & \ldots & \phi_{\nu_A}(x_1) \\ \vdots & & \vdots \\ \phi_{\nu_1}(x_A) & \ldots & \phi_{\nu_A}(x_A) \end{vmatrix} \ . \tag{5.7}$$

The Slater determinant is eigenstate of an IPM many-body hamiltonian which is sum of single-particle hamiltonians:

$$\sum_\nu \hat{h}_\nu \Phi = \hat{H}_0 \Phi = \mathcal{E}_0 \Phi \ . \tag{5.8}$$

5.3 Creation and Destruction Operators

The Slater determinant contains redundant information. From a physics perspective, it is crucial to know whether the state ϕ_ν is occupied or not. However, the specific identity of the fermion occupying this state is irrelevant, as all the fermions in the system are identical. The information contained in the Slater determinant can be summarised by a set of ordered numbers that indicate which single-particle states are occupied. It is possible to establish a correspondence between the Slater determinant and this set of ordered numbers.

$$|\Phi\rangle \longrightarrow |\nu_A, \nu_{A-1} \dots \nu_1\rangle \ . \tag{5.9}$$

In a commonly used convention, the order of the indexes is arranged in decreasing order of energy values. The state denoted as ν_A corresponds to the highest energy, followed by the others in descending order. This representation of a Slater determinant is usually called Occupation Number Representation (ONR).

It is useful to relate states with different particle numbers. For example, the single-particle state

$$|\phi_\nu\rangle \equiv |\nu\rangle, \tag{5.10}$$

can be imagined as obtained by adding a particle to the vacuum state $|0\rangle$. In symbols

$$|\nu\rangle = \hat{a}^+_\nu|0\rangle, \tag{5.11}$$

where $\hat{a}^+_\nu$ represents the operator that adds a particle to the single-particle state characterised by the ν quantum numbers.

In general, we write

$$\hat{a}^+_\nu|\nu_A \dots \nu_1\rangle = |\nu\nu_A \dots \nu_1\rangle. \tag{5.12}$$

The state with $A + 1$ particles is not necessarily ordered. In this operation, we recognise that the normalisation constant has been appropriately adjusted to ensure that all many-body states remain normalised to unity.

The newly added particle, ν, must be positioned correctly. This is achieved by considering that any change in position introduces a multiplicative negative phase, as we are dealing with a determinant that describes a set of fermions. The potential ambiguities regarding the global sign of the state can be resolved by stipulating that the sign of the determinant remains unchanged if the new particle occupies the first position, as we have assumed in the formulation of Eq. (5.12).

For example:

$$\hat{a}^+_{\nu_1}|\nu_3\nu_2\rangle = |\nu_1\nu_3\nu_2\rangle = -|\nu_3\nu_1\nu_2\rangle = |\nu_3\nu_2\nu_1\rangle \ . \tag{5.13}$$

It follows from Eqs. (5.11) and (5.12) that:

$$|\nu_A \dots \nu_1\rangle = \hat{a}^+_{\nu_A}\hat{a}^+_{\nu_{A-1}} \dots \hat{a}^+_{\nu_1}|0\rangle \ . \tag{5.14}$$

For the properties of the Slater determinant we have that:

$$\hat{a}^+_\nu|\nu_A \dots \nu \dots \nu_1\rangle = 0 \ . \tag{5.15}$$

This operation involves inserting a row or a column into a determinant that is identical to an existing row or column. From a physics perspective, Eq. (5.15) expresses the Pauli exclusion principle.

Based on what we have presented so far, it possible to deduce that the creation operators exhibit the following anti-commutation properties

$$\hat{a}^+_{\nu_1}\hat{a}^+_{\nu_2} = -\hat{a}^+_{\nu_2}\hat{a}^+_{\nu_1} \ . \tag{5.16}$$

Let us demonstrate Eq. (5.16). We assume $\nu > \nu'$

$$\begin{aligned}
\hat{a}^+_{\nu}\hat{a}^+_{\nu'}\Phi^A(A\ldots \quad \ldots 1) &= \theta^I_{\nu'}\hat{a}^+_{\nu}\Phi^{A+1}(A\ldots\nu'\ldots 1) \\
&= \theta^I_{\nu}\theta^I_{\nu'}\Phi^{A+2}(A\ldots\nu\ldots\nu'\ldots 1) \ , \\
\hat{a}^+_{\nu'}\hat{a}^+_{\nu}\Phi^A(A\ldots \quad \ldots 1) &= \theta^{II}_{\nu}\hat{a}^+_{\nu'}\Phi^{A+1}(A\ldots\nu\ldots 1) \\
&= \theta^{II}_{\nu}\theta^{II}_{\nu'}\Phi^{A+2}(A\ldots\nu\ldots\nu'\ldots 1) \ .
\end{aligned}$$

We denote the phase acquired from the necessary permutations to position the particle ν as $\theta^I_{\nu} = (-)^p \equiv (-1)^p$. Similarly, we define $\theta^I_{\nu'} = (-)^{p'}$. The phases in the second case are labeled as θ^{II}_{ν} and $\theta^{II}_{\nu'}$. Since $\nu > \nu'$, we have $\theta^I_{\nu} = \theta^{II}_{\nu}$ because the number of permutations required to order ν is the same in both scenarios. Conversely, $\theta^I_{\nu'} = -\theta^{II}_{\nu'}$ because, in the second case, an additional permutation is necessary to arrange ν'. Therefore, we conclude that

$$\hat{a}^+_{\nu}\hat{a}^+_{\nu'}\Phi = -\hat{a}^+_{\nu'}\hat{a}^+_{\nu}\Phi \ ,$$

from where the relation (5.16) between operators is derived.

Equation (5.16) implies

$$\left(a^+_{\nu}\right)^2 = 0 \ , \tag{5.17}$$

which is another way to express Eq. (5.15), i.e. the Pauli exclusion principle.

It is possible to define the adjoint operator of $\hat{a}^+$ whose action can be understood by taking the adjoint of Eq. (5.11).

$$(\hat{a}^+_{\nu})^+ \quad = \quad \hat{a}_{\nu} \ , \tag{5.18}$$

$$|\nu\rangle = \hat{a}^+_{\nu}|0\rangle \quad \Leftrightarrow \quad \langle\nu| = \langle 0|\hat{a}_{\nu} \ , \tag{5.19}$$

$$\langle\nu_1\nu_2\ldots\nu_N| \quad = \quad \langle 0|\hat{a}_{\nu_1}\ldots\hat{a}_{\nu_{N-1}}\hat{a}_{\nu_N} \ , \tag{5.20}$$

since $\langle\nu|\nu\rangle = 1$ we obtain

$$\langle 0\,\underbrace{|\hat{a}_{\nu}\hat{a}^+_{\nu}|0\rangle}_{|0\rangle} = 1 \ , \tag{5.21}$$

and therefore,

$$|0\rangle = \hat{a}_{\nu}\hat{a}^+_{\nu}|0\rangle = \hat{a}_{\nu}|\nu\rangle = |0\rangle \ . \tag{5.22}$$

The effect of applying $\hat{a}_\nu$ on the ket state is to remove a particle that occupy the $|\nu\rangle$ state. Let us analyse the behaviour of $\hat{a}_{\nu_r}$

$$\hat{a}_{\nu_r}|\nu_A \dots \nu_{r+1}\nu_r\nu_{r-1}\dots\nu_1\rangle = |\nu_A \dots \nu_{r+1}\nu_{r-1}\dots\nu_1\rangle(-)^{r-1} \ . \tag{5.23}$$

Since, by definition, in the vacuum all the single-particle states are empty, we can write

$$\hat{a}_\nu|0\rangle = 0 \ . \tag{5.24}$$

If the vector $|\nu_N \dots \nu_1\rangle$ does not contain the ν state, then

$$\hat{a}_\nu|\nu_A \dots \nu_1\rangle = 0 \ . \tag{5.25}$$

Also in the case of the destruction operators it is possible to show that

$$\hat{a}_{\nu_1}\hat{a}_{\nu_2} = -\hat{a}_{\nu_2}\hat{a}_{\nu_1}, \quad \text{and therefore} \quad (\hat{a}_\nu)^2 = 0 \ . \tag{5.26}$$

Similarly, it can be shown that for $\nu_1 \neq \nu_2$, the following relation holds:

$$\hat{a}_{\nu_1}\hat{a}^+_{\nu_2} = -\hat{a}^+_{\nu_2}\hat{a}_{\nu_1} \ . \tag{5.27}$$

By applying the properties we discussed earlier, we obtain:

$$\hat{a}_\nu\hat{a}^+_\nu|\nu_A \dots \nu_1\rangle = \begin{cases} 0 & \text{if} \quad \nu \in S \\ |\nu_A \dots \nu_1\rangle & \text{if} \quad \nu \notin S \end{cases} \qquad S = \{\nu_A \dots \nu_1\} \tag{5.28}$$

$$\hat{a}^+_\nu\hat{a}_\nu|\nu_A \dots \nu_1\rangle = \begin{cases} |\nu_A \dots \nu_1\rangle & \text{if} \quad \nu \in S \\ 0 & \text{if} \quad \nu \notin S \end{cases} \tag{5.29}$$

Equations (5.28) and (5.29) leads to the relation

$$\left(\hat{a}_\nu\hat{a}^+_\nu + \hat{a}^+_\nu\hat{a}_\nu\right)|\nu_A \dots \nu_1\rangle = |\nu_A \dots \nu_1\rangle \ . \tag{5.30}$$

This relation holds for any vector $|\nu_A \dots \nu_1\rangle$ because one of the two terms yields a zero result, while the other reproduces the original vector.

Equation (5.29) defines the occupation number operator

$$\hat{n}_\nu = \hat{a}^+_\nu\hat{a}_\nu \ , \tag{5.31}$$

whose eigenvalues are either 1 or 0, depending on whether the state to which is applied contains the single-particle state ν, or not.

The relations between creation and destruction operators given by Eqs. (5.28), (5.29) and (5.30) is:

$$\hat{a}_\nu\hat{a}^+_\nu + \hat{a}^+_\nu\hat{a}_\nu = 1 \ . \tag{5.32}$$

The relations (5.16), (5.26), (5.27), (5.32) defining the properties of the creation and destruction operators, can be summarised as:

$$\{\hat{a}_\nu, \hat{a}^+_{\nu'}\} = \delta_{\nu\nu'}, \qquad \{\hat{a}_\nu, \hat{a}_{\nu'}\} = 0, \qquad \{\hat{a}^+_\nu, \hat{a}^+_{\nu'}\} = 0 \ , \tag{5.33}$$

where the symbol $\{,\}$ indicates the anti-commutator operator.

When a many-body system exhibits rotational symmetry, it is advantageous to work with operators that possess the characteristics of irreducible spherical tensors. For this reason, it is beneficial to utilise creation and annihilation operators defined in a slightly different manner, as outlined in Appendix B.

5.4 One and Two-Body Operators

We used the ONR to describe state vectors in the form of Slater determinants, but the approach can be extended to describe many-body operators.

We indicated with $|S\rangle$ and $|S'\rangle$ two vectors described in ONR, and with $|\Phi_S\rangle$ and $|\Phi_{S'}\rangle$ the two corresponding Slater determinants. Let us consider, in the configuration space, a generic many-body operator $\hat{O}_S(x_1 \ldots x_A)$. Since the physical quantities which can be measured are related to the expectation value of operators between states, the two representations must provide the same value

$$\langle \Phi_{S'}|\hat{O}_S|\Phi_S\rangle = \langle S'|\hat{O}|S\rangle \ , \tag{5.34}$$

where $\hat{O}$ is the many-body operator expressed in ONR.

In the study of many-body systems, the most common operators related to observables are the one- and two-body operators. In the coordinate space, a one-body operator is given by the sum of operators acting on a single coordinate at the time:

$$\hat{O}^I_S(x_1 \ldots x_A) = \sum_{i=1}^{A} \hat{o}^I(x_i) \ . \tag{5.35}$$

The kinetic energy is a typical one-body operator. For simplicity let us use a basis of single-particle wave functions where $\hat{O}^I$ is diagonal

$$\hat{o}^I(x)\phi_\nu(x) = \omega_\nu \phi_\nu(x). \tag{5.36}$$

By using the Slater determinant built on the ϕ_ν states we obtain:

$$\hat{O}_S^I(x_1 \dots x_A)\Phi_S(x_1 \dots x_A) =$$
$$= \frac{1}{\sqrt{A!}} \sum_P (-)^P \sum_{i=1}^{A} \hat{o}^I(x_i) \hat{P} \phi_{\nu_1}(x_1) \dots \phi_{\nu_A}(x_A)$$
$$= \frac{1}{\sqrt{A!}} \sum_P (-)^P \sum_{i=1}^{A} \hat{P} \hat{o}^I(x_i) \phi_{\nu_1}(x_1) \dots \phi_{\nu_A}(x_A)$$
$$= \frac{1}{\sqrt{A!}} \sum_P (-)^P \sum_{i=1}^{A} (\omega_{\nu_i} \hat{P} \phi_{\nu_1}(x_1) \dots \phi_{\nu_A}(x_A)$$
$$= \sum_{i=1}^{A} \omega_{\nu_i} \Phi_S = \sum_{\nu \text{ occupied}} \omega_\nu \Phi_S = \sum_\nu \omega_\nu n_\nu \Phi_S \ , \tag{5.37}$$

where we have indicated with $\hat{P}$ the operator which makes all the possible P permutations. In the last sum of the previous equations $n_\nu = 1$ for all the occupied levels and $n_\nu = 0$ for the empty ones. The evaluation of the matrix element is:

$$\langle \Phi_{S'} | \hat{O}_S^I | \Phi_S \rangle = \langle \Phi_{S'} | \sum_\nu \omega_\nu n_\nu | \Phi_S \rangle$$
$$= \sum_\nu \omega_\nu \langle S' | \hat{n}_\nu | S \rangle = \langle S' | O^I | S \rangle \ , \tag{5.38}$$

where we used the occupation number operator defined in Eq. (5.31) whose eigenvalue is n_ν. By substituting we obtain

$$\langle S' | O^I | S \rangle = \langle S' | \sum_\nu \omega_\nu \hat{a}_\nu^+ \hat{a}_\nu | S \rangle \ , \tag{5.39}$$

and then:

$$\hat{O}^I = \sum_{\nu\nu} \omega_{\nu\nu} \hat{a}_\nu^+ \hat{a}_\nu \ , \tag{5.40}$$

where

$$\omega_{\nu\nu} = \int d^3 r \phi_\nu^*(\mathbf{r}) \hat{o}^I(\mathbf{r}) \phi_\nu(\mathbf{r}) \ . \tag{5.41}$$

The expression of the one-body operator for a generic single-particle basis is:

$$\hat{O}^I = \sum_{\nu\nu'} O_{\nu\nu'}^I \hat{a}_\nu^+ \hat{a}_{\nu'} \ , \tag{5.42}$$

$$O_{\nu\nu'}^I = \int dx \phi_\nu^*(x) \hat{o}^I(x) \phi_{\nu'}(x) \equiv \langle \nu | o^I | \nu' \rangle \ , \tag{5.43}$$

where x is a generalised coordinate, containing space, spin and, eventually, isospin. This expression shows that the one-body operator removes a particle in ν' and adds a particle in the ν state. Clearly, ν' must be occupied, while ν must be unoccupied. This action is called creation of a particle-hole pair. In the case where $|S\rangle = |S'\rangle$ the only contributions different from zero are those where $\nu' = \nu$.

A similar procedure can be used to obtain the ONR expression for the two-body operators, defined as sum of operators acting on two-coordinates at the time:

$$\hat{O}_S^{II}(x_1 \dots x_A) = \sum_{i<j} \hat{o}^{II}(x_i, x_j) = \frac{1}{2} \sum_{i \neq j} \hat{o}^{II}(x_i, x_j) \ . \tag{5.44}$$

A typical example of two-body operator is the interaction potential

$$\hat{V} = \frac{1}{2} \sum_{i \neq j} V(x_i, x_j).$$

In ONR, operators of this type are represented as:

$$\hat{O}^{II} = \frac{1}{2} \sum_{\nu\nu'\mu\mu'} O^{II}_{\nu\mu\nu'\mu'} \hat{a}^+_\nu \hat{a}^+_\mu \hat{a}_{\mu'} \hat{a}_{\nu'} \ , \tag{5.45}$$

where $\hat{O}^{II}_{\nu\mu\nu'\mu'}$, which is a number, is defined as:

$$\begin{aligned} \hat{O}^{II}_{\nu\mu\nu'\mu'} &= \langle \nu\mu | O^{II} | \nu'\mu' \rangle \\ &= \int dx dx' \phi^*_\nu(x) \phi^*_\mu(x') \hat{o}^{II}(x, x') \phi_{\nu'}(x) \phi_{\mu'}(x') \ . \end{aligned} \tag{5.46}$$

It is worth to remark that the order of the two sub-indexes are inverted in the sequence of the creation and destruction operators with respect to that of the matrix element.

In this case, the two-body operator removes two particles from occupied states and adds them in two empty states.

5.5 Field Operators

The creation and annihilation operators $\hat{a}^+_\nu$ and $\hat{a}_\nu$ presented here follow the same algebra defined by the anti-commutation relations (5.33) as the analogous operators introduced in field theory. However, the physical interpretation of the actions of these operators differs between the two theories. In field theories, these operators describe the creation and annihilation of particles, such as in the formation of particle-antiparticle pairs. In many-body theories, the MF problem has already been solved, providing a basis of single-particle states. The creation operator $\hat{a}^+_\nu$ adds a particle to the state characterised by the quantum numbers ν, while $\hat{a}_\nu$ removes a particle from this state.

In field theories, there are field operators $\hat{\psi}^+(\mathbf{r})$ that create a particle at the point $\mathbf{r}$ and operators $\hat{\psi}(\mathbf{r})$ that annihilate a particle located at $\mathbf{r}$. It is also possible to define similar operators in many-body theories. These field operators are related to the creation and annihilation operators through the single-particle wave functions $\phi_\nu(\mathbf{r})$, which are generated by the solution of the MF problem:

$$\hat{\psi}(\mathbf{r}) = \sum_\nu \hat{a}_\nu \phi_\nu(\mathbf{r}) \ , \tag{5.47}$$

$$\hat{\psi}^+(\mathbf{r}) = \sum_\nu \hat{a}_\nu^+ \phi_\nu^*(\mathbf{r}) \ . \tag{5.48}$$

Because of the orthonormality of the ϕ, these two equations can be inverted to express the creation and destruction operators in terms of field operator:

$$\hat{a}_\nu = \int d^3r \phi_\nu^*(\mathbf{r})\hat{\psi}(\mathbf{r}) \quad \text{and} \quad \hat{a}_\nu^+ = \int d^3r \phi_\nu(\mathbf{r})\hat{\psi}^+(\mathbf{r}) \ . \tag{5.49}$$

By using the anticommutation relations (5.33) in the above definitions, it is possible to obtain analogous relations for the field operators:

$$\left\{\hat{\psi}^+(\mathbf{r}), \hat{\psi}^+(\mathbf{r}')\right\} = 0 \ , \ \left\{\hat{\psi}(\mathbf{r}), \hat{\psi}(\mathbf{r}')\right\} = 0 \ , \ \left\{\hat{\psi}^+(\mathbf{r}), \hat{\psi}(\mathbf{r}')\right\} = \delta(\mathbf{r} - \mathbf{r}') \ . \tag{5.50}$$

The one- and two-body operators can be expressed in terms of field operators as:

$$\hat{O}^I = \int d^3r \ \hat{\psi}^+(\mathbf{r})\hat{o}^I(\mathbf{r})\hat{\psi}(\mathbf{r}) \ , \tag{5.51}$$

$$\hat{O}^{II} = \frac{1}{2}\int d^3r d^3r' \ \hat{\psi}^+(\mathbf{r})\hat{\psi}^+(\mathbf{r}')\hat{o}^{II}(\mathbf{r}, \mathbf{r}')\hat{\psi}(\mathbf{r}')\hat{\psi}(\mathbf{r}) \ . \tag{5.52}$$

The hamiltonian operator can be expressed in terms of creation and destruction operators or in terms of field operators as:

$$\hat{H} = \sum_{\nu\nu'} \langle\nu|\hat{T}|\nu'\rangle \hat{a}_\nu^+ \hat{a}_{\nu'} + \frac{1}{2}\sum_{\nu\nu'\mu\mu'} \langle\nu\mu|\hat{V}|\nu'\mu'\rangle \hat{a}_\nu^+ \hat{a}_\mu^+ \hat{a}_{\mu'} \hat{a}_{\nu'} \ ,$$

$$= \int d^3r \hat{\psi}^+(\mathbf{r}) \left(-\frac{\hbar^2}{2m}\nabla^2\right) \hat{\psi}(\mathbf{r}) \tag{5.53}$$

$$+ \frac{1}{2}\int d^3r d^3r' \hat{\psi}^+(\mathbf{r})\hat{\psi}^+(\mathbf{r}')V(\mathbf{r}, \mathbf{r}')\hat{\psi}(\mathbf{r}')\hat{\psi}(\mathbf{r}) \ . \tag{5.54}$$

Chapter 6
Perturbation Theory of Many-Body Systems

6.1 Pictures

The values of the observable quantities, in Quantum Mechanics, are obtained by calculating expectation values of operators between states. These quantities are invariant if the same unitary transformation is applied to the states and to the operators. These unitary transformations can contain the time and, for example, they can transform states which depend on time in time-independent states. In this latter case, the dependence on the time is inserted in the operator expressions. The expressions of states and operators obtained by applying these time-dependent unitary transformations are called *pictures* of the Quantum Mechanics.

6.1.1 Schrödinger Picture

The most used picture is that of Schrödinger where the states describing the system depend on the time, while the operators are time-independent. In this picture, the time evolution of the system is described by the equation:

$$i\hbar\frac{\partial}{\partial t}|\Psi_S(t)\rangle = \hat{H}|\Psi_S(t)\rangle \ , \tag{6.1}$$

which is the well known time-dependent Schrödinger equation.

For those systems where the energy is conserved, the hamiltonian $\hat{H}$ does not depend on the time. In this case, the formal solution of Eq. (6.1) is:

$$|\Psi_S(t)\rangle = e^{-i\frac{\hat{H}(t-t_0)}{\hbar}}|\Psi_S(t_0)\rangle \ . \tag{6.2}$$

In this equation the exponential function of an operator is present. The action of this function on the state $|\Psi_S(t_0)\rangle$ can be expressed in terms of a power expansion of the

© The Author(s), under exclusive license to Springer Nature Switzerland AG 2026

G. Co', *Concepts in Quantum Many-Body Physics*, UNITEXT for Physics,
https://doi.org/10.1007/978-3-032-08920-5_6

exponential. In addition, since $\hat{H}$ is an hermitian operator, $\exp(i\hat{H}t/\hbar)$ is a unitary operator.

The Eq. (6.2) allows us to know the solution of the Schrödinger equation at any time t once the state of the system is known at a specific time value t_0.

6.1.2 Heisenberg Picture

In this picture, the states are time-independent, while the operators have an explicit time dependence. The states $|\Psi_H\rangle$ in the Heisenberg picture are related to those of the Scrödinger picture $|\Psi_S\rangle$ by the relation

$$|\Psi_H(t)\rangle \equiv e^{i\frac{\hat{H}t}{\hbar}}|\Psi_S(t)\rangle \ . \tag{6.3}$$

The time evolution of these states in the Heisenberg picture is:

$$\begin{aligned} i\hbar\frac{\partial}{\partial t}|\Psi_H(t)\rangle &= -\hat{H}e^{i\frac{\hat{H}t}{\hbar}}|\Psi_S(t)\rangle + e^{i\frac{\hat{H}t}{\hbar}}i\hbar\frac{\partial}{\partial t}|\Psi_S(t)\rangle = \text{for (6.1)} \\ &= -\hat{H}e^{i\frac{\hat{H}t}{\hbar}}|\Psi_S(t)\rangle + \hat{H}e^{i\frac{\hat{H}t}{\hbar}}|\Psi_S(t)\rangle = 0 \ . \end{aligned} \tag{6.4}$$

This clearly shows that $|\Psi_H\rangle$ does not depend on time.

The relation between the expressions of the operators in the two pictures is obtained by considering that the expectation values of the operators must be the same in both pictures:

$$\langle\Psi_S(t)|\hat{O}_S|\Psi_S(t)\rangle = \langle\Psi_H|e^{i\frac{\hat{H}t}{\hbar}}\hat{O}_S e^{-i\frac{\hat{H}t}{\hbar}}|\Psi_H\rangle \ . \tag{6.5}$$

Consequently we obtain

$$\hat{O}_H \equiv e^{i\frac{\hat{H}t}{\hbar}}\hat{O}_S e^{-i\frac{\hat{H}t}{\hbar}} \ . \tag{6.6}$$

The time evolution of the operator provides the equation of motion:

$$\begin{aligned} i\hbar\frac{\partial}{\partial t}\hat{O}_H(t) &= -\hat{H}e^{i\frac{\hat{H}t}{\hbar}}\hat{O}_S e^{-i\frac{\hat{H}t}{\hbar}} + e^{i\frac{\hat{H}t}{\hbar}}\hat{O}_S\hat{H}e^{-i\frac{\hat{H}t}{\hbar}} = \\ &= -\hat{H}\hat{O}_H + \hat{O}_H\hat{H} = [\hat{O}_H, \hat{H}], \end{aligned} \tag{6.7}$$

where we used the fact that $\hat{H}$ and $\exp[i\hat{H}t/\hbar]$ commute. In general, $\hat{O}_H$ and $\hat{H}$ do not commute. In case of commutation, Eq. (6.7) expresses the fact that $\hat{O}_H$ describes a constant of motion.

6.1.3 Interaction Picture

This picture, intermediate between that of Heisenberg and that of Schrödinger, is that of main interest in the description of many-body systems.

Let us consider a time-independent hamiltonian in the Schrödinger picture, and let us separate it in two terms

$$\hat{H} = \hat{H}_0 + \hat{H}_1 \ . \tag{6.8}$$

The state vector in the interaction picture is defined as:

$$|\Psi_I(t)\rangle \equiv e^{i\frac{\hat{H}_0 t}{\hbar}}|\Psi_S(t)\rangle \ . \tag{6.9}$$

The time evolution of the state $|\Psi_I(t)\rangle$ is given by:

$$\begin{aligned} i\hbar\frac{\partial}{\partial t}|\Psi_I(t)\rangle &= -\hat{H}_0 e^{i\frac{\hat{H}_0 t}{\hbar}}|\Psi_S(t)\rangle + e^{i\frac{\hat{H}_0 t}{\hbar}} i\hbar\frac{\partial}{\partial t}|\Psi_S(t)\rangle \\ &= e^{i\frac{\hat{H}_0 t}{\hbar}}\left[-\hat{H}_0 + \hat{H}_0 + \hat{H}_1\right] e^{-i\frac{\hat{H}_0 t}{\hbar}} e^{i\frac{\hat{H}_0 t}{\hbar}}|\Psi_S(t)\rangle \\ &= e^{i\frac{\hat{H}_0 t}{\hbar}}\hat{H}_1 e^{-i\frac{\hat{H}_0 t}{\hbar}}|\Psi_I(t)\rangle = \hat{H}_{1,I}(t)|\Psi_I(t)\rangle \ , \end{aligned} \tag{6.10}$$

where in the second step we used Eq. (6.1). In general, $\hat{H}_1$ and $\hat{H}_0$ do not commute.

Also in this case, the expression of the operators in the interaction picture, in terms of the operator in Schrödinger picture, is obtained by considering that the expectation values must be the same in both pictures:

$$\langle\Psi_S(t)|\hat{O}_S|\Psi_S(t)\rangle = \langle\Psi_I|e^{+i\frac{\hat{H}_0 t}{\hbar}}\hat{O}_S e^{-i\frac{\hat{H}_0 t}{\hbar}}|\Psi_I\rangle \ , \tag{6.11}$$

from where we obtain the definition:

$$\hat{O}_I(t) = e^{i\frac{\hat{H}_0 t}{\hbar}}\hat{O}_S e^{-i\frac{\hat{H}_0 t}{\hbar}} \ . \tag{6.12}$$

From the above equations, it appears clear that, in the interaction picture, both states and operators depend on time. The equation of motion in the interaction picture is:

$$\begin{aligned} i\hbar\frac{\partial}{\partial t}\hat{O}_I(t) &= -\hat{H}_0 e^{i\frac{\hat{H}_0 t}{\hbar}}\hat{O}_S e^{-i\frac{\hat{H}_0 t}{\hbar}} + e^{i\frac{\hat{H}_0 t}{\hbar}}\hat{O}_S\hat{H}_0 e^{-i\frac{\hat{H}_0 t}{\hbar}} \\ &= e^{i\frac{\hat{H}_0 t}{\hbar}}\hat{O}_S e^{-i\frac{\hat{H}_0 t}{\hbar}}\hat{H}_0 - \hat{H}_0 e^{i\frac{\hat{H}_0 t}{\hbar}}\hat{O}_S e^{-i\frac{\hat{H}_0 t}{\hbar}} \\ &= \hat{O}_I\hat{H}_0 - \hat{H}_0\hat{O}_I = \left[\hat{O}_I(t), \hat{H}_0\right] \ . \end{aligned} \tag{6.13}$$

Since in the ONR all the operators can be expressed in terms of creation and destruction operators, it is convenient to obtain the expressions of these operators in the interaction picture. In order to simplify the calculation we consider a special situation where $\hat{H}_0$ is a one-body operator diagonal in the chosen single-particle basis:

$$\hat{H}_0 = \sum_k \hbar\omega_k \hat{a}_k^+ \hat{a}_k \quad . \tag{6.14}$$

The equation of motion for an operator in the interaction picture is

$$\begin{aligned} i\hbar\frac{d}{dt}\hat{a}_{I,k}(t) &= e^{i\frac{\hat{H}_0 t}{\hbar}}\left[\hat{a}_{S,k}, \hat{H}_0\right]e^{-i\frac{\hat{H}_0 t}{\hbar}} \\ &= e^{i\frac{\hat{H}_0 t}{\hbar}}\left[\hat{a}_{S,k}, \sum_{k'} \hbar\omega_{k'}\hat{a}_{S,k'}^+\hat{a}_{S,k'}\right]e^{-i\frac{\hat{H}_0 t}{\hbar}} \\ &= e^{i\frac{\hat{H}_0 t}{\hbar}}\sum_{k'}\left[\hat{a}_{S,k}, \hat{a}_{S,k'}^+\hat{a}_{S,k'}\right]e^{-i\frac{\hat{H}_0 t}{\hbar}}\hbar\omega_{k'} \quad , \end{aligned}$$

where I and S indicate the interaction and Schrödinger pictures respectively. The term in parentheses becomes:

$$\begin{aligned} &[\hat{a}_{S,k}\hat{a}_{S,k'}^+\hat{a}_{S,k'} - \hat{a}_{S,k'}^+(-\hat{a}_{S,k}\hat{a}_{S,k'})] \\ &= [\hat{a}_{S,k}\hat{a}_{S,k'}^+\hat{a}_{S,k'} - (-\delta_{kk'}\hat{a}_{S,k'} + \hat{a}_{S,k}\hat{a}_{S,k'}^+\hat{a}_{S,k'})] = \hat{a}_{S,k} \quad , \end{aligned}$$

where we used the anti-commutation properties (5.33). We have then:

$$i\hbar\frac{d}{dt}\hat{a}_{k,I}(t) = \hbar\omega_k\hat{a}_{k,I}(t) \quad , \tag{6.15}$$

and, by assuming $\hat{a}_I(t=0) = \hat{a}_S$,

$$\hat{a}_{I,k}(t) = \hat{a}_{S,k}e^{-i\omega_k t} \quad . \tag{6.16}$$

For the adjoint operator, we obtain:

$$\hat{a}_{I,k}^+(t) = \hat{a}_{S,k}^+ e^{i\omega_k t} \quad . \tag{6.17}$$

Since the time dependence is present only in the complex phase, the anti-commutation properties of the creation and destruction operators in the interaction picture are the same as those in the Schrödinger picture.

To obtain the expression for any operator in the interaction picture, it is sufficient to substitute the annihilation operator $\hat{a}_k$ with $\hat{a}_{I,k}(t)$, and similarly for the creation operators. Additionally, the field operators can be expressed in terms of $\hat{a}_{S,k}$ and $\hat{a}_{S,k}^+$; therefore, their expressions in the interaction picture can be derived using the same procedure.

6.2 Time-Evolution Operator

In the interaction picture, we define an operator $\hat{U}(t, t_0)$, the time evolution operator, which determines the state vector at the time t once the state vector is known at the time t_0:

$$|\Psi_I(t)\rangle = \hat{U}(t, t_0)|\Psi_I(t_0)\rangle \ . \tag{6.18}$$

From the definition of state in the interaction picture, and from Eq. (6.2) we can write:

$$\begin{aligned} |\Psi_I(t)\rangle &= e^{i\frac{\hat{H}_0 t}{\hbar}}|\Psi_S(t)\rangle = e^{i\frac{\hat{H}_0 t}{\hbar}} e^{-i\frac{\hat{H}}{\hbar}(t-t_0)}|\Psi_S(t_0)\rangle \\ &= e^{i\frac{\hat{H}_0 t}{\hbar}} e^{-i\frac{\hat{H}}{\hbar}(t-t_0)} e^{-i\frac{\hat{H}_0 t_0}{\hbar}}|\Psi_I(t_0)\rangle \ , \end{aligned}$$

and therefore

$$\hat{U}(t, t_0) \equiv e^{i\frac{\hat{H}_0 t}{\hbar}} e^{-i\frac{\hat{H}(t-t_0)}{\hbar}} e^{-i\frac{\hat{H}_0 t_0}{\hbar}} \ . \tag{6.19}$$

The properties of $\hat{U}(t, t_0)$ can be deduced by the above equation. For example:

$$\hat{U}(t_0, t_0) = \hat{\mathbb{I}} \ , \tag{6.20}$$

where we indicated with $\hat{\mathbb{I}}$ the identity operator, and

$$\hat{U}^+(t, t_0)\hat{U}(t, t_0) = \hat{U}(t, t_0)\hat{U}^+(t, t_0) = \hat{\mathbb{I}} \ , \tag{6.21}$$

which implies

$$\hat{U}^+(t, t_0) = \hat{U}^{-1}(t, t_0) \ . \tag{6.22}$$

Furthermore, we have

$$\hat{U}(t_1, t_2)\hat{U}(t_2, t_3) = \hat{U}(t_1, t_3) \ , \tag{6.23}$$

and

$$\hat{U}(t, t_0)\hat{U}(t_0, t) = \hat{\mathbb{I}} \ , \tag{6.24}$$

which implies

$$\hat{U}(t_0, t) = \hat{U}^+(t, t_0) \ , \tag{6.25}$$

The time evolution of the state can be determined by considering Eq. (6.18) in conjunction with the expression for $\hat{U}$ provided in Eq. (6.19). This approach is equivalent to solving the full Schrödinger equation and presents similar challenges. The advantage of using $\hat{U}$ lies in the possibility to perform a perturbation expansion on it.

Let us rewrite Eq. (6.10) as

$$i\hbar\frac{\partial}{\partial t}|\Psi_I(t)\rangle = \hat{H}_1(t)|\Psi_I(t)\rangle \quad , \tag{6.26}$$

where $\hat{H}_1(t)$ is the term of the hamiltonian in the interaction picture. To simplify the writing, we have omitted the subscript I in the Hamiltonian term, which will be understood henceforth. By using Eq. (6.18) we obtain:

$$i\hbar\frac{\partial}{\partial t}\hat{U}(t,t_0)|\Psi_I(t_0)\rangle = \hat{H}_1(t)\hat{U}(t,t_0)|\Psi_I(t_0)\rangle.$$

We divide this expression by $|\Psi_I(t_0)\rangle$ which is, by construction, different from zero, and we obtain a relation between operators:

$$i\hbar\frac{\partial}{\partial t}\hat{U}(t,t_0) = \hat{H}_1(t)\hat{U}(t,t_0) \ . \tag{6.27}$$

By integrating from t_0 to t we obtain

$$\int_{t_0}^{t} i\hbar\frac{\partial}{\partial t'}\hat{U}(t',t_0)dt' = \int_{t_0}^{t} \hat{H}_1(t')\hat{U}(t',t_0)dt'$$
$$= i\hbar\left[\hat{U}(t,t_0) - \hat{U}(t_0,t_0)\right] = \int_{t_0}^{t} dt'\hat{H}_1(t')\hat{U}(t',t_0).$$

By considering Eq. (6.20) we have:

$$\hat{U}(t,t_0) = 1 - \frac{i}{\hbar}\int_{t_0}^{t} dt'\hat{H}_1(t')\hat{U}(t',t_0). \tag{6.28}$$

We obtain a formal solution Eq. (6.28) by inserting its expression of $\hat{U}(t',t_0)$ in the right hand side:

$$\hat{U}(t,t_0) = 1 - \frac{i}{\hbar}\int_{t_0}^{t} dt'\hat{H}_1(t')\left[1 - \frac{i}{\hbar}\int_{t_0}^{t'} dt''\hat{H}_1(t'')[1-\ldots]\right]$$
$$\hat{U}(t,t_0) = 1 - \frac{i}{\hbar}\int_{t_0}^{t} dt'\hat{H}_1(t') + \left(\frac{-i}{\hbar}\right)^2 \underbrace{\int_{t_0}^{t} dt' \int_{t_0}^{t'} dt''\hat{H}_1(t')\hat{H}_1(t'')}_{t>t'}$$
$$+ \ldots \ . \tag{6.29}$$

Let us consider the third term in the right hand side of the above equation:

$$\int_{t_0}^{t} dt' \int_{t_0}^{t'} dt'' \hat{H}_1(t')\hat{H}_1(t'') = \frac{1}{2}\underbrace{\int_{t_0}^{t} dt' \int_{t_0}^{t'} dt'' \hat{H}_1(t')\hat{H}_1(t'')}_{t'>t''} + \frac{1}{2}\underbrace{\int_{t_0}^{t} dt'' \int_{t_0}^{t''} dt' \hat{H}_1(t'')\hat{H}_1(t')}_{t''>t'} \ . \tag{6.30}$$

In this expression the dummy indexes of the integration variables have been interchanged. In general, one has

$$\int_a^b dy \int_y^b dx f(x) f(y) = \int_a^b dx \int_a^x dy f(x) f(y) \ , \tag{6.31}$$

therefore Eq. (6.30) can be written as:

$$\begin{aligned}
&\int_{t_0}^{t} dt' \int_{t_0}^{t'} dt'' \hat{H}_1(t')\hat{H}_1(t'') = \\
&\frac{1}{2}\int_{t_0}^{t} dt' \int_{t_0}^{t'} dt'' \hat{H}_1(t')\hat{H}_1(t'') + \frac{1}{2}\int_{t_0}^{t} dt' \int_{t'}^{t} dt'' \hat{H}_1(t'')\hat{H}_1(t') = \\
&\frac{1}{2}\int_{t_0}^{t} dt' \int_{t_0}^{t} dt'' \left[\hat{H}_1(t')\hat{H}_1(t'')\theta(t'-t'') + \hat{H}_1(t'')\hat{H}_1(t')\theta(t''-t')\right] = \\
&\frac{1}{2}\int_{t_0}^{t} dt' \int_{t_0}^{t} dt'' \hat{\mathbb{T}}\left[\hat{H}_1(t')\hat{H}_1(t'')\right] \ ,
\end{aligned} \tag{6.32}$$

where we have introduced a **time-ordering operator** $\hat{\mathbb{T}}$ whose action consists in ordering a product of operators in decreasing time from the left to the right.

By generalising the result in Eq. (6.32) for each term, we derive the expression for Eq. (6.29).

$$\hat{U}(t, t_0) = \sum_{n=0}^{\infty} \left(\frac{-i}{\hbar}\right)^n \frac{1}{n!} \int_{t_0}^{t} dt_1 \ldots \int_{t_0}^{t} dt_n \hat{\mathbb{T}}\left[\hat{H}_1(t_1)\ldots\hat{H}_1(t_n)\right] . \tag{6.33}$$

A simple test of the validity of Eq. (6.31).

Let us define

$$\frac{dF(x)}{dx} = f(x) \ .$$

We consider the first term of Eq. (6.31).

$$\int_a^b dy \int_y^b dx f(x) f(y) = \int_a^b dy f(y) \left[F(b) - F(y)\right]$$

$$= F(b)\left[F(b) - F(a)\right] - \int_a^b dy f(y) F(y)$$
$$= F(b)\left[F(b) - F(a)\right] - \int_a^b dy \frac{1}{2}\frac{d}{dy}\left[F(y)\right]^2$$
$$= F^2(b) - F(b)F(a) - \frac{1}{2}\left[F^2(b) - F^2(a)\right]$$
$$= \frac{1}{2}F^2(b) + \frac{1}{2}F^2(a) - F(b)F(a) = \frac{1}{2}\left[F(b) - F(a)\right]^2 \ .$$

The second term of Eq. (6.31) is

$$\int_a^b dx \int_a^x dy f(x) f(y) = \int_a^b dx f(x)\left[F(x) - F(a)\right]$$
$$= \int_a^b dx f(x) F(x) - F(a)\left[F(b) - F(a)\right]$$
$$= \int_a^b dx \frac{1}{2}\frac{d}{dx}\left[F(x)\right]^2 - F(a)F(b) + F^2(a)$$
$$= \frac{1}{2}\left[F^2(b) - F^2(a)\right] - F(a)F(b) + F^2(a)$$
$$= \frac{1}{2}F^2(b) + \frac{1}{2}F^2(a) - F(b)F(a) = \frac{1}{2}\left[F(b) - F(a)\right]^2 \ .$$

6.3 Wick's Theorem

Before we delve into the use of the time-evolution operator in the perturbation expansion, we would like to present a theorem that is highly beneficial for calculating the expectation values of operators expressed in ONR.

6.3.1 Time-Ordering Operator

We have already introduced in Eq. (6.32) the time-ordering operator

$$\hat{\mathbb{T}}[\hat{A}\hat{B}\hat{C}\ldots]. \tag{6.34}$$

This involves ordering a sequence of time-dependent operators, such as those expressed in the interaction picture, with the latest times positioned to the left. Since these operators can be represented in terms of creation and annihilation operators, it is important to consider the anti-commutation relations when performing this ordering. Specifically, a -1 sign must be added for each interchange of fermionic operators.

For example, assuming $t_{n+1} < t_n$, we have that:

$$\hat{\mathbb{T}}\left[\hat{a}(t_3)\hat{a}^+(t_1)\hat{a}^+(t_2)\right] = \hat{a}^+(t_1)\hat{a}^+(t_2)\hat{a}(t_3). \tag{6.35}$$

Another example:

$$\hat{\mathbb{T}}\left[\hat{a}(t_2)\hat{a}^+(t_1)\hat{a}^+(t_3)\right] = -\hat{a}^+(t_1)\hat{a}(t_2)\hat{a}^+(t_3) \ . \tag{6.36}$$

6.3.2 Normal Order Product

By definition, the expectation value of a normally ordered set of creation and annihilation operators with respect to the ground state of the system is zero. This necessitates the construction of an operator $\hat{\mathbb{N}}$ that organises the sequence of creation and annihilation operators to ensure a zero expectation value. The action of $\hat{\mathbb{N}}$ is contingent upon the definition of the system ground state.

In Field Theory, the ground state of the system is referred to as the physical vacuum, denoted as $|0\rangle$, which is a state devoid of particles. In this context, the action of $\hat{\mathbb{N}}$ involves moving the annihilation operators to the right and the creation operators to the left [1]. This because $\hat{a}|0\rangle = 0$, and $\langle 0|\hat{a}^+ = 0$. For example:

$$\hat{\mathbb{N}}[\hat{a}_1\hat{a}_2^+\hat{a}_3\hat{a}_4^+] = -\hat{a}_2^+\hat{a}_4^+\hat{a}_1\hat{a}_3 \ , \quad \text{Field Theory}. \tag{6.37}$$

In the context of many-body systems, it is often more convenient to consider a ground state represented by the solution to the MF problem, denoted as $|\Phi_0\rangle$. In this state, all the single particle states below the Fermi surface are occupied, referred to as hole states, while those above the Fermi surface are unoccupied and represent particle states. The action of the operator $\hat{\mathbb{N}}$ on the creation and annihilation operators is more complex than in the previous case. Following conventional terminology, we use the letters i, j, k, l to denote hole states, and m, n, p, q, r to represent particle states. Greek letters are employed to indicate a generic state, which can be either a hole or a particle. The action of the creation and destruction operators on the MF ground state is:

$$\begin{aligned} \hat{a}_j|\Phi_0\rangle \neq 0 \quad &; \quad \hat{a}_m|\Phi_0\rangle = 0 \ ; \\ \hat{a}_j^+|\Phi_0\rangle = 0 \quad &; \quad \hat{a}_m^+|\Phi_0\rangle \neq 0 \ . \end{aligned} \tag{6.38}$$

In this case, the action of $\hat{\mathbb{N}}$ involves moving the $\hat{a}_m$ and $\hat{a}_j^+$ operators to the right, while shifting the $\hat{a}_j$ and $\hat{a}_m^+$ operators to the left. Additionally, each time the position of an operator is changed, a minus sign is introduced. The following example illustrates this, containing two minus signs:

$$\hat{\mathbb{N}}[\hat{a}_m\hat{a}_j^+\hat{a}_j\hat{a}_m^+] = (-\hat{a}_j\hat{a}_m^+)(-\hat{a}_m\hat{a}_j^+) = \hat{a}_j\hat{a}_m^+\hat{a}_m\hat{a}_j^+ \ . \tag{6.39}$$

6.3.3 Contraction

The contraction is defined as the difference between two time-ordered and normal-ordered operators.

$$\overline{\hat{A}\hat{B}} \equiv \hat{\mathbb{T}}[\hat{A}\hat{B}] - \hat{\mathbb{N}}[\hat{A}\hat{B}] \ . \tag{6.40}$$

We identify the two operators to be contracted by a line that connect them.

If the operators are time-independent, or if they are defined at the same time, the action of $\hat{\mathbb{T}}$ is that of the identity operator: $\hat{\mathbb{T}}[\hat{A}\hat{B}] = \hat{A}\hat{B}$. For example,

$$\overline{\hat{a}_m^+\hat{a}_i} = \hat{\mathbb{T}}[\hat{a}_m^+\hat{a}_i] - \hat{\mathbb{N}}[\hat{a}_m^+\hat{a}_i] = \hat{a}_m^+\hat{a}_i - \hat{a}_m^+\hat{a}_i = 0 \ .$$

The result of the contraction is not an operator, or an operator sequence, but it is a number. This is a consequence of the anti-commutation rules (5.33) of the creation and destruction operators.

The contraction is the expectation value of the two operators on the ground state:

$$\langle\Phi_0|\hat{A}\hat{B}|\Phi_0\rangle = \langle\Phi_0|\overline{\hat{A}\hat{B}}|\Phi_0\rangle + \langle\Phi_0|\hat{\mathbb{N}}[\hat{A}\hat{B}]|\Phi_0\rangle = \overline{\hat{A}\hat{B}}\langle\Phi_0|\Phi_0\rangle \ , \tag{6.41}$$

where we used the fact that $\langle\Phi_0|\hat{\mathbb{N}}[\hat{A}\hat{B}]|\Phi_0\rangle = 0$ by definition.

In the case of a sequence of operators where contractions need to be performed, it is essential for the involved operators to be in close proximity. This requirement is fulfilled by taking into account the anti-commutation rules, which introduce a factor of minus one with each exchange of operators. For example:

$$\hat{A}\hat{B}\hat{C}\hat{D}\hat{E}\hat{F} = -\overline{\hat{A}\hat{C}}\,\overline{\hat{B}\hat{F}}\hat{D}\hat{E} \ . \tag{6.42}$$

6.3.4 Formulation of the Wick's Theorem

The Wick's theorem [2] states that a time-ordered product of operators can be expressed as a sum of normal-ordered products, with all possible contractions performed.

$$
\begin{aligned}
\hat{\mathbb{T}}[\hat A\hat B\hat C\dots\hat Z] &= \hat{\mathbb{N}}[\hat A\hat B\hat C\dots\hat Z] + \hat{\mathbb{N}}[\overline{\hat A\hat B}\dots\hat Z] + \hat{\mathbb{N}}[\hat A^{\bullet}\hat B\hat C^{\bullet}\dots\hat Z] \\
&\quad + \hat{\mathbb{N}}[\overline{\hat A\hat B}\,\overline{\hat C\hat D}\dots\hat Z] + \hat{\mathbb{N}}[\hat A^{\bullet}\hat B^{\circ}\hat C^{\bullet}\hat D^{\circ}\dots\hat Z] \\
&\quad + \hat{\mathbb{N}}[\hat A^{\bullet}\overline{\hat B\hat C}\hat D^{\bullet}\dots\hat Z] + \dots\ .
\end{aligned}
$$

Since the result of a contraction is a number, all the contracted operators are not involved by the action of $\hat{\mathbb{N}}$. For example, the Wick's theorem applied to the product of four operators gives:

$$
\begin{aligned}
\hat A\hat B\hat C\hat D &= \hat{\mathbb{N}}[\hat A\hat B\hat C\hat D] + \hat{\mathbb{N}}[\overline{\hat A\hat B}\hat C\hat D] + \hat{\mathbb{N}}[\hat A^{\bullet}\hat B\hat C^{\bullet}\hat D] + \hat{\mathbb{N}}[\hat A^{\bullet}\hat B\hat C\hat D^{\bullet}] \\
&\quad + \hat{\mathbb{N}}[\hat A\overline{\hat B\hat C}\hat D] + \hat{\mathbb{N}}[\hat A\hat B^{\bullet}\hat C\hat D^{\bullet}] + \hat{\mathbb{N}}[\hat A\hat B\overline{\hat C\hat D}] \\
&\quad + \hat{\mathbb{N}}[\overline{\hat A\hat B}\,\overline{\hat C\hat D}] + \hat{\mathbb{N}}[\hat A^{\bullet}\hat B^{\circ}\hat C^{\bullet}\hat D^{\circ}] + \hat{\mathbb{N}}[\hat A^{\bullet}\hat B^{\circ}\hat C^{\circ}\hat D^{\bullet}] \\
&= \hat{\mathbb{N}}[\hat A\hat B\hat C\hat D] + \overline{\hat A\hat B}\,\hat{\mathbb{N}}[\hat C\hat D] - \overline{\hat A\hat C}\,\hat{\mathbb{N}}[\hat B\hat D] + \overline{\hat A\hat D}\,\hat{\mathbb{N}}[\hat B\hat C] \\
&\quad + \overline{\hat B\hat C}\,\hat{\mathbb{N}}[\hat A\hat D] - \overline{\hat B\hat D}\,\hat{\mathbb{N}}[\hat A\hat C] + \overline{\hat C\hat D}\,\hat{\mathbb{N}}[\hat A\hat B] \\
&\quad + \overline{\hat A\hat B}\,\overline{\hat C\hat D} - \overline{\hat A\hat C}\,\overline{\hat B\hat D} + \overline{\hat A\hat D}\,\overline{\hat B\hat C}\ .
\end{aligned}
\tag{6.43}
$$

This expression indicates that calculating the expectation value of these operators with respect to the ground state simplifies to evaluating only the terms that contain contractions. This is because the terms involving $\hat{\mathbb{N}}$ are zero by definition.

6.4 Adiabatic Switching on of the Interaction

Here, we employ the adiabatic switching on of the interaction as a mathematical technique to express the eigenstates of a system of interacting particles in terms of the states of non-interacting particles, specifically MF states. This approach is particularly advantageous because the MF problem can be solved with relative ease.

Let us consider a hamiltonian of the form

$$
\hat H = \hat H_0 + e^{-\epsilon|t|}\hat H_1\ , \tag{6.44}
$$

where ϵ is a real, and positive, number. Obviously:

$$
\lim_{t\to\pm\infty} \hat H = \hat H_0\ , \tag{6.45}
$$

and at the time $t = 0$ the hamiltonian (6.44) is the full hamiltonian:

$$\lim_{t \to 0} \hat{H} = \hat{H}_0 + \hat{H}_1 \ . \tag{6.46}$$

The value of the parameter ϵ can be chosen to gradually turn the perturbation on and off as desired. The final results must be independent of the value of ϵ. In interaction picture, the eigenstates of the hamiltonian (6.44) can be expressed as:

$$|\Psi_I(t)\rangle = \hat{U}_\epsilon(t, t_0)|\Psi_I(t_0)\rangle \ . \tag{6.47}$$

The subscript ϵ has been added to the time evolution operator $\hat{U}$ since in its definition, Eq. (6.19), instead of $\hat{H}_1(t)$ we use $e^{-\epsilon|t|}\hat{H}_1(t)$. The expression (6.33) of the time-evolution operator becomes:

$$\hat{U}_\epsilon(t, t_0) = \sum_{n=0}^{\infty} \left(\frac{-i}{\hbar}\right)^n \frac{1}{n!} \int_{t_0}^{t} dt_1 \ldots \int_{t_0}^{t} dt_n \\ e^{-\epsilon[|t_1|+|t_2|+\ldots]}\hat{\mathbb{T}}[\hat{H}_1(t_1)\ldots\hat{H}_1(t_n)] \ , \tag{6.48}$$

where the exponential terms have been extracted from $\hat{\mathbb{T}}$ since they commute with $\hat{H}_1$. The equation of motion (6.10) for a system described by the hamiltonian (6.44) can be expressed as:

$$i\hbar\frac{\partial}{\partial t}|\Psi_I(t)\rangle = e^{-\epsilon|t|}\hat{H}_1(t)|\Psi_I(t)\rangle \underset{t\to\pm\infty}{\longrightarrow} 0 \ , \tag{6.49}$$

therefore, in the limit $t \to \pm\infty$ we have that $|\Psi_I(\pm\infty)\rangle$ is time independent. For $t \to \pm\infty$ the hamiltonian reduces to $\hat{H}_0$, therefore we have

$$|\Psi_I(t)\rangle = \hat{U}_\epsilon(t, -\infty)|\Phi_0\rangle \ ,$$

where $|\Phi_0\rangle$ is eigenstate of $\hat{H}_0$.

If the interaction term $\hat{H}_1$ were not present, then $|\Psi_I\rangle$ would always be equal to $|\Phi_0\rangle$. As time progresses, the interaction is gradually turned on until $t = 0$, when it is fully active. After this moment, the interaction is slowly turned off until it becomes zero as t approaches $+\infty$.

The definitions of the state in the Heisenberg, Schrödinger and interaction pictures determine the states at $t = 0$:

$$|\Psi_H(t)\rangle_{t\to 0} = \lim_{t\to 0} e^{i\frac{\hat{H}t}{\hbar}}|\Psi_S(t)\rangle = |\Psi_S(0)\rangle, \tag{6.50}$$

$$|\Psi_I(t)\rangle_{t\to 0} = \lim_{t\to 0} e^{i\frac{\hat{H}_0 t}{\hbar}}|\Psi_S(t)\rangle = |\Psi_S(0)\rangle, \tag{6.51}$$

$$|\Psi_H(0)\rangle = |\Psi_I(0)\rangle = |\Psi_S(0)\rangle \ , \tag{6.52}$$

and then:

$$|\Psi_S(0)\rangle = |\Psi_H(0)\rangle = |\Psi_I(0)\rangle = \hat{U}_\epsilon(0, -\infty)|\Phi_0\rangle \ . \tag{6.53}$$

The equation above represents the eigenstate of an interacting Hamiltonian $\hat{H}$ in terms of an eigenstate of a non-interacting, MF Hamiltonian $\hat{H}_0$. This result is physically meaningful if all quantities calculated in the limit as $\epsilon \to 0$ remain finite.

The **Gell-Mann and Low theorem** [3] answer to this question. Let us consider a perturbation expansion of the following mathematical entity

$$\frac{\hat{U}_\epsilon(0, -\infty)|\Phi_0\rangle}{\langle\Phi_0|\hat{U}_\epsilon(0, -\infty)|\Phi_0\rangle} \equiv \frac{|\Psi_0\rangle}{\langle\Phi_0|\Psi_0\rangle} = \sum_{n=0}^{\infty} \left|\xi_\epsilon^{(n)}\right\rangle g^n \ , \tag{6.54}$$

where g^n is a generic expansion parameter. If, for each n the $\lim_{\epsilon\to 0}\left|\xi_\epsilon^{(n)}\right\rangle$ exists, i.e. it is finite, then, the full quantity defined in Eq. (6.54) is eigenstate of the hamiltonian $\hat{H}$:

$$\hat{H}\frac{|\Psi_0\rangle}{\langle\Phi_0|\Psi_0\rangle} = E_0\frac{|\Psi_0\rangle}{\langle\Phi_0|\Psi_0\rangle} \ . \tag{6.55}$$

Multiplying on the left both terms of the above equation by $\langle\Phi_0|$ we obtain:

$$\begin{aligned}\frac{\langle\Phi_0|\hat{H}|\Psi_0\rangle}{\langle\Phi_0|\Psi_0\rangle} = E_0 &= \frac{\langle\Phi_0|\hat{H}_0|\Psi_0\rangle}{\langle\Phi_0|\Psi_0\rangle} + \frac{\langle\Phi_0|\hat{H}_1|\Psi_0\rangle}{\langle\Phi_0|\Psi_0\rangle} \\ &= \mathcal{E}_0 + \frac{\langle\Phi_0|\hat{H}_1|\Psi_0\rangle}{\langle\Phi_0|\Psi_0\rangle} \ ,\end{aligned}$$

therefore:

$$E_0 - \mathcal{E}_0 = \frac{\langle\Phi_0|\hat{H}_1|\Psi_0\rangle}{\langle\Phi_0|\Psi_0\rangle} = \frac{\langle\Phi_0|\hat{H}_1\,\hat{U}(0, -\infty)|\Phi_0\rangle}{\langle\Phi_0|\hat{U}(0, -\infty)|\Phi_0\rangle}. \tag{6.56}$$

Detailed proofs of the Gell-Mann and Low theorem can be found in various textbooks, such as [4, 5]. Here, we would like to emphasis the key point of the theorem: in the limit as $\epsilon \to 0$, both the numerator and denominator of Eq. (6.54) do not exist separately; however, the limit of their ratio is finite. The numerator contains divergences that are canceled out by corresponding divergences in the denominator.

The result in Eq. (6.56) is significant because it highlights the difference in energy between a system of interacting particles and that of non-interacting particles as described by a MF model. All the necessary components to calculate this quantity are well understood. Our approach is based on the fundamental assumption that the MF problem has been solved, which means that the state $|\Phi_0\rangle$ is known. The time-evolution operator $\hat{U}$ is determined using the perturbation expansion outlined in Eq. (6.33). Here, Wick's theorem facilitates the evaluation of the time-ordering operator $\hat{\mathbb{T}}$ applied to the various $\hat{H}_1$ operators acting at different times.

References

1. M.E. Peskin, D.V. Schroeder, *An Introduction to Quantum Field Theory* (Westwiew Press, Boulder, 1995)
2. G.C. Wick, The evaluation of the collision matrix. Phys. Rev. **80**, 268 (1950)
3. M. Gell-Mann, F. Low, Bound states in quantum field theory. Phys. Rev. **84**, 350 (1951)
4. A.L. Fetter, J.D. Walecka, *Quantum Theory of Many-Particle Systems* (Mc Graw-Hill, New York, 1971)
5. E.K.U. Gross, E. Runge, O. Heinonen, *Many-Particle Theory* (Adam Hilger, Bristol, 1991)

Chapter 7
Goldstone Theorem

7.1 Goldstone Diagrams

The analysis of the different terms in the perturbative expansion is made easier through the use of graphical techniques. These techniques involve associating a graphical symbol, or diagram, with each term of the perturbative expansion.

In the relativistic framework of Field Theory, these diagrams are known as **Feynman diagrams**. There is a specific set of rules that enables the precise reconstruction of the mathematical expression corresponding to the perturbative term depicted graphically.

We work in the field of non-relativistic quantum mechanics, and therefore we prefer to refer to these graphical symbols as **Goldstone diagrams**. Additionally, we do not establish rules for mapping the diagrams to precise mathematical expressions. Instead, we use these diagrams to identify the terms in the perturbative expansion that share similar characteristics.

In Fig. 7.1, various elements that make up the diagram are presented. In this representation, we envision an ideal time arrow oriented from the bottom to the top of the diagram. This line is illustrated in Fig. 7.1, and it will be understood in all subsequent figures.

An oriented line extending from a point and moving in the direction of positive time (case A in the figure) represents the creation of a particle. This diagram is associated with the creation operator $\hat{a}_m^+$. We use the convention of denoting states above the Fermi surface with the letters m, n, p, q, r, and states below it with the letters i, j, k, l. In the case of coordinate representation, this symbol corresponds to the field operator $\hat{\psi}^+(\mathbf{r})$, which indicates the creation of a particle at the point $\mathbf{r}$.

Conversely, an oriented line moving in the positive time direction and terminating at a point (case B in the figure) represents the annihilation of a particle. The operators associated with this process are $\hat{a}_m$ or $\hat{\psi}(\mathbf{r})$.

In relativistic Field Theory, lines oriented in opposite directions with respect to the time arrow (negative times) represent antiparticles. In the context of Goldstone

© The Author(s), under exclusive license to Springer Nature Switzerland AG 2026

G. Co', *Concepts in Quantum Many-Body Physics*, UNITEXT for Physics,
https://doi.org/10.1007/978-3-032-08920-5_7

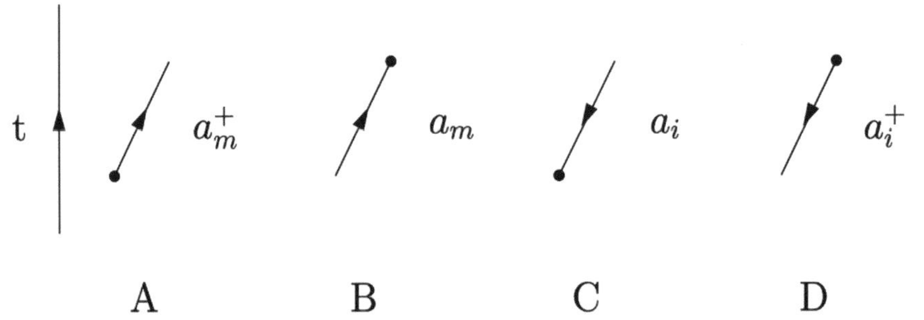

Fig. 7.1 Basic graphical symbols used in Goldstone diagrams

diagrams, however, these lines represent hole states, which are single-particle states located below the Fermi surface.

A disappearing line (case D in the figure) indicates the creation of a hole state, with the associated operators being $\hat{a}_i^+$ or $\hat{\psi}(\mathbf{r})$. Conversely, an oriented line that moves backward in time and disappears at a point (case C in the figure) signifies the destruction of a hole state. In this scenario, the relevant operators are $\hat{a}_i$ o $\hat{\psi}^+(\mathbf{r})$.

The final graphic element needed to complete the diagrams is related to the interaction $\hat{V}(\mathbf{r}_1, \mathbf{r}_2)$. We have chosen a dashed line connecting two points to represent the action of two-body interactions. Since the framework is non-relativistic, the effect of the interaction is instantaneous; thus, in the Goldstone diagrams, the dashed lines are always horizontal. In contrast, this is not the case in Feynman diagrams, where the interaction propagates at a finite velocity.

As an example of Goldstone diagrams, we present two diagrams in Fig. 7.2, for which we will calculate the explicit expressions. The diagrams represent the expectation values of the type:

$$\langle \Phi_a | \hat{V} | \Phi_b \rangle \ , \tag{7.1}$$

where $|\Phi_a\rangle$ indicates the initial state and $|\Phi_b\rangle$ the final one.

Let us consider the A diagram of Fig. 7.2. The initial state is formed by a particle state n and by a hole state i:

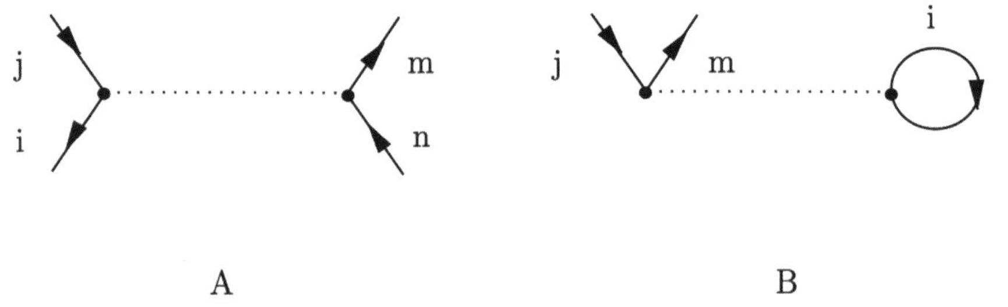

Fig. 7.2 Example of Goldstone diagrams

$$|\Phi_a\rangle = \hat{a}_n^+ \hat{a}_i |\Phi_0\rangle \Longrightarrow \langle\Phi_a| = \langle\Phi_0|\hat{a}_i^+ \hat{a}_n \ . \tag{7.2}$$

In this diagram also the final state is composed by one hole and one particle states which are different from those of the initial state:

$$|\Phi_b\rangle = \hat{a}_m^+ \hat{a}_j |\Phi_0\rangle. \tag{7.3}$$

The expression of the A diagram of Fig. 7.2 is:

$$\langle\Phi_0|\hat{a}_i^+ \hat{a}_n \hat{V} \hat{a}_m^+ \hat{a}_j |\Phi_0\rangle. \tag{7.4}$$

At this point we insert the ONR expression of the interaction $\hat{V}$

$$\hat{V} = \frac{1}{2} \sum_{\nu\nu'\mu\mu'} V_{\nu\mu\nu'\mu'} \hat{a}_\nu^+ \hat{a}_\mu^+ \hat{a}_{\mu'} \hat{a}_{\nu'}, \tag{7.5}$$

and, with the help of the Wick's theorem, we can evaluate the expectation value indicated by the diagram.

The evaluation of the B diagram of Fig. 7.2 proceeds in similar way. The difference with the previous case is that the hole states of the initial and final states are the same. The expression of the diagram is:

$$\langle\Phi_0|\hat{a}_i^+ \hat{a}_j \hat{V} \hat{a}_i \hat{a}_m^+ |\Phi_0\rangle. \tag{7.6}$$

7.2 Goldstone Theorem

The Goldstone theorem [1] states that the difference between the energy of system of interacting particles and that of a system of non-interacting particles can be expressed as:

$$E_0 - \mathcal{E}_0 = \langle\Phi_0|\hat{H}_1 \sum_{n=0}^{\infty} \left(\frac{1}{\mathcal{E}_0 - \hat{H}_0} \hat{H}_1 \right)^n |\Phi_0\rangle_c \ , \tag{7.7}$$

where $\hat{H}_0$ and $\hat{H}_1$ are time-independent operators in the Schrödinger picture, and $|\Phi_0\rangle$ is the eigenstate of $\hat{H}_0$, the MF hamiltonian. The meaning of the important subscript c will be clarified below.

Let us write explicitly the first terms of Eq. (7.7):

$$\begin{aligned} E_0 - \mathcal{E}_0 &= \langle\Phi_0|\hat{H}_1|\Phi_0\rangle + \langle\Phi_0|\hat{H}_1 \frac{1}{\mathcal{E}_0 - \hat{H}_0} \hat{H}_1|\Phi_0\rangle_c \\ &+ \langle\Phi_0|\hat{H}_1 \frac{1}{\mathcal{E}_0 - \hat{H}_0} \hat{H}_1 \frac{1}{\mathcal{E}_0 - \hat{H}_0} \hat{H}_1|\Phi_0\rangle_c + \cdots \ . \end{aligned} \tag{7.8}$$

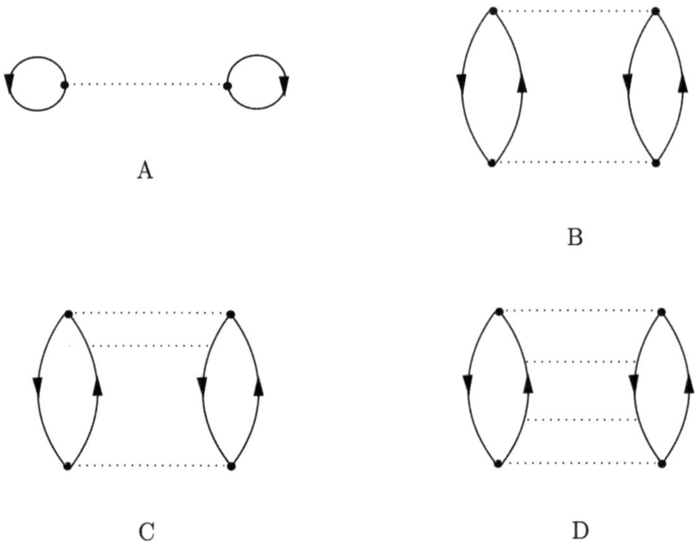

Fig. 7.3 Some diagrams related to the terms of Eq. (7.8)

Some of the diagrams related to the expansion terms are shown in Fig. 7.3. The diagram A is contained in the first term of Eq. (7.8). A single dashed line indicates the presence of only one interaction $\hat{H}_1$. There are not open lines of particle or holes since the expectation value is evaluated on the MF ground state. The other diagram present in the first expansion term is that called exchange term. In this diagram the particle and hole lines are interchanged between the various points. (This is the diagram B of Fig. 7.4).

The diagram B in Fig. 7.3 represents one of the diagrams associated with the second term of Eq. (7.8), as it includes two interaction lines of the operator $\hat{H}_1$. Additionally, this diagram features an exchange term. It illustrates that $\hat{H}_1$ creates an intermediate state characterised by two particle states and two hole states (denoted as $2p-2h$). This occurs because $\hat{H}_1$ is a two-body operator. The intermediate state propagates, with the term $(\mathcal{E}_0 - \hat{H}_0)^{-1}$ which is the Fourier transform of the time propagation operator, and it is subsequently de-excited by the action of another $\hat{H}_1$ term.

The traditional expression of the perturbative expansion can be obtained by inserting a completeness of $\hat{H}_0$ eigenstates $\hat{\mathbb{I}} = \sum_{n\neq 0} |\Phi_n\rangle\langle\Phi_n|$, (where $\hat{\mathbb{I}}$ is the identity operator):

$$E_0 - \mathcal{E}_0 = \langle\Phi_0|\hat{H}_1|\Phi_0\rangle + \sum_{n\neq 0} \frac{\langle\Phi_0|\hat{H}_1|\Phi_n\rangle\langle\Phi_n|\hat{H}_1|\Phi_0\rangle}{\mathcal{E}_0 - \mathcal{E}_n} + \dots \ . \tag{7.9}$$

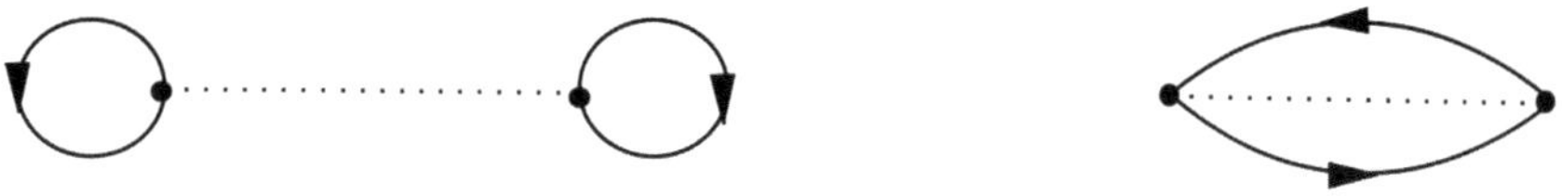

A B

Fig. 7.4 Diagrams representing the (7.14) and (7.15) terms

The absence of the $n = 0$ terms is a subtle point that we clarify below, where we demonstrate how to derive Eq. (7.7) from the expression of the Gell-Mann and Low theorem.

We consider Eq. (6.56)

$$E_0 - \mathcal{E}_0 = \frac{\langle \Phi_0 | \hat{H}_1 | \Psi_0 \rangle}{\langle \Phi_0 | \Psi_0 \rangle} = \frac{\langle \Phi_0 | \hat{H}_1 \hat{U}(0, -\infty) | \Phi_0 \rangle}{\langle \Phi_0 | \hat{U}(0, -\infty) | \Phi_0 \rangle} . \tag{7.10}$$

By using the expression (6.33) of the time evolution operator, we obtain, for the numerator:

$$\langle \Phi_0 | \hat{H}_1 \hat{U}(0, -\infty) | \Phi_0 \rangle = \sum_{\nu=0} \left(\frac{-i}{\hbar} \right)^{\nu} \frac{1}{\nu!} \int_{-\infty}^{0} dt_1 \ldots \int_{-\infty}^{0} dt_{\nu}$$
$$\langle \Phi_0 | \hat{\mathbb{T}}[\hat{H}_1, \hat{H}_1(t_1) \ldots, \hat{H}_1(t_{\nu})] | \Phi_0 \rangle , \tag{7.11}$$

where we used the $\lim \varepsilon \to 0$ of Eq. (6.48). The operator $\hat{H}_1$, which in (6.56) is outside the action of $\hat{\mathbb{T}}$, has been placed in the first position among the operators that need to be ordered by $\hat{\mathbb{T}}$. This adjustment does not affect the results, as $\hat{H}_1$ is defined at $t = 0$, which is the latest time considered.

The different terms of the perturbative expansion can be calculated using Wick's theorem. Analysing the generated diagrams reveals that they can be classified into two categories: *linked* and *unlinked*.

The linked diagrams illustrate the arrangement of contractions between the creation and destruction operators, ensuring that all terms in the diagrams, each defined at a different time, are interconnected.

Let us consider, for example, the term with $\nu = 1$:

$$\langle \Phi_0 | \hat{H}_1(0) \hat{H}_1(t_1) | \Phi_0 \rangle \sim$$
$$\langle \Phi_0 | \hat{V}_{\mu\nu\mu'\nu'} \hat{a}^+_{\mu}(0) \hat{a}^+_{\nu}(0) \hat{a}_{\nu'}(0) \hat{a}_{\mu'}(0)$$
$$\hat{V}_{\eta\xi\eta'\xi'} \hat{a}^+_{\eta}(t_1) \hat{a}^+_{\xi}(t_1) \hat{a}_{\xi'}(t_1) \hat{a}_{\eta'}(t_1) | \Phi_0 \rangle , \tag{7.12}$$

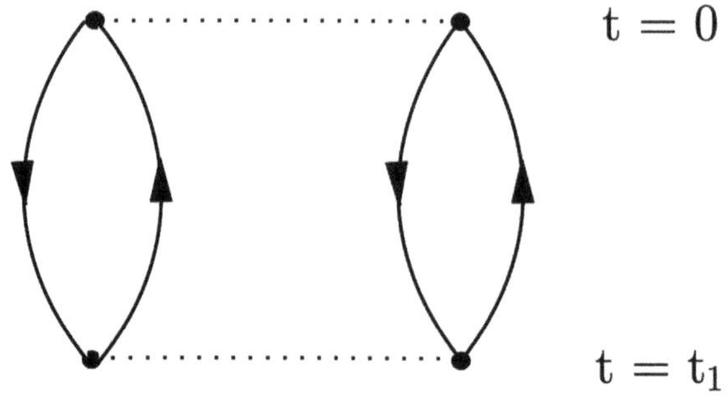

Fig. 7.5 Diagram representing Eq. (7.16)

where we have explicitly written $\hat{H}_1$ in terms of creation and destruction operators in interaction picture, and this is the reason why they are time-dependent. For simplicity, we will assume that sums over repeated Greek indexes are understood.

In absence of contraction between operators defined at the time t_1 and those defined at $t = 0$, the two terms can be separated by inserting an identity operator $|\Phi_0\rangle\langle\Phi_0| = \hat{\mathbb{I}}$:

$$\langle\Phi_0|V_{\mu\nu\mu'\nu'}a_\mu^+ a_\nu^+ a_{\nu'} a_{\mu'}|\Phi_0\rangle_{t=0}\langle\Phi_0|V_{\eta\xi\eta'\xi'}a_\eta^+ a_\xi^+ a_{\xi'} a_{\eta'}|\Phi_0\rangle_{t=t_1} \ . \tag{7.13}$$

Here the subscripts $t = 0$ and $t = t_1$ indicate the time dependence of the creation and destruction operators. The contractions different from zero in these two terms are:

$$\hat{a}_\mu^+(t)\hat{a}_\nu^+(t)\hat{a}_{\nu'}(t)\hat{a}_{\mu'}(t) \ , \tag{7.14}$$

and

$$\hat{a}_\mu^+(t)\hat{a}_\nu^+(t)\hat{a}_{\nu'}(t)\hat{a}_{\mu'}(t) \ . \tag{7.15}$$

The first of these terms is illustrated in Diagram A of Fig. 7.4, while the second term is represented in Diagram B of the same figure.

By utilizing this result, we can see that the term in Eq. (7.13) generates four diagrams. Each diagram is formed by coupling two of the diagrams from Fig. 7.4: one at time $t = 0$ and the other at time $t = t_1$. It is evident that these diagrams are unlinked.

Conversely, if we apply the Wick theorem by utilizing contractions that connect operators defined at $t = 0$ with those defined at $t = t_1$, we obtain linked diagrams. For example:

$$\langle\Phi_0|V_{\mu\nu\mu'\nu'}\hat{a}_\mu^+(0)\hat{a}_\nu^+(0)\hat{a}_{\nu'}(0)\hat{a}_{\mu'}(0)V_{\eta\xi\eta'\xi'}\hat{a}_\eta^+(t)\hat{a}_\xi^+(t)\hat{a}_{\xi'}(t)\hat{a}_{\eta'}(t)|\Phi_0\rangle \ . \tag{7.16}$$

The diagram representing this term is shown in Fig. 7.5.

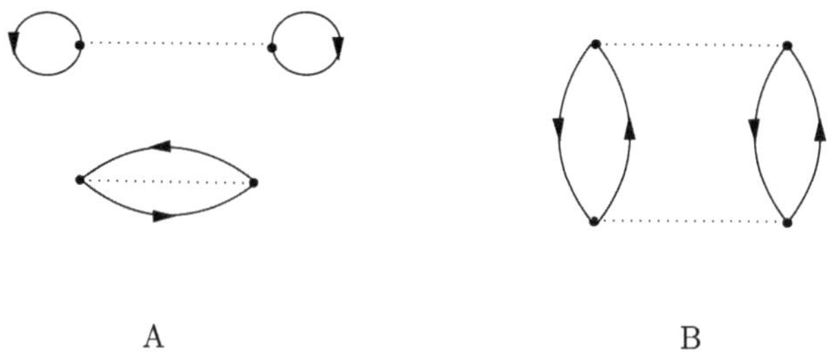

Fig. 7.6 Linked and unlinked terms of Eq. (7.12)

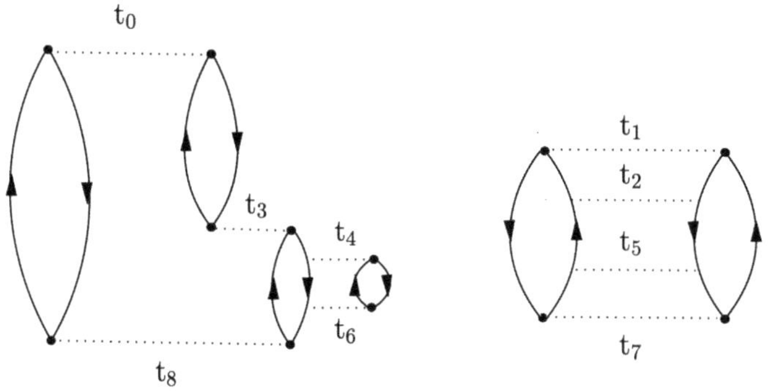

Fig. 7.7 An unliked diagram composed by two parts

The term with $\nu = 1$ in Eq. (7.12) produces both unlinked diagrams, as the diagram A of Fig. 7.6, and linked diagrams, as the diagram B of the same figure.

Following our discussion on the structure of the various terms in the expansion, we will now derive Eq. (7.7). Let us consider one term from the sum in Eq. (7.11) and assume it consists of two unlinked components, similar to the diagram illustrated in Fig. 7.7.

The contribution of a diagram of this type is:

$$\langle\Phi_0|\hat{H}_1\hat{U}(0,-\infty)|\Phi_0\rangle_\nu = \left(\frac{-i}{\hbar}\right)^\nu \frac{1}{\nu!}\int_{-\infty}^0 dt_1 \dots \int_{-\infty}^0 dt_n \langle\Phi_0|\hat{\mathbb{T}}[\hat{H}_1(0)\dots\hat{H}_1(t_n)]|\Phi_0\rangle_c$$
$$\int_{-\infty}^0 dt_{n+1}\dots\int_{-\infty}^0 dt_{n+m} \quad \langle\Phi_0|\hat{\mathbb{T}}[\hat{H}_1(t_{n+1})\dots\hat{H}_1(t_{n+m})]|\Phi_0\rangle_c \ . \quad (7.17)$$

We factorised the two linked parts identified by the subscript c. The first term contains n interaction terms $\hat{H}_1$ and the second one m interactions.

The total contribution in Eq. (7.11) from all the terms illustrated in Fig. 7.7 is derived by considering all possible exchanges of two $\hat{H}_1$ operators defined at different times. Since each $\hat{H}_1$ operator comprises four creation and annihilation operators, the total phase generated by this exchange is always positive. The number of possible permutations is given by $\nu!$.

Conversely, two $\hat{H}_1$ operators that belong to the same partition do not yield a new diagram, as $\hat{\mathbb{T}}$ reorders the various terms. The number of permutations that do not result in any new diagram is $n!\,m!$.

The total contribution to Eq. (7.11) of diagrams such as that of Fig. 7.7 is:

$$\sum_n \sum_m \left(\frac{-i}{\hbar}\right)^{n+m} \frac{\nu!}{n!m!}\frac{1}{\nu!}\int_{-\infty}^{0} dt_1 \ldots \int_{-\infty}^{0} dt_n \langle \Phi_0|\hat{\mathbb{T}}[\hat{H}_1, \hat{H}_1(t_1)\ldots \hat{H}_n(t_n)]|\Phi_0\rangle_c$$
$$\int_{-\infty}^{0} dt_{n+1}\ldots \int_{-\infty}^{0} dt_{n+m}\langle \Phi_0|\hat{\mathbb{T}}[\hat{H}_1(t_{n+1})\ldots \hat{H}_1(t_{n+m})]|\Phi_0\rangle_c. \tag{7.18}$$

The second term of this expression appears in the denominator of Eq. (7.10). We note that the difference between the numerator and denominator is the absence of $\hat{H}_1(0)$ in the latter. Equation (7.10) can be rewritten as:

$$\begin{aligned}
&E_0 - \mathcal{E}_0 \\
&= \frac{\left[\sum_{\nu=0}^{\infty}\left(\frac{-i}{\hbar}\right)^{\nu}\frac{1}{\nu!}\int_{-\infty}^{0} dt_1\ldots\int_{-\infty}^{0} dt_\nu \langle\Phi_0|\hat{\mathbb{T}}[\hat{H}_1(0), \hat{H}_1(t_1)\ldots, \hat{H}_1(t_\nu)]|\Phi_0\rangle\right]_c \left[\sum_{\mu=0}^{\infty}\left(\frac{-i}{\hbar}\right)^{\mu}\frac{1}{\mu!}\int_{-\infty}^{0} dt_1\ldots\int_{-\infty}^{0} dt_\mu \langle\Phi_0|\hat{\mathbb{T}}[\hat{H}_1(t_1)\ldots, \hat{H}_1(t_\mu)]|\Phi_0\rangle\right]}{\langle\Phi_0|\hat{U}(0,-\infty)|\Phi_0\rangle} \\
&= \left[\sum_{\nu=0}^{\infty}\left(\frac{-i}{\hbar}\right)^{\nu}\frac{1}{\nu!}\int_{-\infty}^{0} dt_1\ldots\int_{-\infty}^{0} dt_\nu \langle\Phi_0|\hat{\mathbb{T}}[\hat{H}_1, \hat{H}_1(t_1)\ldots, \hat{H}_1(t_\nu)]|\Phi_0\rangle\right]_c ,
\end{aligned}$$

where we used the fact that ν, μ, and the times t, are dummy integrations indexes. All the diagrams not linked to $\hat{H}_1(0)$ are eliminated by the denominator.

After clarifying why only connected diagrams appear in the expression of the Goldstone theorem, we will calculate the time integrals. Let us consider the order n of the expansion and utilise the explicit expression of the $\hat{H}_1$ operator in the interaction picture:

$$\hat{H}_{I,1}(t) = e^{i\frac{\hat{H}_0 t}{\hbar}}\hat{H}_1 e^{-i\frac{\hat{H}_0 t}{\hbar}}. \tag{7.19}$$

The term of order n can be written as:

$$[E_0 - \mathcal{E}_0]_n = \left(\frac{-i}{\hbar}\right)^n \int_{-\infty}^{0} dt_1 \int_{-\infty}^{t_1} dt_2 \ldots \int_{-\infty}^{t_{n-1}} dt_n e^{\epsilon(t_1+\ldots t_n)}$$
$$\langle\Phi_0|\hat{H}_1 e^{i\frac{\hat{H}_0 t_1}{\hbar}}\hat{H}_1 e^{-i\frac{\hat{H}_0 t_1}{\hbar}} e^{i\frac{\hat{H}_0 t_2}{\hbar}}\hat{H}_1 \ldots e^{-i\frac{\hat{H}_0}{\hbar}t_{n-1}} e^{i\frac{\hat{H}_0}{\hbar}t_n}\hat{H}_1 e^{-i\frac{\hat{H}_0}{\hbar}t_n}|\Phi_0\rangle_c.$$

In the above expression $\hat{\mathbb{T}}$ has been eliminated since we have explicitly written the integration limits and we inserted the factor $e^{\epsilon t}$. We define a new set of integration variables which we denote as x such as

$$x_1 = t_1 \quad , \quad x_2 = t_2 - t_1 \quad , \quad x_3 = t_3 - t_2 \quad , \quad \dots \quad , \quad x_n = t_n - t_{n-1} \,.$$

The time variables are related to these new variables x as

$$t_1 = x_1 \quad , \quad t_2 = x_2 + x_1 \quad , \quad t_3 = x_3 + x_2 + x_1 \quad , \quad \dots \quad , \quad t_n = \sum_n x_n \,.$$

By considering $\hat{H}_0|\Phi_0\rangle = \mathcal{E}_0|\Phi_0\rangle$ we obtain:

$$[E_0 - \mathcal{E}_0]_n = \left(\frac{-i}{\hbar}\right)^n \langle\Phi_0| \int_{-\infty}^{0} dx_1 \int_{-\infty}^{0} dx_2 \dots \int_{-\infty}^{0} dx_n$$
$$e^{\epsilon(x_1+(x_2+x_1)+(x_3+x_2+x_1)+\dots(x_n+x_{n-1}\dots+x_2+x_1)}$$
$$\hat{H}_1 e^{i\frac{\hat{H}_0 x_1}{\hbar}} \hat{H}_1 e^{i\frac{\hat{H}_0 x_2}{\hbar}} \hat{H}_1 \dots e^{i\frac{\hat{H}_0 x_n}{\hbar}} \hat{H}_1 e^{-i\frac{\hat{H}_0 t_n}{\hbar}} |\Phi_0\rangle_c.$$

Let us consider the last term on the right,

$$e^{-i\frac{\hat{H}_0 t_n}{\hbar}} |\Phi_0\rangle = e^{-i\frac{\mathcal{E}_0}{\hbar}t_n} |\Phi_0\rangle = e^{-i\frac{\mathcal{E}_0}{\hbar}(x_1+\dots x_n)} |\Phi_0\rangle,$$

and insert it in the integral:

$$[E_0 - \mathcal{E}_0]_n = \left(\frac{-i}{\hbar}\right)^n \langle\Phi_0|\hat{H}_1 \int_{-\infty}^{0} dx_1\, e^{n\epsilon x_1} e^{i\frac{(\hat{H}_0-\mathcal{E}_0)}{\hbar}x_1} \cdot \hat{H}_1$$
$$\int_{-\infty}^{0} dx_2\, e^{(n-1)\epsilon x_2} e^{i\frac{(\hat{H}_0-\mathcal{E}_0)}{\hbar}x_2} \cdot \hat{H}_1 \dots \int_{-\infty}^{0} dx_N\, e^{\epsilon x_n} e^{i\frac{(\hat{H}_0-\mathcal{E}_0)}{\hbar}x_n} \hat{H}_1|\Phi_0\rangle_c.$$

The nth term is expressed as product of integrals of the type:

$$\int_{-\infty}^{0} dx_1 e^{\frac{i}{\hbar}(\hat{H}_0-\mathcal{E}_0-in\epsilon\hbar)x_1} = \frac{\hbar}{-i}\frac{1}{[\mathcal{E}_0 - \hat{H}_0 + in\varepsilon\hbar]}.$$

Since there are n equal terms, there is a factor $(\hbar/-i)^n$ which can be factorised, therefore:

$$[E - \mathcal{E}_0]_n =$$
$$\langle\Phi_0|\hat{H}_1 \frac{1}{\mathcal{E}_0 - \hat{H}_0 + i\epsilon n\hbar} \hat{H}_1 \frac{1}{\mathcal{E}_0 - \hat{H}_0 + i\epsilon(n-1)\hbar} \dots \hat{H}_1 \frac{1}{\mathcal{E}_0 - \hat{H}_0 + i\epsilon\hbar} \hat{H}_1|\Phi_0\rangle_c.$$

By executing the $\lim \epsilon \to 0$ we obtain the expression of the Goldstone theorem.

Since Eq. (7.7) is valid only for linked diagrams, as indicated by the subscript c, the insertion of terms like $|\Phi_0\rangle\,\langle\Phi_0|$ is prohibited. Consequently, there are no divergences, as the contribution from the denominator $\mathcal{E}_0 - \hat{H}_0$ is always different from zero.

Reference

1. J. Goldstone, Derivation of the Brueckner many-body theory. Proc. Roy. Soc. (London) **A239** 267 (1957)

Chapter 8
Brueckner Theory

8.1 Introduction

The evaluation of the ground state energy for a system of interacting particles is based on the Goldstone equation, which we can rewrite as follows:

$$E_0 - \mathcal{E}_0 = \langle \Phi_0 | \hat{H}_1 \sum_{n=0}^{\infty} \left(\frac{1}{\mathcal{E}_0 - \hat{H}_0} \hat{H}_1 \right)^n | \Phi_0 \rangle_c \ , \tag{8.1}$$

where E_0 represents the energy of the system of interacting particles, and $\mathcal{E}_0$ denotes the smallest eigenvalue of H_0, which corresponds to the ground state energy of the MF solution to the many-body problem. The subscript c indicates that only connected diagrams must be considered in the expansion formula. This last point avoids the divergencies produced by zeros of the denominator.

The computational scheme is, in principle, well-defined. After obtaining $\hat{V}$, which is included in $\hat{H}_1$, from the analysis of two-particle systems as described in Chap. 3, it is necessary to insert it into the Goldstone expansion (8.1) to determine the ground state energy. The challenge arises from the behaviour of the two-particle potential at short distances. When the relative distance between the two particles is small, the microscopic interaction can become extremely large, and may even diverge. This empirical observation has been noted across a wide range of two-particle systems, from nucleons to molecules. A similar issue also occurs in the electron gas, where the Coulomb interaction diverges at zero relative distances.

This issue, related to divergence or strong repulsion at short distances, is illustrated in Fig. 8.1. In this figure, we denote the wave function describing the relative motion of the two interacting particles as ψ. Since ψ is an eigenstate of the full Hamiltonian, the product $\hat{V}(r)\ \psi(r)$ remains finite for every value of r. At short distances, where the potential becomes very large, the wave function $\psi(r)$ is quite small. In the limit where the potential approaches infinity, the wave function tends to zero more rapidly than the divergence of the potential, ensuring that the product $\hat{V}(r)\ \psi(r)$ remains

© The Author(s), under exclusive license to Springer Nature Switzerland AG 2026

G. Co', *Concepts in Quantum Many-Body Physics*, UNITEXT for Physics,
https://doi.org/10.1007/978-3-032-08920-5_8

Fig. 8.1 Microscopic potential between two particles and their relative wave functions. We denote the wave function describing the relative motion of the two interacting particles as ψ (represented by the dashed-dotted line), while the wave function for two non-interacting particles is indicated as ϕ (represented by the dashed line). The units used are arbitrary

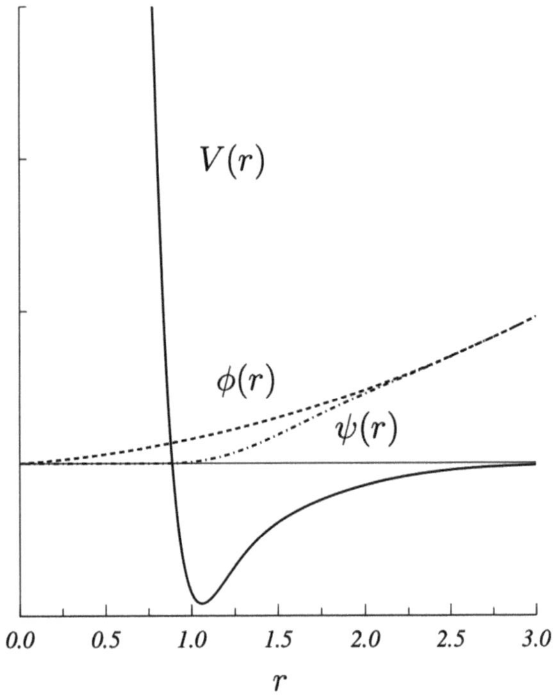

finite. However, this is not the case when we multiply the potential by the relative wave functions ϕ of two free, non-interacting particles. Here, ϕ is an eigenstate of the Hamiltonian $\hat{H}_0$.

In the Goldstone equation (8.1), one must calculate the matrix elements of the interaction between the eigenstates $|\Phi_0\rangle$ of $\hat{H}_0$. The challenges that arise in this process become evident when we assume that the potential $\hat{V}(r)$ approaches infinity at short distances. Each term in the expansion diverges, leading to a situation where the solution to the Goldstone equation involves summing and subtracting infinities to obtain a finite result. Even if we consider finite but large values of the potential at short distances, the issue persists. Each term in the expansion would still be significantly larger than the expected energy value. This situation is fundamentally at odds with the basic principles that make perturbative theory useful. Clearly, microscopic potentials are not perturbative, and therefore, they should not be directly employed in Goldstone expansions.

The idea of the Brueckner theory [1] is that of using in the Goldstone expansion (8.1) an effective interaction which behaves well at short distances in order to be perturbative. This implies the definition of a new interaction $\hat{\mathcal{G}}$ such that

$$\hat{\mathcal{G}}|\Phi_0\rangle = \hat{H}_1|\Psi_0\rangle \ , \tag{8.2}$$

with

$$\hat{H}_0|\Phi_0\rangle = \mathcal{E}_0|\Phi_0\rangle \ , \tag{8.3}$$

and

$$\hat{H}|\Psi_0\rangle = (\hat{H}_0 + \hat{H}_1)|\Psi_0\rangle = E_0|\Psi_0\rangle \ . \tag{8.4}$$

8.2 The Bethe-Goldstone Equation

The relationship between $\hat{H}_0$ and $\hat{H}_1$, which satisfies the definition of effective interaction given in equation (8.2), is referred to as the **Bethe-Goldstone equation**. We derived this equation starting from the definition in (8.2) and utilizing the equations

$$\hat{H}_0|\Phi_n\rangle = \mathcal{E}_n|\Phi_n\rangle \ , \tag{8.5}$$

$$\hat{H}|\Psi_n\rangle = (\hat{H}_0 + \hat{H}_1)|\Psi_n\rangle = E_n|\Psi_n\rangle \ . \tag{8.6}$$

Since the eigenstates of Eq. (8.5) form a complete basis, it is possible to expand $|\Psi_0\rangle$ on this basis. Neglecting the role of a global normalisation constant, we obtain

$$|\Psi_0\rangle \simeq |\Phi_0\rangle + \sum_{n\neq 0} a_n|\Phi_n\rangle \ . \tag{8.7}$$

where the coefficients a_n are numbers. Considering Eq. (8.6), we can write for the ground state

$$(\hat{H}_0 + \hat{H}_1)\left(|\Phi_0\rangle + \sum_{n\neq 0} a_n|\Phi_n\rangle\right) = E_0\left(|\Phi_0\rangle + \sum_{n\neq 0} a_n|\Phi_n\rangle\right) \ , \tag{8.8}$$

$$(\hat{H}_0 - E_0)\left(|\Phi_0\rangle + \sum_{n\neq 0} a_n|\Phi_n\rangle\right) + \hat{H}_1|\Psi_0\rangle = 0 \ . \tag{8.9}$$

By multiplying on the left by $\langle\Phi_0|$, and considering the orthogonality of the $|\Phi_n\rangle$ states, we obtain:

$$\langle\Phi_0|(\hat{H}_0 - E_0)\Big(|\Phi_0\rangle + \sum_{n\neq 0} a_n|\Phi_n\rangle\Big) + \langle\Phi_0|\hat{H}_1|\Psi_0\rangle = 0 \ ,$$

$$\mathcal{E}_0 - E_0 + \langle\Phi_0|\hat{H}_1|\Psi_0\rangle = 0 \ . \tag{8.10}$$

Repeating an analogous operation on Eq. (8.9), and multiplying by $\langle\Phi_n|$ with $n > 0$, we have

$$\langle\Phi_n|(\hat{H}_0 - E_0)\Big(|\Phi_0\rangle + \sum_{n'\neq 0} a_{n'}|\Phi_{n'}\rangle\Big) + \langle\Phi_n|\hat{H}_1|\Psi_0\rangle = 0 \ , \tag{8.11}$$

and, since $\langle\Phi_n|\Phi_{n'}\rangle = \delta_{n,n'}$, we obtain,

$$(\mathcal{E}_n - E_0)a_n + \langle\Phi_n|\hat{H}_1|\Psi_0\rangle = 0 \ . \tag{8.12}$$

By using the expression of a_n taken from the above equation, and inserting it in (8.7) we obtain

$$\begin{aligned}|\Psi_0\rangle &\simeq |\Phi_0\rangle + \sum_{n\neq 0}\frac{1}{E_0 - \mathcal{E}_n}|\Phi_n\rangle\langle\Phi_n|\hat{H}_1|\Psi_0\rangle \\ &= |\Phi_0\rangle + \hat{Q}\frac{1}{E_0 - \hat{H}_0}\hat{H}_1|\Psi_0\rangle \ ,\end{aligned} \tag{8.13}$$

where we defined the operator $\hat{Q}$ as

$$\hat{Q} = \sum_{n\neq 0}|\Phi_n\rangle\langle\Phi_n| \ . \tag{8.14}$$

Multiplying on the left by $\hat{H}_1$ we get the expression

$$\hat{H}_1|\Psi_0\rangle = \hat{H}_1|\Phi_0\rangle + \hat{H}_1\frac{\hat{Q}}{E_0 - \hat{H}_0}\hat{H}_1|\Psi_0\rangle \ , \tag{8.15}$$

and considering the definition (8.2) of the operator $\hat{\mathcal{G}}$ we obtain the Bethe-Goldstone equation

$$\hat{\mathcal{G}}|\Phi_0\rangle = \hat{H}_1|\Psi_0\rangle = \left(\hat{H}_1 + \hat{H}_1\frac{\hat{Q}}{E_0 - \hat{H}_0}\hat{\mathcal{G}}\right)|\Phi_0\rangle \ . \tag{8.16}$$

8.3 The Sum of the Ladder Diagrams

In this section, we show how the Bethe-Goldstone equation can be derived using the diagrammatic techniques introduced in Chap. 7. This alternative derivation offers valuable insights that enhance our understanding of the physical significance of the equation.

We consider a hamiltonian containing only two-body interactions,

$$\hat{H} = \sum_i \hat{t}_i + \sum_{i<j}\hat{v}_{ij} = \sum_i(\hat{t}_i + \hat{u}_i) + \sum_{i<j}\hat{v}_{ij} - \sum_i \hat{u}_i \ , \tag{8.17}$$

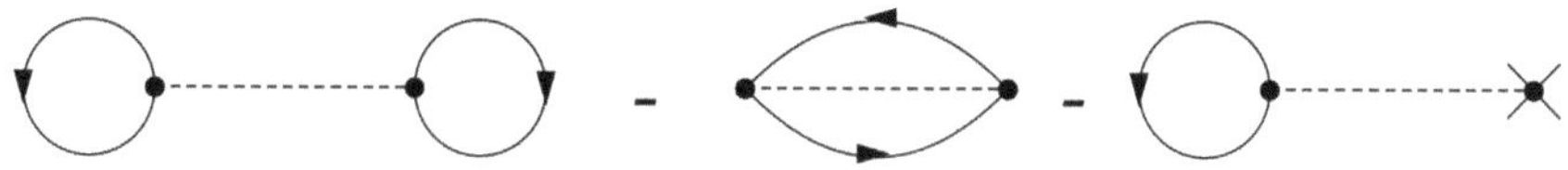

Fig. 8.2 Diagrams representing the terms of Eq. (8.25)

where the MF potential $\hat{u}_i$ has been summed and subtracted. We define

$$\hat{H}_0|\Phi_0\rangle = (\hat{T} + \hat{U})|\Phi_0\rangle = \sum_i (\hat{t}_i + \hat{u}_i)|\Phi_0\rangle = \sum_i \hat{h}_i|\Phi_0\rangle \ , \quad (8.18)$$

where Φ_0 is the ground-state Slater determinant

$$\Phi_0(\mathbf{r}_1, \mathbf{r}_2, \cdots, \mathbf{r}_A) = \frac{1}{\sqrt{A!}} \det |\phi_1(\mathbf{r}_1)\phi_2(\mathbf{r}_2)\cdots\phi_A(\mathbf{r}_A)| \ , \quad (8.19)$$

and the single-particle wave functions are defined as:

$$\hat{h}_i|\phi_i\rangle = \epsilon_i|\phi_i\rangle \ , \quad (8.20)$$

and

$$\mathcal{E}_0 = \sum_i \epsilon_i \ . \quad (8.21)$$

In ONR the two terms of the hamiltonian under consideration can be expressed as:

$$\hat{H}_0 = \sum_\nu \left(\langle\nu|\hat{t}|\nu\rangle + \langle\nu|\hat{u}|\nu\rangle\right)\hat{a}^+_\nu \hat{a}_\nu \ , \quad (8.22)$$

and

$$\hat{H}_1 = \frac{1}{2} \sum_{\nu\nu'\mu\mu'} \langle\nu\mu|\hat{v}|\nu'\mu'\rangle \hat{a}^+_\nu \hat{a}^+_\mu \hat{a}_{\mu'} \hat{a}_{\nu'} - \sum_\nu \langle\nu|\hat{u}|\nu\rangle \hat{a}^+_\nu \hat{a}_\nu \ . \quad (8.23)$$

By definition, we have that

$$\mathcal{E}_0 = \langle\Phi_0|\hat{H}_0|\Phi_0\rangle = \langle\Phi_0|\hat{T}|\Phi_0\rangle + \langle\Phi_0|\hat{U}|\Phi_0\rangle \ . \quad (8.24)$$

The first term of the expansion (8.1) is that with $n = 0$,

$$
\begin{aligned}
(E_0 - \mathcal{E}_0)_{(n=0)} &= \langle \Phi_0 | \hat{H}_1 | \Phi_0 \rangle = \langle \Phi_0 | \hat{V} - \hat{U} | \Phi_0 \rangle \\
&= \frac{1}{2} \sum_{\nu\nu'\mu\mu'} \langle \nu\mu | \hat{v} | \nu'\mu' \rangle \langle \Phi_0 | \hat{a}^+_\nu \hat{a}^+_\mu \hat{a}_{\mu'} \hat{a}_{\nu'} | \Phi_0 \rangle \\
&\quad - \sum_\nu \langle \nu | \hat{u} | \nu' \rangle \langle \Phi_0 | \hat{a}^+_\nu \hat{a}_{\nu'} | \Phi_0 \rangle \\
&= \frac{1}{2} \sum_{ij} \left(\langle ij | \hat{v} | ij \rangle - \langle ij | \hat{v} | ji \rangle \right) - \sum_i \langle i | \hat{u} | i \rangle \ . \qquad (8.25)
\end{aligned}
$$

The diagrams corresponding to the three terms of the above equation are shown in Fig. 8.2 where we inserted a new symbol to indicate $\langle i|\hat{u}|i\rangle$. We use the common convention of indicating with the i, j, k, l letters the single-particle states of hole type and the with m, n, p, q, r the particle states.

Let us consider now the term $n = 1$ of the expansion (8.1):

$$
\begin{aligned}
&(E_0 - \mathcal{E}_0)_{(n=1)} = \langle \Phi_0 | \hat{H}_1 (\mathcal{E}_0 - \hat{H}_0)^{-1} \hat{H}_1 | \Phi_0 \rangle_c \\
&= \langle \Phi_0 | \hat{V} (\mathcal{E}_0 - \hat{H}_0)^{-1} \hat{V} | \Phi_0 \rangle_c - \langle \Phi_0 | \hat{U} (\mathcal{E}_0 - \hat{H}_0)^{-1} \hat{V} | \Phi_0 \rangle_c \\
&- \langle \Phi_0 | \hat{V} (\mathcal{E}_0 - \hat{H}_0)^{-1} \hat{U} | \Phi_0 \rangle_c + \langle \Phi_0 | \hat{U} (\mathcal{E}_0 - \hat{H}_0)^{-1} \hat{U} | \Phi_0 \rangle_c \ . \qquad (8.26)
\end{aligned}
$$

The first term of the above expression can be written as

$$
\begin{aligned}
&\langle \Phi_0 | \hat{V} (\mathcal{E}_0 - \hat{H}_0)^{-1} \hat{V} | \Phi_0 \rangle_c \\
&= \frac{1}{4} \langle \Phi_0 | \sum_{\nu\nu'\mu\mu'} \langle \nu\mu | \hat{v} | \nu'\mu' \rangle \hat{a}^+_\nu \hat{a}^+_\mu \hat{a}_{\mu'} \hat{a}_{\nu'} (\mathcal{E}_0 - \hat{H}_0)^{-1} \\
&\sum_{\alpha\beta\alpha'\beta'} \langle \alpha\beta | \hat{v} | \alpha'\beta' \rangle \hat{a}^+_\alpha \hat{a}^+_\beta \hat{a}_{\beta'} \hat{a}_{\alpha'} | \Phi_0 \rangle_c \ . \qquad (8.27)
\end{aligned}
$$

Let us insert a projection operator $|\Phi_n\rangle\langle\Phi_n| = \hat{\mathbb{I}}$ on both the left and right sides of the denominator $(E_0 - \hat{H}_0)^{-1}$, ensuring that $n \neq 0$. This is important to avoid unwanted diagrams that would violate the constraints imposed by the subscript c. The Slater determinants $|\Phi_n\rangle$ represent excited states of the system of non-interacting particles, specifically those formed by 2-particle, 2-hole $(2p - 2h)$ excitations,

$$
|\Phi_n\rangle = \hat{a}^+_m \hat{a}^+_n \hat{a}_j \hat{a}_i |\Phi_0\rangle \ , \qquad (8.28)
$$

whose energy is given by

$$
\mathcal{E}_0 + \epsilon_m + \epsilon_n - \epsilon_i - \epsilon_j \ , \qquad (8.29)
$$

where the ϵ indicate the energies of the single-particle states.

Since $|\Phi_n\rangle$ is an eigenstate of $\hat{H}_0$, the denominator of Eq. (8.27) is diagonal with respect to these states. This implies that only contributions with the same state on both

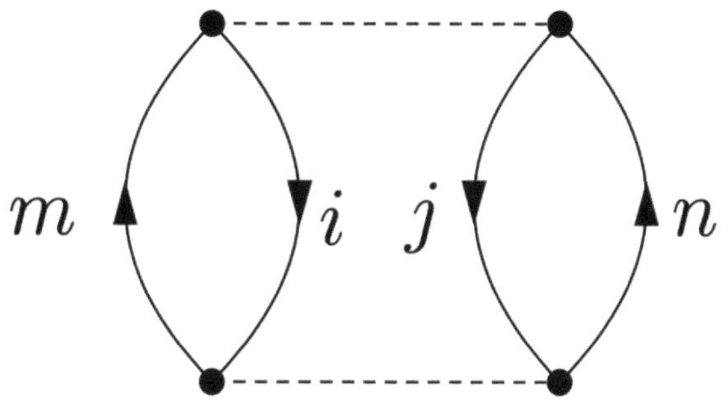

Fig. 8.3 Diagram corresponding to the (8.30) term

the left and right sides of the denominator are non-zero. Consequently, the particle and hole lines created by the interaction on the left side are connected to those that are annihilated by the interaction on the right side. The corresponding direct diagram for this term is illustrated in Fig. 8.3. The contribution of this diagram to Eq. (8.1) is:

$$
\begin{aligned}
&\langle \Phi_0 | \hat{V} (\mathcal{E}_0 - \hat{H}_0)^{-1} \hat{V} | \Phi_0 \rangle_c \\
&= \frac{1}{4} \sum_{ijmn} \Big(\langle ij | \hat{v} | mn \rangle (\mathcal{E}_0 - (\mathcal{E}_0 + \epsilon_m + \epsilon_n - \epsilon_i - \epsilon_j))^{-1} \langle mn | \hat{v} | ij \rangle \Big) \\
&= \frac{1}{4} \sum_{ijmn} \Big(\langle ij | \hat{v} | mn \rangle (\epsilon_i + \epsilon_j - \epsilon_m - \epsilon_n)^{-1} \langle mn | \hat{v} | ij \rangle \Big) \ . \qquad (8.30)
\end{aligned}
$$

The other terms of Eq. (8.26) can be calculated in analogous manner. For example:

$$
\langle \Phi_0 | \hat{V} (\mathcal{E}_0 - \hat{H}_0)^{-1} \hat{U} | \Phi_0 \rangle_c = \frac{1}{2} \sum_{ijm} \Big(\langle ij | \hat{v} | mj \rangle (\epsilon_i - \epsilon_m)^{-1} \langle m | \hat{u} | i \rangle \Big) \ . \qquad (8.31)
$$

Now, let us consider a generic diagram and focus our attention on an interaction affecting a particle line. This helps us avoid the issues related to Pauli blocking. The interaction transforms the state $|\phi_p\rangle$ into the state $|\phi_n\rangle$. Since all particle states are unoccupied, the transition $|\phi_p\rangle \rightarrow |\phi_n\rangle$ is not restricted by the Pauli exclusion principle

Let us consider the diagram of Fig. 8.4 and insert interaction lines in the part of the p e n line. The expression of the diagram with one additional interaction between p and n is

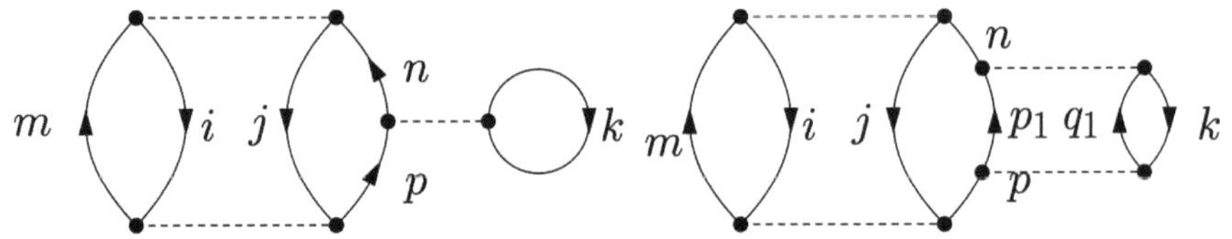

Fig. 8.4 Insertion of interactions on the n particle line of Fig. 8.3

$$
\begin{aligned}
&\langle \Phi_0 | \hat{V} (\mathcal{E}_0 - \hat{H}_0)^{-1} \hat{V} (\mathcal{E}_0 - \hat{H}_0)^{-1} \hat{V} | \Phi_0 \rangle_c \\
&= \frac{1}{8} \sum_{ijmp} \langle ij | \hat{v} | mp \rangle (\epsilon_i + \epsilon_j - \epsilon_m - \epsilon_p)^{-1} \\
&\quad \sum_{kn} \langle pk | \hat{v} | nk \rangle (\epsilon_i + \epsilon_j - \epsilon_m - \epsilon_n)^{-1} \langle mn | \hat{v} | ij \rangle \ . \qquad (8.32)
\end{aligned}
$$

The expression of the diagram with two additional interaction lines between p and n is

$$
\begin{aligned}
&\langle \Phi_0 | \hat{V} (\mathcal{E}_0 - \hat{H}_0)^{-1} \hat{V} (\mathcal{E}_0 - \hat{H}_0)^{-1} \hat{V} (\mathcal{E}_0 - \hat{H}_0)^{-1} \hat{V} | \Phi_0 \rangle_c \\
&= \frac{1}{16} \sum_{ijmp} \langle ij | \hat{v} | mp \rangle (\epsilon_i + \epsilon_j - \epsilon_m - \epsilon_p)^{-1} \\
&\quad \sum_{kn} \left[\sum_{p_1 q_1} \langle pk | \hat{v} | p_1 q_1 \rangle (\epsilon_i + \epsilon_j + \epsilon_k - \epsilon_m - \epsilon_{p_1} - \epsilon_{q_1})^{-1} \langle p_1 q_1 | \hat{v} | nk \rangle \right] \\
&\quad (\epsilon_i + \epsilon_j - \epsilon_m - \epsilon_n)^{-1} \langle mn | \hat{v} | ij \rangle \ . \qquad (8.33)
\end{aligned}
$$

In the equation above, we can identify terms that also appear in Eq. (8.32) and remain unchanged. These are the first and last terms of the equation. The modified portion is enclosed in square brackets. Additionally, within this modified term, there is a part in the denominator that is identical to that in Eq. (8.32), representing the energy differences between the components of the diagram that did not change after the inclusion of a new interaction line. We call

$$
W = \epsilon_i + \epsilon_j + \epsilon_k - \epsilon_m \ . \qquad (8.34)
$$

this term of the energy denominator.

The terms of the diagram which are modified by inserting the interaction lines are:

1 line $\quad \frac{1}{2} \langle pk | \hat{v} | nk \rangle \equiv \langle pk | \hat{V} | nk \rangle$
(Diagram A in Fig. 8.5),

2 lines $\frac{1}{4}\sum_{p'q'}\langle pk|\hat{v}|p_1q_1\rangle(W-\epsilon_{p_1}-\epsilon_{q_1})^{-1}\langle p_1q_1|\hat{v}|nk\rangle$
$\equiv \langle pk|\hat{V}|p_1q_1\rangle(W-\epsilon_{p_1}-\epsilon_{q_1})^{-1}\langle p_1q_1|\hat{V}|nk\rangle$
(Diagram B in Fig. 8.5),

3 lines $\frac{1}{8}\sum_{p_1q_1p_2q_2}\langle pk|\hat{v}|p_1q_1\rangle(W-\epsilon_{p_1}-\epsilon_{q_1})^{-1}$
$\langle p_1q_1|\hat{v}|p_2q_2\rangle(W-\epsilon_{p_2}-\epsilon_{q_2})^{-1}\langle p_2q_2|\hat{v}|nk\rangle$
$\equiv \sum_{p_1q_1p_2q_2}\langle pk|\hat{V}|p_1q_1\rangle(W-\epsilon_{p_1}-\epsilon_{q_1})^{-1}$
$\langle p_1q_1|\hat{V}|p_2q_2\rangle(W-\epsilon_{p_2}-\epsilon_{q_2})^{-1}\langle p_2q_2|\hat{V}|nk\rangle$
(Diagram C in Fig. 8.5),

n lines $\cdots$.

We construct an operator $\hat{\mathcal{G}}$ whose action consists in inserting interaction lines between the states $\langle pk|$ and $|nk\rangle$. To accomplish this, we define an operator $\hat{Q}$ such that:

$$\hat{Q}|\alpha\beta\rangle = |\alpha\beta\rangle \text{ , if } \epsilon_\alpha, \epsilon_\beta > \epsilon_{\mathrm{F}}, \tag{8.35}$$

$$\hat{Q}|\alpha\beta\rangle = 0 \text{ , (if } \epsilon_\alpha, \epsilon_\beta < \epsilon_{\mathrm{F}}) \text{ .} \tag{8.36}$$

This operator selects only particle lines in the diagram. It is useful to consider also the operator $\hat{\mathcal{W}}$ whose action is

$$\hat{\mathcal{W}}|pq\rangle = (W-\epsilon_p-\epsilon_q)|pq\rangle \ , \tag{8.37}$$

where W has been defined in Eq. (8.34). In the example we are discussing, the expectation value of $\hat{\mathcal{G}}$ is given by

$$\begin{aligned}
\langle pk|\hat{\mathcal{G}}|nk\rangle &= \langle pk|\hat{V}|nk\rangle \\
&+ \sum_{p_1q_1}\langle pk|\hat{V}|p_1q_1\rangle(W-\epsilon_{p_1}-\epsilon_{q_1})^{-1}\langle p_1q_1|\hat{V}|nk\rangle \\
&+ \sum_{p_1q_1p_2q_2}\langle pk|\hat{V}|p_2q_2\rangle(W-\epsilon_{p_2}-\epsilon_{q_2})^{-1}\langle p_2q_2|\hat{V}|p_1q_1\rangle \\
&\quad (W-\epsilon_{p_1}-\epsilon_{q_1})^{-1}\langle p_1q_1|\hat{V}|nk\rangle \\
&+ \cdots \\
&= \langle pk|\hat{V} + \hat{V}\frac{\hat{Q}}{\hat{\mathcal{W}}}\hat{V} + \hat{V}\frac{\hat{Q}}{\hat{\mathcal{W}}}\hat{V}\frac{\hat{Q}}{\hat{\mathcal{W}}}\hat{V} + \cdots|nk\rangle
\end{aligned}$$

$$= \langle pk|\hat{V} + \hat{V}\frac{\hat{Q}}{\hat{\mathcal{W}}}\hat{\mathcal{G}}|nk\rangle \ . \tag{8.38}$$

From the operator's perspective the above equation is the Bethe-Goldstone equation (8.16)

$$\hat{\mathcal{G}} = \hat{V} + \hat{V}\frac{\hat{Q}}{\hat{\mathcal{W}}}\hat{\mathcal{G}} \ . \tag{8.39}$$

This equation provides a guideline for constructing an effective interaction that does not produce divergences at short distances. We have already mentioned that the idea is to use $\hat{\mathcal{G}}$ instead of $\hat{V}$ in the Goldstone expansion (8.1) (Fig. 8.5).

Since the effective interaction $\hat{\mathcal{G}}$ has been derived by considering an infinite set of ladder diagrams, it is important to address the possibility that some of the diagrams in the Goldstone expansion may already be included in the definition of $\hat{\mathcal{G}}$. To avoid the issue of double counting, it is essential to carefully select the Goldstone diagrams to be calculated. For instance, in the diagram shown in Fig. 8.6, the two upper lines represent a case of double counting.

8.4 Comparison with the Lippmann-Schwinger Equation

The Bethe-Goldstone equation (8.39) shares strong similarities with the Lippmann-Schwinger equation [2], which provides a formal description of the scattering of two particles. In this section, we will revisit the key physical assumptions that lead to the

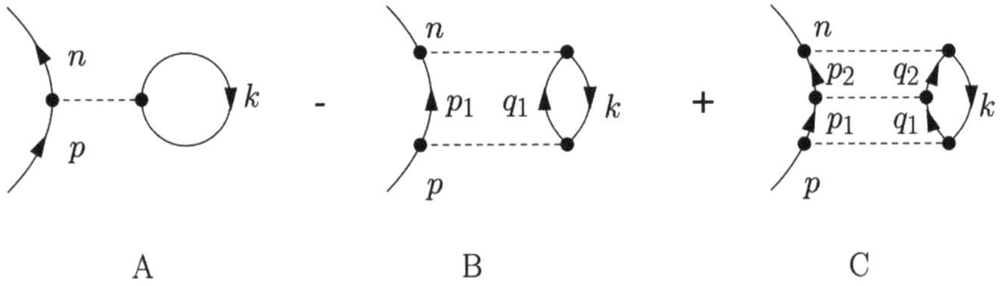

Fig. 8.5 Insertion of interactions in a particle line

Fig. 8.6 Diagram showing a double counting

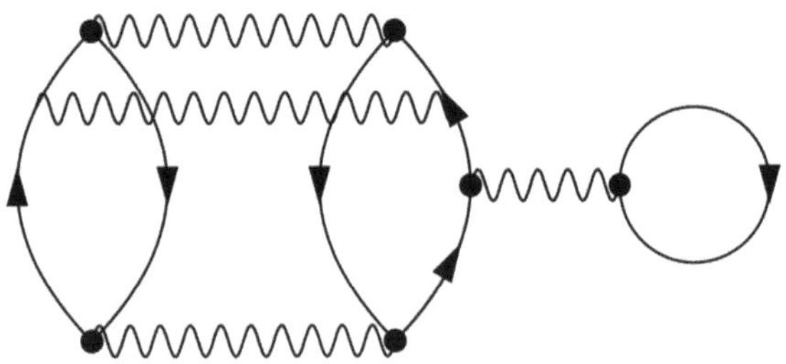

Lippmann-Schwinger equation, in order to clarify the distinctions between these two closely related equations. For more comprehensive discussions and derivations of the Lippmann-Schwinger equation, readers can refer to standard Quantum Mechanics textbooks, such as [3].

Let us consider the scattering of two particles in a vacuum that interact through a central potential. The asymptotic behaviour of the wave function describing the relative motion of the two particles can be expressed as follows

$$\lim_{r\to\infty} \psi_{k_a}(\mathbf{r}) = e^{i\mathbf{k}_a\cdot\mathbf{r}} + f_{k_a}(\Omega)\frac{e^{ik_a r}}{r} \; . \tag{8.40}$$

The cross section is related to the transition amplitude by the relation

$$\frac{d\sigma}{d\Omega} = \left|f_{k_a}(\Omega)\right|^2 \; . \tag{8.41}$$

For a hamiltonian $\hat{H} = \hat{T} + \hat{V}$ the scattering amplitude is defined by the relation [3]

$$\langle\phi_b|\hat{V}|\psi_a\rangle = -\frac{2\pi\hbar^2}{m} f_a(\Omega) \; , \tag{8.42}$$

where the writing has been simplified by using $a \equiv k_a$ and analogously for b. Strictly speaking, Eq. (8.42) does not represent a matrix element since the two states of the expectation value are not eigenstates of the same hamiltonian. We indicated with ϕ the eigenfunction of $\hat{H}_0 = \hat{T}$, and with ψ that of he full hamiltonian $\hat{H}$.

We define here the Green's function as a resolvent of the hamiltonian. For the Green's function describing the motion of a free particle we have the expression

$$\frac{\hbar^2}{2m}\left[\nabla^2 + k^2\right] G^0(\mathbf{r},\mathbf{r}') = \delta(\mathbf{r}-\mathbf{r}') \; , \tag{8.43}$$

implying

$$G^0(\mathbf{r},\mathbf{r}') = -\frac{m}{2\pi\hbar^2}\frac{e^{ik|\mathbf{r}-\mathbf{r}'|}}{|\mathbf{r}-\mathbf{r}'|} \; , \tag{8.44}$$

where the energy of the particle is

$$E = \frac{\hbar^2 k^2}{2m} \; . \tag{8.45}$$

The validity of the expression (8.44) can be verified by substitution and remembering that

$$\nabla^2 \frac{1}{|\mathbf{r}-\mathbf{r}'|} = -4\pi\delta(\mathbf{r}-\mathbf{r}') \; .$$

The solution of the Schrödinger equation for a particle under the effect of a potential $\hat{V}(\mathbf{r})$ is

$$\psi(\mathbf{r}) = e^{i\mathbf{k}\cdot\mathbf{r}} + \int d^3r'\, G^0(\mathbf{r}, \mathbf{r}')\hat{V}(\mathbf{r}')\psi(\mathbf{r}') \ , \tag{8.46}$$

in fact, by inserting the above expression in the equation

$$\frac{\hbar^2}{2m}\left[\nabla^2 + k^2\right]\psi(\mathbf{r}) = \hat{V}(\mathbf{r})\psi(\mathbf{r}) \ , \tag{8.47}$$

we obtain

$$\begin{aligned} &\frac{\hbar^2}{2m}\left[\nabla^2 + k^2\right]e^{i\mathbf{k}\cdot\mathbf{r}} + \int d^3r'\,\frac{\hbar^2}{2m}\left[\nabla^2 + k^2\right]G^0(\mathbf{r}, \mathbf{r}')\hat{V}(\mathbf{r}')\psi(\mathbf{r}') \\ &= \frac{\hbar^2}{2m}\left[-k^2 + k^2\right]e^{i\mathbf{k}\cdot\mathbf{r}} + \int d^3r'\,\delta(\mathbf{r} - \mathbf{r}')\hat{V}(\mathbf{r}')\psi(\mathbf{r}') \\ &= 0 + \hat{V}(\mathbf{r})\psi(\mathbf{r}) \ . \end{aligned}$$

Let us define an operator $\hat{\mathcal{T}}$ such as

$$\langle\phi_b|\hat{\mathcal{T}}|\phi_a\rangle = \langle\phi_b|\hat{V}|\psi_a\rangle \ . \tag{8.48}$$

which is the analogous of the definition (8.2) of $\hat{\mathcal{G}}$ but for two particles in vacuum.

We can consider Eq. (8.43) as an equality between operators

$$(E - \hat{H}_0)G^0 = \hat{\mathbb{I}} \ ; \ G^0 = \frac{\hat{\mathbb{I}}}{(E - \hat{H}_0)} \ , \tag{8.49}$$

and Eq. (8.46) can be written as

$$|\psi_a\rangle = |\phi_a\rangle + G^0\hat{V}|\psi_a\rangle = |\phi_a\rangle + \frac{\hat{\mathbb{I}}}{E - \hat{H}_0 + i\eta}\hat{V}|\psi_a\rangle \ , \tag{8.50}$$

where the $i\eta$ term has been inserted to avoid divergences. Multiplying on the left by $\hat{V}$ and considering the definition (8.48) of $\hat{\mathcal{T}}$ we obtain

$$\hat{V}|\psi_a\rangle = \hat{V}|\phi_a\rangle + \hat{V}\frac{\hat{\mathbb{I}}}{E - \hat{H}_0 + i\eta}\hat{V}|\psi_a\rangle,$$

or equivalently

$$\hat{\mathcal{T}}|\phi_a\rangle = \left(\hat{V} + \hat{V}\frac{\hat{\mathbb{I}}}{E - \hat{H}_0 + i\eta}\hat{\mathcal{T}}\right)|\phi_a\rangle \ , \tag{8.51}$$

which is the Lippmann-Schwinger equation.

The analogy with the Bethe-Goldstone equation (8.16) is clear. The Bethe-Goldstone equation describes the interaction between two particles that scatter within a medium. The most significant difference lies in the presence of the operator $\hat{Q}$, which accounts for the Pauli exclusion principle. In the medium, only reactions that populate states above the Fermi surface are permitted. In contrast, in a vacuum, there are no restrictions on the final states.

Another difference pertains to the energy denominator. In the Lippmann-Schwinger equation, the denominator includes only the kinetic energies of the colliding particles. The presence of the imaginary term is essential because there is a possibility that the denominator could become zero. In contrast, the Bethe-Goldstone equation ensures that the denominator never equals zero, as it does not involve unlinked diagrams. Additionally, the single-particle energies in this context are not merely the kinetic energies of the particles; they also incorporate the MF energy term. Lastly, there is the W term, defined by Eq. (8.34), which introduces a dependence on the single-particle energies in the denominator for the diagram terms that are not directly influenced by the interaction between the two particles

8.5 Application to Nuclear Matter

The most notable application of Brueckner theory to physical systems has been in the study of nuclear matter. This is a theoretical system composed of an infinite number of nucleons, exhibiting translational symmetry, with the Coulomb interaction turned off. These characteristics simplify the MF problems, as the potential remains constant and the single-particle wave functions take the form of plane waves. We focus on the scenario where the number of protons equals the number of neutrons, known as symmetric nuclear matter. The single-particle wave functions are characterised by the wave number $\mathbf{k} = \mathbf{p}/\hbar c$, where $\mathbf{p}$ represents the momentum of the particle (see Sect. 2.3).

The empirical evidence from elastic electron scattering off nuclei indicates that the charge distributions at the center of the nucleus exhibit similar values across different nuclei on the nuclide chart [4]. To simplify the analysis, we can model the nucleus as a rigid sphere with a constant density and a radius R. The density can be calculated by dividing the number of nucleons by the volume of the sphere:

$$\rho = \frac{A}{\frac{4}{3}\pi R^3} = \frac{A}{\frac{4}{3}\pi r_0^3 A} = \frac{3}{4\pi r_0^3} = 0.17 \pm 0.02 \text{ fm}^{-3} , \qquad (8.52)$$

where the empirical relation $R = r_0 A^{1/3}$ with r_0=1.12 fm has been used.

We estimate the binding energy of symmetric nuclear matter by considering the following expression from the semi-empirical mass formula [4]

$$B(A, Z) = a_v A + a_s A^{\frac{2}{3}} + a_c \frac{Z^2}{A^{\frac{1}{3}}} + a_i \frac{(N - Z)^2}{A} + \delta(A) \ , \qquad (8.53)$$

and examine its behaviour under the aforementioned assumptions. Since we assume that the number of protons and neutrons is equal, the asymmetry term, which is the fourth term, becomes zero. This approximation is made here to simplify the analysis, but it is not essential for the stability of the system. There are studies on asymmetric nuclear matter, particularly focusing on pure neutron matter. Currently, neutron stars, which are associated with pulsars, represent the physical systems most similar to nuclear matter.

The other assumption is much more relevant: the Coulomb interaction is turned off. This assumption is essential for the stability of the infinite system. Since the Coulomb interaction has an infinite range, all protons interact with one another. In the case of an infinite number of protons, this leads to infinite repulsion.

With these assumptions the expression of the binding energy per nucleon of nuclear matter is:

$$\frac{B(A, Z)}{A} = a_v + a_s A^{-\frac{1}{3}} + \frac{\delta(A)}{A} \ , \qquad (8.54)$$

and neglecting the last term, which is the pairing term, in the limit as A approaches infinity, only the volume term remains significant: $a_v = 16.0 \pm 1.0$ MeV. A proper description of nuclear matter should provide an equation of state that relates the binding energy per nucleon to the density of the system. The minimum of this function should occur in the empirical region where, for $\rho = 0.17 \pm 0.02$ fm^{-3}, the binding energy per nucleon $B(A, Z)/A = a_v$.

The MF treatment of a system exhibiting translational invariance is discussed in Sect. 2.3, where Eq. (2.52) expresses the system density as a function of the Fermi momentum

$$\rho(\mathbf{r}) = \frac{2}{3\pi^2} k_F{}^3. \qquad (8.55)$$

By utilizing the empirical value of the density, we determine that the numerical value of the Fermi momentum is $k_F = 1.36$ fm^{-1}, which corresponds to 250 MeV/c. Using Eq. (2.56), we can calculate the kinetic energy per particle of the system

$$\frac{T}{A} = \frac{3}{5} \epsilon_F, \qquad (8.56)$$

where the Fermi energy is given by

$$\epsilon_F = \frac{\hbar^2}{2m} k_F{}^2 \ . \qquad (8.57)$$

We present here below how the Brueckner theory is applied to this system. We define the relative coordinates of the two interacting particles p and q

$$\mathbf{R} = \frac{1}{2}(\mathbf{r}_p + \mathbf{r}_q) \ ; \ \ \mathbf{r} = \mathbf{r}_p - \mathbf{r}_q , \tag{8.58}$$

and the relative momenta

$$\mathbf{K}_{pq} = \mathbf{k}_p + \mathbf{k}_q \ ; \ \ \mathbf{k}_{pq} = \mathbf{k}_p - \mathbf{k}_q \ . \tag{8.59}$$

The unperturbed wave function describing the motion of the two particles is

$$\Phi_{pq}(\mathbf{r}_p, \mathbf{r}_q) = \frac{1}{\mathcal{V}} e^{i\mathbf{k}_p \cdot \mathbf{r}_p} e^{i\mathbf{k}_q \cdot \mathbf{r}_q} = \frac{1}{\mathcal{V}} e^{i\mathbf{K}_{pq} \cdot \mathbf{R}} e^{i\mathbf{k}_{pq} \cdot \mathbf{r}} = \frac{1}{\mathcal{V}} e^{i\mathbf{K}_{pq} \cdot \mathbf{R}} \phi_{pq}(\mathbf{r}) \ , \tag{8.60}$$

here we denote the volume by $\mathcal{V}$, which will approach infinity at the end of the calculation, and neglect the contributions from spin and isospin terms.

The operators $\hat{Q}$ and $\hat{\mathcal{W}}$ defined in Eqs. (8.35, 8.36) and Eq. (8.37), operate on this wave function. The action of $\hat{Q}$ on the wave function

$$\hat{Q}|\Phi_{pq}\rangle = |\Phi_{pq}\rangle \ , \tag{8.61}$$

is different from zero only when both $|\mathbf{k}_p|$ and $|\mathbf{k}_q|$ are larger than $|\mathbf{k}_\mathrm{F}|$, and the action of $\hat{\mathcal{W}}$ is

$$\hat{\mathcal{W}}|\Phi_{pq}\rangle = \left[W - \epsilon(\mathbf{k}_p) - \epsilon(\mathbf{k}_q)\right]|\Phi_{pq}\rangle = e(\mathbf{k}_p, \mathbf{k}_q)|\Phi_{pq}\rangle \ . \tag{8.62}$$

The wave function of the two interacting particles can be expressed by factoring in the free motion of the center of mass of the nucleon pair, while considering their relative motion

$$\Psi_{pq}(\mathbf{r}_p, \mathbf{r}_q) = \frac{1}{\mathcal{V}} e^{i\mathbf{K}_{pq} \cdot \mathbf{R}} \psi_{pq}(\mathbf{r}) \ . \tag{8.63}$$

Using the definition (8.2) we write

$$\begin{aligned} \langle\Phi_{pq}|\hat{\mathcal{G}}|\Phi_{rs}\rangle &= \int d^3r_p\, d^3r_q\ \Phi_{pq}(\mathbf{r}_p, \mathbf{r}_q)\hat{V}(\mathbf{r})\Psi_{rs}(\mathbf{r}_p, \mathbf{r}_q) \\ &= \frac{1}{(2\pi)^3}\delta(\mathbf{K}_{pq} - \mathbf{K}_{sr}) \int d^3r e^{i\mathbf{k}_{pq} \cdot \mathbf{r}} \hat{V}(\mathbf{r})\psi_{rs}(\mathbf{r}) \ . \end{aligned} \tag{8.64}$$

The part related to the center of mass coordinate can be factorized, therefore, by considering Eq. (8.63), we obtain

$$\psi_{pq}(\mathbf{r}) = \phi_{pq}(\mathbf{r}) + \int d^3r'\, \mathcal{K}_{pq}(\mathbf{r}, \mathbf{r}')\hat{V}(r')\psi_{pq}(\mathbf{r}') \ , \tag{8.65}$$

with

$$\mathcal{K}_{pq}(\mathbf{r}, \mathbf{r}') = \frac{1}{(2\pi)^3} \int d^3k \frac{\hat{Q}(\mathbf{K}_{pq}, \mathbf{k})}{e(\mathbf{K}_{pq}, \mathbf{k})} e^{i\mathbf{k}\cdot(\mathbf{r}-\mathbf{r}')} \ . \tag{8.66}$$

In solving the Goldstone equation (8.1), it is essential to choose the type of diagram for calculation. This diagram will certainly include a matrix element of the type described in (8.64). The procedure used to determine the value of this matrix element consists of the following steps.

1. Choice of the MF potential $\hat{U}$ to select the single particle energies of the denominator.
2. Numerical solution of Eq. (8.66).
3. Insertion of the kernel (8.66) in Eq. (8.65).
4. Insertion of Eq. (8.65) in the matrix element (8.64).

8.6 Final Considerations

Using Brueckner theory, we define an effective interaction $\hat{\mathcal{G}}$ that acts on the mean-field wave functions $|\Phi\rangle$. This effective interaction yields results that are quantitatively consistent with the action of the microscopic interaction $\hat{V}$ on the wave function $|\Psi\rangle$, which describes the interacting system (see Eq. (8.2)). The relationship between $\hat{\mathcal{G}}$ and $\hat{V}$ is established by the Bethe-Goldstone equation, (8.16) or (8.39). Notably, there is a strong similarity between the Bethe-Goldstone equation and the Lippmann-Schwinger equation, which describes the scattering of two particles in a vacuum.

1. The denominator of the Lippmann-Schwinger equation (8.51) contains an imaginary part. This indicates that, asymptotically, there is a phase shift between the scattering wave functions and the free wave functions. In contrast, the Bethe-Goldstone equation (8.16) does not include an imaginary term, which means there is no phase shift between the relative wave functions of the interacting particles, ψ_{pq}, and those of the non-interacting particles, ϕ_{pq}. These two wave functions differ only at short distances, where the finite-range potential plays a significant role, particularly in the region of the strongly repulsive core.
2. The key parameter in the Goldstone expansion is not the number of interacting lines, but rather the number of hole lines. Diagrams that differ by a single interaction line yield contributions of the same order of magnitude. In contrast, diagrams that include an additional hole line produce contributions that are one order of magnitude smaller [5]. The hole line expansion is an expansion in powers of density. In many-body physics, it is essential to consider the relative density, which refers to the number of point-like particles present within the volume defined by the strongly repulsive core of the interaction (see Sect. 3.4).
3. In principle, the interaction $\hat{\mathcal{G}}$ should be independent of the choice of the mean-field potential $\hat{U}$. However, in practice, this is not the case due to the truncation of the Goldstone expansion (8.1). The convergence of the expansion using $\hat{\mathcal{G}}$, calculated with $\hat{H}_1 = \hat{V}$, is quite slow. By incorporating a one-body term into the perturbative Hamiltonian, specifically $\hat{H}_1 = \hat{V} - \hat{U}$, we can enhance the rate of convergence. A self-consistent choice of $\hat{U}$ is

$$\hat{U} = \sum_{\alpha} \langle \phi_\alpha | \hat{\mathcal{G}}(W) | \phi_\alpha \rangle. \tag{8.67}$$

The so-called *normal* choice involves considering only the states below the Fermi surface in the sum mentioned above. This approach introduces a discontinuity in the single-particle energies: the states below the Fermi energy include the potential term, while those above it consist solely of kinetic energies. Although this relatively straightforward choice may seem appealing, it does not ensure the stability of the results, which are highly sensitive to the selection of $\hat{U}$.
The advancements in computing capabilities have enabled a more accurate approximation by taking into account the entire space in the sum of Eq. (8.67). These calculations, conducted using the so-called *continuous* choice, have demonstrated a notable independence from the selection of $\hat{U}$ [6].

References

1. K.A. Brueckner, Many-body Problem for srongly interacting Particles. Phys. Rev. **100**, 36 (1955)
2. B.A. Lippmann, J. Schwinger, Variational Principles for Scattering Processes. I. Phys. Rev. **79**, 469 (1950)
3. A. Messiah, *Quantum Mechanics* (North Holland, Amsterdam, 1961)
4. K. Krane, *Introductory Nuclear Physics* (John Wiley & sons, New York, 1988)
5. B.D. Day, Elements of the Brueckner-Goldstone theory of nuclear matter. Rev. Mod. Phys. **39**, 719 (1967)
6. O. Benhar, S. Fantoni, *Nuclear Matter Theory* (CRC Press, Boca Raton, 2021)

Chapter 9
Green's Function

9.1 Introduction

Until now, we have refrained from using a theoretical concept introduced in Quantum Field Theory and Many-Body Theories, known as the Green's function. In fact, we briefly mentioned the Green's function in Sect. 8.4 as the resolvent of a differential equation. While the mathematical expressions for the Green's function presented in this chapter and in Sect. 8.4 are identical, their interpretations are entirely different.

The Green's function is highly valuable in the description of many-body systems, particularly for calculating observable quantities, including the energy of the system. It encapsulates general features of the many-body system, regardless of the specific observable being investigated.

The Green's function is not uniquely defined, as its definition depends on the number of field operators considered. In this chapter, we first introduce the one-body Green's function, provide its physical interpretation, demonstrate how to use it to calculate observables, and relate it to the Green's function discussed in Sect. 8.4. Following that, we define the two-body Green's function. At the end of the chapter, we present a set of interconnected equations that establish relationships between Green's functions defined for any number of particles. The challenges associated with solving this set of coupled equations motivate us to develop an alternative approach for calculating Green's functions, based on perturbative theory. This topic will be addressed in Chap. 10.

9.2 One-Body Green's Functions

The fermionic field operator in Heisenberg representation is defined as

$$\hat{\psi}_{\mathrm{H},\alpha}(\mathbf{x}, t) = e^{\frac{i}{\hbar}\hat{H}t}\hat{\psi}_{\alpha}(\mathbf{x})e^{-\frac{i}{\hbar}\hat{H}t} \quad , \tag{9.1}$$

© The Author(s), under exclusive license to Springer Nature Switzerland AG 2026

G. Co', *Concepts in Quantum Many-Body Physics*, UNITEXT for Physics,
https://doi.org/10.1007/978-3-032-08920-5_9

where the subscript H indicates Heisenberg while $\hat{H}$ is the system hamiltonian. With α we indicated all the quantum numbers, other than space and time, which characterise the particle, for example spin and isospin.

The **one-body Green's function** for a fermion system is defined as

$$i\,G_{\alpha\beta}(\mathbf{x},t,\mathbf{x}',t') = \frac{\langle\Psi_0|\hat{\mathbb{T}}\left[\hat{\psi}_{\mathrm{H},\alpha}(\mathbf{x},t)\hat{\psi}^+_{\mathrm{H},\beta}(\mathbf{x}',t')\right]|\Psi_0\rangle}{\langle\Psi_0|\Psi_0\rangle} \quad . \tag{9.2}$$

In the above expression $|\Psi_0\rangle$ indicates the ground state of the system

$$\hat{H}|\Psi_0\rangle = E|\Psi_0\rangle, \tag{9.3}$$

and $\hat{\mathbb{T}}$ indicates the time-ordering operator

$$\hat{\mathbb{T}}\left[\hat{\psi}_{\mathrm{H},\alpha}(\mathbf{x},t)\hat{\psi}^+_{\mathrm{H},\beta}(\mathbf{x}',t')\right] = \begin{cases} \hat{\psi}_{\mathrm{H},\alpha}(\mathbf{x},t)\hat{\psi}^+_{\mathrm{H},\beta}(\mathbf{x}',t') & t > t' \\ -\,\hat{\psi}^+_{\mathrm{H},\beta}(\mathbf{x}',t')\hat{\psi}_{\mathrm{H},\alpha}(\mathbf{x},t) & t < t' \, . \end{cases} \tag{9.4}$$

The one-body Green's functions can be related to the following observable quantities:

- expectation value of a one-body operator with respect to the system ground state,
- energy of the ground state,
- excitation spectrum of the system for single particle excitations.

From now on, to simplify the notation, we will not explicitly write the quantum numbers α and β, assuming their sum whenever we indicate an integration over the coordinates.

For a time-independent hamiltonian $\hat{H}$ we can write the Green's function (9.2) as

$$i\,G(\mathbf{x},t,\mathbf{x}',t') = \begin{cases} e^{\frac{i}{\hbar}E(t-t')}\dfrac{\langle\Psi_0|\hat{\psi}(\mathbf{x})e^{-\frac{i}{\hbar}\hat{H}(t-t')}\hat{\psi}^+(\mathbf{x}')|\Psi_0\rangle}{\langle\Psi_0|\Psi_0\rangle} & t > t' \, , \\ \\ -\,e^{-\frac{i}{\hbar}E(t-t')}\dfrac{\langle\Psi_0|\hat{\psi}^+(\mathbf{x}')e^{\frac{i}{\hbar}\hat{H}(t-t')}\hat{\psi}(\mathbf{x})|\Psi_0\rangle}{\langle\Psi_0|\Psi_0\rangle} & t < t' \, . \end{cases} \tag{9.5}$$

where we used the explicit expression (9.1) of the field operator in Heisenberg picture.

For the calculation of the expectation value of a one-body operator on the ground state of the system, we use the expression of one-body operator (5.51) in terms of field operators

$$\hat{\mathbb{O}}^I = \int d^3x\,\hat{\psi}^+(\mathbf{x})\hat{\mathbb{O}}(\mathbf{x})\hat{\psi}(\mathbf{x}) \quad . \tag{9.6}$$

We can write its expectation value with respect to the ground state as

$$\langle \hat{\mathbb{O}}^I \rangle = \int d^3x \, \frac{\langle \Psi_0 | \hat{\psi}^+(\mathbf{x}) \hat{\mathbb{O}}(\mathbf{x}) \hat{\psi}(\mathbf{x}) | \Psi_0 \rangle}{\langle \Psi_0 | \Psi_0 \rangle}$$
$$= \int d^3x \lim_{\mathbf{x}' \to \mathbf{x}} \hat{\mathbb{O}}(\mathbf{x}) \frac{\langle \Psi_0 | \hat{\psi}^+(\mathbf{x}') \hat{\psi}(\mathbf{x}) | \Psi_0 \rangle}{\langle \Psi_0 | \Psi_0 \rangle} \;, \tag{9.7}$$

where the limit $\mathbf{x}' \to \mathbf{x}$ has been inserted to allow the commutation of $\hat{\mathbb{O}}(\mathbf{x})$ with $\hat{\psi}^+(\mathbf{x}')$. By using the second Eq. (9.5), and denoting t^+ as a time greater than t, we obtain

$$\langle \hat{\mathbb{O}}^I \rangle = -i \lim_{t' \to t^+ \to t} \int d^3x \lim_{\mathbf{x} \to \mathbf{x}'} \hat{\mathbb{O}}(\mathbf{x}) G(\mathbf{x}, t, \mathbf{x}', t') \;. \tag{9.8}$$

Let us illustrate this type of calculation by considering the number density operator. The operator associated with the number of particles in the system can be expressed as a function of the density operator

$$\hat{\rho}(\mathbf{x}) = \delta(\mathbf{x} - \mathbf{x}'). \tag{9.9}$$

The expectation value of this operator with respect to the ground state of the system is:

$$\langle \hat{\rho} \rangle = -i \lim_{t' \to t^+} \int d^3x \lim_{\mathbf{x} \to \mathbf{x}'} \hat{\rho}(\mathbf{x}) G(\mathbf{x}, t, \mathbf{x}', t')$$
$$= -i \lim_{t' \to t^+} \lim_{\mathbf{x} \to \mathbf{x}'} G(\mathbf{x}, t, \mathbf{x}', t') = \frac{\langle \Psi_0 | \hat{\psi}^+(\mathbf{x}) \hat{\psi}(\mathbf{x}) | \Psi_0 \rangle}{\langle \Psi_0 | \Psi_0 \rangle} \;. \tag{9.10}$$

In the case of non-interacting fermions we have $|\Psi_0\rangle = |\Phi_0\rangle$ and we obtain

$$\langle \hat{\rho} \rangle = \frac{\langle \Phi_0 | \hat{\psi}^+(\mathbf{x}) \hat{\psi}(\mathbf{x}) | \Phi_0 \rangle}{\langle \Phi_0 | \Phi_0 \rangle}. \tag{9.11}$$

By using the representation of the field operators in terms of creation and destruction operators, Eqs. (5.47) and (5.48), we have

$$\langle \hat{\rho} \rangle = \sum_{\alpha,\alpha'} \phi^*_\alpha(\mathbf{x}) \phi_{\alpha'}(\mathbf{x}) \langle \Phi_0 | \hat{a}^+_\alpha \hat{a}_{\alpha'} | \Phi_0 \rangle$$
$$= \sum_{\alpha,\alpha'=1}^{\epsilon_F} \phi^*_\alpha(\mathbf{x}) \phi_{\alpha'}(\mathbf{x}) \delta_{\alpha,\alpha'} = \sum_{\alpha=1}^{\epsilon_F} |\phi_\alpha(x)|^2 \;, \tag{9.12}$$

where we have assumed the orthonormality of the single particle wave functions ϕ_α and that the many-body state $|\Phi_0\rangle$ is normalized to 1. We indicated with ϵ_F the

Fermi energy. This is the traditional expression of the density for a system of non interacting particles.

The ground state energy of the system is related to the one-body Green's function by the expression

$$E_0 = \frac{\langle\Psi_0|\hat{T}|\Psi_0\rangle}{\langle\Psi_0|\Psi_0\rangle} + \frac{\langle\Psi_0|\hat{W}|\Psi_0\rangle}{\langle\Psi_0|\Psi_0\rangle}$$
$$= -\frac{i}{2}\int d^3r \lim_{\mathbf{r}'\to\mathbf{r}}\lim_{t'\to t}\left[i\hbar\frac{\partial}{\partial t} + \left(-\hbar^2\frac{\nabla_{\mathbf{r}}^2}{2m}\right)\right] G(\mathbf{r}, t, \mathbf{r}', t') \; . \tag{9.13}$$

Below, we demonstrate how to derive the expression mentioned above.

We use the following expressions of the commutators of 3 operators:

$$[\hat{A}, \hat{B}\hat{C}] = \hat{A}\hat{B}\hat{C} - \hat{B}\hat{C}\hat{A} = \hat{A}\hat{B}\hat{C} - \hat{B}\hat{C}\hat{A} + \hat{B}\hat{A}\hat{C} - \hat{B}\hat{A}\hat{C} \tag{9.14}$$
$$\left[\hat{A}, \hat{B}\hat{C}\right] = \{\hat{A}, \hat{B}\}\hat{C} - \hat{B}\{\hat{A}, \hat{C}\} \tag{9.15}$$
$$\left[\hat{A}, \hat{B}\hat{C}\right] = [\hat{A}, \hat{B}]\hat{C} - \hat{B}[\hat{C}, \hat{A}] \; , \tag{9.16}$$

and also

$$\{\hat{A}, \hat{B}\hat{C}\} = \{\hat{A}, \hat{B}\}\hat{C} - \hat{B}[\hat{A}, \hat{C}] \; . \tag{9.17}$$

We consider the commutator of the operator $\hat{\psi}(\mathbf{r})$ with the hamiltonian

$$\hat{H} = \hat{T} + \hat{W} = \int d^3x\, \hat{\psi}^+(\mathbf{x})\left(-\hbar^2\frac{\nabla_{\mathbf{x}}^2}{2m}\right)\hat{\psi}(\mathbf{x})$$
$$+ \frac{1}{2}\int d^3x\, d^3y\, \hat{\psi}^+(\mathbf{x})\hat{\psi}^+(\mathbf{y})\hat{V}(\mathbf{x}, \mathbf{y})\hat{\psi}(\mathbf{y})\hat{\psi}(\mathbf{x}) \; , \tag{9.18}$$

Let us first evaluate the commutator with the kinetic energy term

$$[\psi(\mathbf{r}), \hat{T}] = \left[\underbrace{\hat{\psi}(\mathbf{r})}_{\hat{A}}, \int d^3x\, \underbrace{\hat{\psi}^+(\mathbf{x})}_{\hat{B}}\underbrace{\left(-\hbar^2\frac{\nabla_{\mathbf{x}}^2}{2m}\right)\hat{\psi}(\mathbf{x})}_{\hat{C}}\right] \; . \tag{9.19}$$

We apply Eq. (9.15)

$$[\hat{\psi}(\mathbf{r}), \hat{T}] = \int d^3x\, \underbrace{\{\hat{\psi}(\mathbf{r}), \hat{\psi}^+(\mathbf{x})\}}_{\delta(\mathbf{x}-\mathbf{r})}\left(-\hbar^2\frac{\nabla_{\mathbf{x}}^2}{2m}\hat{\psi}(\mathbf{x})\right)$$
$$- \int d^3x\, \hat{\psi}^+(\mathbf{x})\left(-\hbar^2\frac{\nabla_{\mathbf{x}}^2}{2m}\right)\underbrace{\{\hat{\psi}(\mathbf{r}), \hat{\psi}(\mathbf{x})\}}_{0}$$
$$= \int d^3x\, \delta(\mathbf{x} - \mathbf{r})\left(-\hbar^2\frac{\nabla_{\mathbf{x}}^2}{2m}\hat{\psi}(\mathbf{x})\right) = -\frac{\hbar^2\nabla_{\mathbf{r}}^2}{2m}\hat{\psi}(\mathbf{r}) \; . \tag{9.20}$$

For the interaction term we consider

$$[\hat{\psi}(\mathbf{r}), \hat{V}] = \frac{1}{2}\int d^3x\, d^3y \left[\underbrace{\hat{\psi}(\mathbf{r})}_{\hat{A}}, \underbrace{\hat{\psi}^+(\mathbf{x})}_{\hat{B}}\underbrace{\hat{\psi}^+(\mathbf{y})\hat{V}(\mathbf{x},\mathbf{y})\hat{\psi}(\mathbf{y})\hat{\psi}(\mathbf{x})}_{\hat{C}}\right] . \tag{9.21}$$

By applying Eq. (9.15) we obtain

$$\begin{aligned}
&[\hat{\psi}(\mathbf{r}), \hat{V}] \\
&= \frac{1}{2}\int d^3x\, d^3y \left[\underbrace{\{\psi(\mathbf{r}), \psi^+(\mathbf{x})\}}_{\delta(\mathbf{r}-\mathbf{x})}\hat{C} - \hat{\psi}^+(\mathbf{x})\left\{\underbrace{\hat{\psi}(\mathbf{r})}_{\hat{A}'}, \underbrace{\hat{\psi}^+(\mathbf{y})}_{\hat{B}'}\underbrace{\hat{V}(\mathbf{x},\mathbf{y})\hat{\psi}(\mathbf{y})\hat{\psi}(\mathbf{x})}_{\hat{C}'}\right\}\right] \\
&= \frac{1}{2}\int d^3y\, \hat{\psi}^+(\mathbf{y})\hat{V}(\mathbf{r},\mathbf{y})\hat{\psi}(\mathbf{y})\hat{\psi}(\mathbf{r}) \\
&+ \frac{1}{2}\int d^3x\, d^3y \left(-\hat{\psi}^+(\mathbf{x})\right)\underbrace{\left\{\hat{\psi}(\mathbf{r}), \hat{\psi}^+(\mathbf{y})\right\}}_{\delta(\mathbf{r}-\mathbf{y})}\hat{V}(\mathbf{x},\mathbf{y})\underbrace{\hat{\psi}(\mathbf{y})\hat{\psi}(\mathbf{x})}_{-\hat{\psi}(\mathbf{x})\hat{\psi}(\mathbf{y})} \\
&- \frac{1}{2}\int d^3x\, d^3y \left(-\hat{\psi}^+(\mathbf{x})\right)\hat{\psi}^+(\mathbf{y})\hat{V}(\mathbf{x},\mathbf{y})\underbrace{[\hat{\psi}(\mathbf{r}), \hat{\psi}(\mathbf{y})\hat{\psi}(\mathbf{x})]}_{0} \\
&= \frac{1}{2}\int d^3y\, \hat{\psi}^+(\mathbf{y})\hat{V}(\mathbf{r},\mathbf{y})\hat{\psi}(\mathbf{y})\hat{\psi}(\mathbf{r}) + \frac{1}{2}\int d^3x\, \hat{\psi}^+(\mathbf{x})\hat{V}(\mathbf{x},\mathbf{r})\hat{\psi}(\mathbf{x})\hat{\psi}(\mathbf{r}) ,
\end{aligned} \tag{9.22}$$

where we used Eq. (9.17) and the anti-commutation rules of the field operators. For the symmetry of the potential $\hat{V}(\mathbf{x}, \mathbf{y}) = \hat{V}(\mathbf{y}, \mathbf{x})$, the two terms of the last expression are equal, therefore

$$[\hat{\psi}(\mathbf{r}), \hat{V}] = \int d^3x\, \hat{\psi}^+(\mathbf{x})\hat{V}(\mathbf{r},\mathbf{x})\hat{\psi}(\mathbf{x})\hat{\psi}(\mathbf{r}) . \tag{9.23}$$

Putting together Eqs. (9.20) and (9.23) we obtain

$$[\hat{\psi}(\mathbf{r}), \hat{H}] = -\hbar^2\frac{\nabla^2_{\mathbf{r}}}{2m}\hat{\psi}(\mathbf{r}) + \int d^3x\, \hat{\psi}^+(\mathbf{x})\hat{V}(\mathbf{r},\mathbf{x})\hat{\psi}(\mathbf{x})\hat{\psi}(\mathbf{r}) . \tag{9.24}$$

We use this expression in the equation of motion for the field operator in Heisenberg representation, Eq. (6.7),

$$\begin{aligned}
i\hbar\frac{\partial}{\partial t}\hat{\psi}_{\mathrm{H}}(\mathbf{r},t) &= \left[\hat{\psi}_{\mathrm{H}}(\mathbf{r},t), \hat{H}\right] = e^{\frac{i}{\hbar}\hat{H}t}\left[\hat{\psi}(\mathbf{r}), \hat{H}\right]e^{-\frac{i}{\hbar}\hat{H}t} \\
&= -\hbar^2\frac{\nabla^2_{\mathbf{r}}}{2m}\hat{\psi}_{\mathrm{H}}(\mathbf{r},t) + \int d^3x\, \hat{\psi}^+_{\mathrm{H}}(\mathbf{x},t)\hat{V}(\mathbf{r},\mathbf{x})\hat{\psi}_{\mathrm{H}}(\mathbf{x},t)\hat{\psi}_{\mathrm{H}}(\mathbf{r},t) .
\end{aligned} \tag{9.25}$$

We multiply on the left by the operator $\hat{\psi}^+_{\mathrm{H}}(\mathbf{r}', t')$, then we evaluate the expectation value on the ground state $|\Psi_0\rangle$ and divide it by its norm

$$\begin{aligned}
&\left[i\hbar\frac{\partial}{\partial t} - \left(-\hbar^2\frac{\nabla^2_{\mathbf{r}}}{2m}\right)\right]\frac{\langle\Psi_0|\hat{\psi}^+_{\mathrm{H}}(\mathbf{r}',t')\hat{\psi}_{\mathrm{H}}(\mathbf{r},t)|\Psi_0\rangle}{\langle\Psi_0|\Psi_0\rangle} \\
&= \frac{1}{\langle\Psi_0|\Psi_0\rangle}\int d^3x\, \langle\Psi_0|\hat{\psi}^+_{\mathrm{H}}(\mathbf{r}',t')\hat{\psi}^+_{\mathrm{H}}(\mathbf{x},t)\hat{V}(\mathbf{r},\mathbf{x})\hat{\psi}_{\mathrm{H}}(\mathbf{x},t)\hat{\psi}_{\mathrm{H}}(\mathbf{r},t)|\Psi_0\rangle .
\end{aligned} \tag{9.26}$$

We take the limit for $\mathbf{r}' \to \mathbf{r}$ e $t' \to t$ and integrate on d^3r,

$$\int d^3r \lim_{\mathbf{r}'\to\mathbf{r}} \lim_{t'\to t} \left[i\hbar\frac{\partial}{\partial t} - \left(-\hbar^2\frac{\nabla^2_{\mathbf{r}}}{2m}\right)\right] \left[-iG(\mathbf{r},t,\mathbf{r}',t')\right] = 2\frac{\langle\Psi_0|\hat{V}|\Psi_0\rangle}{\langle\Psi_0|\Psi_0\rangle} , \tag{9.27}$$

where the factor 2 is due to the definition of $\hat{W}$ given by Eq. (9.18).

We use Eq. (9.7) to calculate the expectation value of the kinetic energy operator

$$\frac{\langle\Psi_0|\hat{T}|\Psi_0\rangle}{\langle\Psi_0|\Psi_0\rangle} = \int d^3r \lim_{\mathbf{r}'\to\mathbf{r}} \lim_{t'\to t} \left(-\hbar^2\frac{\nabla^2_{\mathbf{r}}}{2m}\right)\left(-i\,G(\mathbf{r},t,\mathbf{r}',t')\right) . \tag{9.28}$$

By using these two equations in (9.13) we obtain the desired result.

9.2.1 System of Non-interacting Fermions

In this section, we derive an expression for the one-body Green's function of an infinite system of non-interacting fermions. This result serves as the foundation for the perturbative calculation of the Green's function that describes a system of interacting particles.

The system of non-interacting particle is defined by the fact that it is described by a MF hamiltonian

$$\hat{H} = \hat{H}_0 = \sum_\alpha \hat{h}_\alpha , \tag{9.29}$$

where $\hat{h}_\alpha$ are single particle hamiltonians of which $|\phi_\alpha\rangle$ are the eigenstates. In this case, all the operators in the Heisenberg picture are equal to those expressed in the interaction picture

$$\hat{\mathbb{O}}_{\mathrm{H}} \equiv e^{\frac{i}{\hbar}\hat{H}t}\hat{\mathbb{O}}e^{-\frac{i}{\hbar}\hat{H}t} = e^{\frac{i}{\hbar}\hat{H}_0 t}\hat{\mathbb{O}}e^{-\frac{i}{\hbar}\hat{H}_0 t} \equiv \hat{\mathbb{O}}_{\mathrm{I}} . \tag{9.30}$$

We simplify the writing by assuming that the many-body states $|\Phi_0\rangle$ are normalised to one. By definition, the one-body Green's function can be expressed as

$$\begin{aligned} i\,G^0(\mathbf{x},t,\mathbf{x}',t') &= \langle\Phi_0|\hat{\mathbb{T}}\left[\hat{\psi}_{\mathrm{I}}(\mathbf{x},t)\hat{\psi}^+_{\mathrm{I}}(\mathbf{x}',t')\right]|\Phi_0\rangle \\ &= \langle\Phi_0|\hat{\psi}_{\mathrm{I}}(\mathbf{x},t)\hat{\psi}^+_{\mathrm{I}}(\mathbf{x}',t')|\Phi_0\rangle\Theta(t-t') \\ &- \langle\Phi_0|\hat{\psi}^+_{\mathrm{I}}(\mathbf{x}',t')\hat{\psi}_{\mathrm{I}}(\mathbf{x},t)|\Phi_0\rangle\Theta(t'-t) , \end{aligned} \tag{9.31}$$

where $\Theta(x)$ is the step function (Fig. 9.1).

Let us consider the field operators in the interaction picture and express them in terms of creation and destruction operators

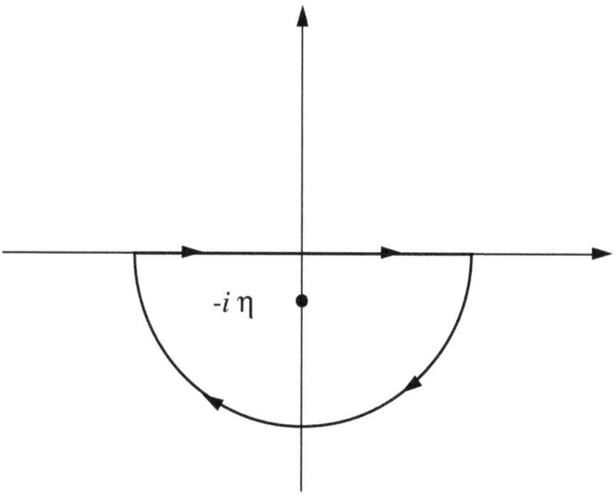

Fig. 9.1 Integration contour for the integral (9.41)

$$\hat{\psi}_{\mathrm{I}}(\mathbf{x},t) = \sum_k \hat{a}_{I,k}(t)\phi_k(\mathbf{x}) \quad \text{and} \quad \hat{\psi}_{\mathrm{I}}^+(\mathbf{x},t) = \sum_k \hat{a}_{I,k}^+(t)\phi_k^*(\mathbf{x}) \ , \tag{9.32}$$

therefore, by using Eqs. (6.15) and (6.16) we have

$$\begin{aligned} & i\, G^0(\mathbf{x},t,\mathbf{x}',t') \\ = \ & \langle\Phi_0|\sum_k \phi_k(\mathbf{x})\hat{a}_k e^{-i\omega_k t}\sum_{k'}\phi_{k'}^*(\mathbf{x}')\hat{a}_{k'}^+ e^{i\omega_{k'}t'}|\Phi_0\rangle\Theta(t-t') \\ - \ & \langle\Phi_0|\sum_{k'}\phi_{k'}^*(\mathbf{x}')\hat{a}_{k'}^+ e^{i\omega_{k'}t'}\sum_k \phi_k(\mathbf{x})\hat{a}_k e^{-i\omega_k t}|\Phi_0\rangle\Theta(t'-t) \\ = \ & \sum_{k\,k'}\phi_k(\mathbf{x})\phi_{k'}^*(\mathbf{x}')e^{-i\omega_k t}e^{i\omega_{k'}t'} \\ & \left[\langle\Phi_0|\hat{a}_k\hat{a}_{k'}^+|\Phi_0\rangle\Theta(t-t') - \langle\Phi_0|\hat{a}_{k'}^+\hat{a}_k|\Phi_0\rangle\Theta(t'-t)\right] \ , \end{aligned} \tag{9.33}$$

where we defined $\omega_k = E_k/\hbar$.

By definition of $|\Phi_0\rangle$ the following relations hold

$$\langle\Phi_0|\hat{a}_k\hat{a}_{k'}^+|\Phi_0\rangle = \delta_{k\,k'}\Theta(k'-k_F) \ , \tag{9.34}$$

and

$$\langle\Phi_0|\hat{a}_{k'}^+\hat{a}_k|\Phi_0\rangle = \delta_{k\,k'}\Theta(k_F-k) \ . \tag{9.35}$$

In an infinite, and homogeneos, system the expression of the single-particle wave function is (see Sect. 2.3)

$$\phi_k(\mathbf{x}) = \frac{1}{\sqrt{\mathcal{V}}}e^{-i\mathbf{k}\cdot\mathbf{x}} \ , \tag{9.36}$$

where spin and isospin dependent terms have been neglected. The unperturbed Green's function (9.33) assumes the expression

$$i\,G^0(\mathbf{x},t,\mathbf{x}',t') = \frac{1}{\mathcal{V}}\sum_k e^{-i\mathbf{k}\cdot(\mathbf{x}-\mathbf{x}')}e^{-i\omega_k(t-t')} \left[\Theta(k-k_F)\Theta(t-t') - \Theta(k_F-k)\Theta(t'-t)\right] \; . \tag{9.37}$$

We remember that, for an infinite system, we have to substitute the sums in integrals with the prescription (see Sect. 2.3)

$$\sum_k \rightarrow \frac{\mathcal{V}}{(2\pi)^3}\int d^3k \; . \tag{9.38}$$

By using the integral representation of the step function

$$\Theta(x) = -\lim_{\eta\to 0}\int_{-\infty}^{\infty} dk \left(\frac{1}{2\pi\, i}\right)\frac{e^{-ikx}}{k+i\eta} \; , \tag{9.39}$$

we can write

$$i\,G^0(\mathbf{x},t,\mathbf{x}',t') = \frac{1}{(2\pi)^3}\int d^3k\, e^{-i\mathbf{k}\cdot(\mathbf{x}-\mathbf{x}')}e^{-i\omega_k(t-t')} \left[-\int_{-\infty}^{\infty}\frac{d\omega'}{2\pi\, i}\frac{e^{-i\omega'(t-t')}}{\omega'+i\eta}\Theta(k-k_F) - \int_{-\infty}^{\infty}\frac{d\omega'}{2\pi\, i}\frac{e^{-i\omega'(t-t')}}{\omega'-i\eta}\Theta(k_F-k)\right] \; . \tag{9.40}$$

Let us obtain Eq. (9.39) using the residue theorem.

$$\int_{-\infty}^{\infty} dk \frac{1}{2\pi\, i}\frac{e^{-ikx}}{k+i\eta} = -\lim_{k\to -i\eta} 2\pi\, i \frac{1}{2\pi\, i}(k+i\eta)\frac{e^{-ikx}}{k+i\eta} = -e^{-\eta x} \; , \tag{9.41}$$

where the overall minus sign is due to the direction of the integration contour for $x > 0$. In the case $x < 0$ the integration contour will be closed without considering the pole, therefore the integral will be zero. We have that

$$\lim_{\eta\to 0} -e^{-\eta x} = -1 \; . \tag{9.42}$$

from which Eq. (9.39).

Putting everything together, and multiplying by $-i$, we obtain

$$G^0(\mathbf{x}, t, \mathbf{x}', t') = \frac{1}{(2\pi)^4} \int d^3k\, e^{-i\mathbf{k}\cdot(\mathbf{x}-\mathbf{x}')} \int_{-\infty}^{\infty} d\omega' e^{-i\omega'(t-t')} e^{-i\omega_k(t-t')} \left[\frac{\Theta(k-k_F)}{\omega'+i\eta} + \frac{\Theta(k_F-k)}{\omega'-i\eta}\right] . \qquad (9.43)$$

We define a new variable $\omega = \omega' + \omega_k$ and rewrite the above equation as

$$G^0(\mathbf{x}, t, \mathbf{x}', t') = \frac{1}{(2\pi)^4} \int d^3k\, e^{-i\mathbf{k}\cdot(\mathbf{x}-\mathbf{x}')} \int_{-\infty}^{\infty} d\omega\, e^{-i\omega(t-t')} \left[\frac{\Theta(k-k_F)}{\omega-\omega_k+i\eta} + \frac{\Theta(k_F-k)}{\omega-\omega_k-i\eta}\right] , \qquad (9.44)$$

From the expression above, we can derive the definition of the unperturbed one-body Green's function, which depends on both energy and momentum

$$\tilde{G}^0(\mathbf{k}, \omega) \equiv \left[\frac{\Theta(k-k_F)}{\omega-\omega_k+i\eta} + \frac{\Theta(k_F-k)}{\omega-\omega_k-i\eta}\right] . \qquad (9.45)$$

9.2.2 Lehmann Representation

Let us consider the complete Green's function for a system of interacting particles. To simplify our notation, we will assume that the ground state is normalised to unity, such that $\langle\Psi_0|\Psi_0\rangle = 1$. We can rewrite the expression for the Green's function by inserting a complete set of eigenstates of the Hamiltonian $\hat{H}$

$$\begin{aligned} i\,G(\mathbf{x}, t, \mathbf{x}', t') = &\ \langle\Psi_0|\hat{\mathbb{T}}\left[\hat{\psi}_{\mathrm{H}}(\mathbf{x}, t)\hat{\psi}^+_{\mathrm{H}}(\mathbf{x}', t')\right]|\Psi_0\rangle \\ = &\ \sum_n \langle\Psi_0|\hat{\psi}_{\mathrm{H}}(\mathbf{x}, t)|\Psi_n\rangle\langle\Psi_n|\hat{\psi}^+_{\mathrm{H}}(\mathbf{x}', t')|\Psi_0\rangle\Theta(t-t') \\ - &\ \sum_{n'} \langle\Psi_0|\hat{\psi}^+_{\mathrm{H}}(\mathbf{x}', t')|\Psi_{n'}\rangle\langle\Psi_{n'}|\hat{\psi}_{\mathrm{H}}(\mathbf{x}, t)|\Psi_0\rangle\Theta(t'-t) . \end{aligned} \qquad (9.46)$$

The eigenstates $|\Psi_n\rangle$ and $|\Psi_{n'}\rangle$ of $\hat{H}$ contain different number of particles. Specifically, the $|\Psi_n\rangle$ multiplying $\Theta(t-t')$ have $A+1$ particles, while the $|\Psi_{n'}\rangle$ multiplying $\Theta(t'-t)$ have $A-1$ particles.

Let us indicate with A the number of particles, eigenvalue of the particle number operator

$$\hat{n} = \int d^3x\, \hat{\psi}^+(\mathbf{x})\hat{\psi}(\mathbf{x}) . \qquad (9.47)$$

We calculate the commutator

$$
\begin{aligned}
[\hat{n}, \hat{\psi}(\mathbf{z})] &= \int d^3x [\hat{\psi}^+(\mathbf{x})\hat{\psi}(\mathbf{x}), \hat{\psi}(\mathbf{z})] \\
&= \int d^3x \, [\hat{\psi}^+(\mathbf{x})\hat{\psi}(\mathbf{x})\hat{\psi}(\mathbf{z}) - \hat{\psi}(\mathbf{z})\hat{\psi}^+(\mathbf{x})\hat{\psi}(\mathbf{x})] \\
&= \int d^3x \, [\hat{\psi}^+(\mathbf{x})\hat{\psi}(\mathbf{x})\hat{\psi}(\mathbf{z}) - \hat{\psi}^+(\mathbf{x})\hat{\psi}(\mathbf{x})\hat{\psi}(\mathbf{z}) - \hat{\psi}(\mathbf{z})\delta(\mathbf{x}-\mathbf{z})] = -\hat{\psi}(\mathbf{z}) \ .
\end{aligned}
$$

This means

$$
[\hat{n}, \hat{\psi}] = -\hat{\psi} \ , \quad \hat{n}\hat{\psi} - \hat{\psi}\hat{n} = -\hat{\psi} \ , \quad \hat{n}\hat{\psi} = \hat{\psi}(\hat{n} - 1) \ , \tag{9.48}
$$

therefore

$$
\hat{n}\hat{\psi}|\Psi_0\rangle = \hat{\psi}(\hat{n} - 1)|\Psi_0\rangle = \hat{\psi}(A - 1)|\Psi_0\rangle = (A - 1)\hat{\psi}|\Psi_0\rangle \ , \tag{9.49}
$$

which indicates that the states of the second term of Eq. (9.46) have a particle less than the states $|\Psi_0\rangle$.

We write Eq. (9.46) by expressing the field operators in the Schrödinger picture,

$$
\hat{\mathbb{O}}_H = e^{\frac{i}{\hbar}\hat{H}t}\hat{\mathbb{O}}_S e^{-\frac{i}{\hbar}\hat{H}t} \ , \tag{9.50}
$$

and we obtain

$$
\begin{aligned}
&i\,G(\mathbf{x}, t, \mathbf{x}', t') \\
&= \sum_n \Theta(t - t')e^{-\frac{i}{\hbar}(E_n - E_0)(t-t')}\langle\Psi_0|\hat{\psi}(\mathbf{x})|\Psi_n\rangle\langle\Psi_n|\hat{\psi}^+(\mathbf{x}')|\Psi_0\rangle \\
&- \sum_{n'} \Theta(t - t')e^{\frac{i}{\hbar}(E_{n'} - E_0)(t-t')}\langle\Psi_0|\hat{\psi}^+(\mathbf{x}')|\Psi_{n'}\rangle\langle\Psi_{n'}|\hat{\psi}(\mathbf{x})|\Psi_0\rangle \ .
\end{aligned} \tag{9.51}
$$

By substituting the expressions of the field operators

$$
\hat{\psi}(\mathbf{x}) = \sum_k \hat{a}_k\phi_k(\mathbf{x}) \quad \text{and} \quad \hat{\psi}^+(\mathbf{x}) = \sum_k \hat{a}_k^+\phi_k^*(\mathbf{x}),
$$

we obtain

$$
\begin{aligned}
&i\,G(\mathbf{x}, t, \mathbf{x}', t') \\
&= \sum_n \Theta(t - t')e^{-\frac{i}{\hbar}(E_n - E_0)(t-t')}\sum_{k,k'}\phi_k(\mathbf{x})\phi_{k'}^*(\mathbf{x}')\langle\Psi_0|\hat{a}_k|\Psi_n\rangle\langle\Psi_n|\hat{a}_{k'}^+|\Psi_0\rangle \\
&- \sum_{n'} \Theta(t' - t)e^{\frac{i}{\hbar}(E_{n'} - E_0)(t-t')}\sum_{k,k'}\phi_k(\mathbf{x})\phi_{k'}^*(\mathbf{x}')\langle\Psi_0|\hat{a}_{k'}^+|\Psi_{n'}\rangle\langle\Psi_{n'}|\hat{a}_k|\Psi_0\rangle \ .
\end{aligned}
$$

By considering the case of an infinite system, we can use Eqs. (9.36) and (9.38) and obtain

$$
\begin{aligned}
& i\,G(\mathbf{x},t,\mathbf{x}',t') = \\
& \sum_n \Theta(t-t') e^{-\frac{i}{\hbar}(E_n-E_0)(t-t')} \frac{1}{(2\pi)^3} \int d^3k\, e^{i\mathbf{k}\cdot(\mathbf{x}-\mathbf{x}')} \langle\Psi_0|\hat{a}_k|\Psi_n\rangle\langle\Psi_n|\hat{a}_k^+|\Psi_0\rangle \\
- & \sum_{n'} \Theta(t'-t) e^{\frac{i}{\hbar}(E_{n'}-E_0)(t-t')} \frac{1}{(2\pi)^3} \int d^3k\, e^{i\mathbf{k}\cdot(\mathbf{x}-\mathbf{x}')} \langle\Psi_0|\hat{a}_k^+|\Psi_{n'}\rangle\langle\Psi_{n'}|\hat{a}_k|\Psi_0\rangle \ .
\end{aligned}
$$

In the above equations, we considered that $|\Psi_n\rangle$ is characterized by the momentum k, therefore

$$
\mathbb{I} = \sum_n |\Psi_n\rangle\langle\Psi_n| \equiv \frac{1}{(2\pi)^3} \int d^3k\, |k\rangle\langle k'| = \delta_{k,k'} \ .
$$

By using the integral expression of the step function (9.39) we obtain

$$
\begin{aligned}
G(\mathbf{x},t,\mathbf{x}',t') = & \frac{1}{(2\pi)^4} \int d^3k\, e^{i\mathbf{k}\cdot(\mathbf{x}-\mathbf{x}')} \int d\omega e^{-i\omega(t-t')} \\
& \left[\sum_n \frac{\langle\Psi_0|\hat{a}_k|\Psi_n\rangle\langle\Psi_n|\hat{a}_k^+|\Psi_0\rangle}{\omega-(E_n-E_0)/\hbar+i\eta} + \sum_{n'} \frac{\langle\Psi_0|\hat{a}_k^+|\Psi_{n'}\rangle\langle\Psi_{n'}|\hat{a}_k|\Psi_0\rangle}{\omega+(E_{n'}-E_0)/\hbar-i\eta} \right] \ . \qquad (9.52)
\end{aligned}
$$

We define the Fourier transform

$$
\begin{aligned}
\tilde{G}(\mathbf{k},\omega) &= \int d^3(x-x')\, e^{-i\mathbf{k}\cdot(\mathbf{x}-\mathbf{x}')} \int d(t-t') e^{i\omega(t-t')} G(\mathbf{x},t,\mathbf{x}',t') \\
&= \sum_n \frac{\langle\Psi_0|\hat{a}_k|\Psi_n\rangle\langle\Psi_n|\hat{a}_k^+|\Psi_0\rangle}{\omega-(E_n-E_0)/\hbar+i\eta} + \sum_{n'} \frac{\langle\Psi_0|\hat{a}_k^+|\Psi_{n'}\rangle\langle\Psi_{n'}|\hat{a}_k|\Psi_0\rangle}{\omega+(E_{n'}-E_0)/\hbar-i\eta} \ . \qquad (9.53)
\end{aligned}
$$

Let us consider the denominator of the first term. The energy E_0 is eigenvalue of $|\Psi_0\rangle$ describing the system with A particles, while E_n is eiegenvalue of $|\Psi_n\rangle$ which has $A+1$ particles. We rewrite the denominator indicating in round brackets the number of particles of the system which the energy refers to.

$$
\begin{aligned}
& \hbar\omega - E_n(A+1) + E_0(A) = \hbar\omega - E_n(A+1) + E_0(A+1) - E_0(A+1) + E_0(A) \\
= \ & \hbar\omega - [E_n(A+1) - E_0(A+1)] - [E_0(A+1) - E_0(A)] \\
= \ & \hbar\omega - \hbar\omega_n(A+1) - \mu \ .
\end{aligned}
$$

In the above equations we call $\omega_n(A+1)$ the excitation energy of the system with $A+1$ particles and with μ the **chemical potential** defined as

$$
\mu = \left(\frac{\partial E}{\partial \mathrm{A}}\right)_V \equiv \left[\frac{E_0(A+1) - E_0(A)}{\Delta 1}\right] \ . \qquad (9.54)
$$

where we indicated with $\Delta 1$ the variation of one unit of the number of fermions.

The Green's function in Lehmann representation is

$$\tilde{G}(\mathbf{k},\omega) = \sum_n \frac{\langle\Psi_0|\hat{a}_k|\Psi_n\rangle\langle\Psi_n|\hat{a}_k^+|\Psi_0\rangle}{\omega - \omega_n - \mu/\hbar + i\eta} + \sum_{n'} \frac{\langle\Psi_0|\hat{a}_k^+|\Psi_{n'}\rangle\langle\Psi_{n'}|\hat{a}_k|\Psi_0\rangle}{\omega + \omega_{n'} - \mu/\hbar - i\eta} \ . \tag{9.55}$$

The sign of the chemical potential in the second denominator is due to the definition (9.54) where the derivative has positive sign when a particle is added to the system. It worth to remark that ω_n in the first term is refers to the system with $A+1$ particles, while $\omega_{n'}$ in the second term to a system with $A-1$ particles.

9.2.3 Physical Interpretation

Let us consider a many-body state in the interaction picture and add a particle in the point $\mathbf{x}'$, at the time t'

$$\hat{\psi}_{\mathrm{I}}^+(\mathbf{x}',t')|\Psi_{\mathrm{I}}(t')\rangle \ . \tag{9.56}$$

This states propagates at the time t as

$$\hat{U}(t,t')\hat{\psi}_{\mathrm{I}}^+(\mathbf{x}',t')|\Psi_{\mathrm{I}}(t')\rangle \ , \tag{9.57}$$

where $\hat{U}$ is the time evolution operator of Sect. 6.2. Let's recall here below some property of this operator:

$$\begin{aligned}
&\hat{U}(t,t')\hat{U}(t',t) = \hat{U}(t,t) = 1 \ , \\
&|\Psi_{\mathrm{H}}\rangle = |\Psi_{\mathrm{I}}(0)\rangle = \hat{U}(0,-\infty)|\Phi_0\rangle = |\Psi_0\rangle \ , \\
&\hat{U}(t,t_0) = e^{\frac{i}{\hbar}\hat{H}_0 t} e^{-\frac{i}{\hbar}(\hat{H}_0+\hat{H}_1)(t-t_0)} e^{-\frac{i}{\hbar}\hat{H}_0 t_0} \ , \\
&\hat{U}(t,0) = e^{-\frac{i}{\hbar}\hat{H}_1 t} \ ; \ \hat{U}(0,t) = e^{\frac{i}{\hbar}\hat{H}_1 t} \ , \\
&\hat{\mathbb{O}}_{\mathrm{I}}(t) = e^{-\frac{i}{\hbar}\hat{H}_1 t}\hat{\mathbb{O}}_{\mathrm{H}}(t)e^{\frac{i}{\hbar}\hat{H}_1 t} = \hat{U}(t,0)\hat{\mathbb{O}}_{\mathrm{H}}\hat{U}(0,t) \ , \\
&\hat{\mathbb{O}}_{\mathrm{H}}(t) = e^{\frac{i}{\hbar}\hat{H}t}\hat{\mathbb{O}}_{\mathrm{S}}e^{-\frac{i}{\hbar}\hat{H}t} \ ; \ \hat{\mathbb{O}}_{\mathrm{I}}(t) = e^{\frac{i}{\hbar}\hat{H}_0 t}\hat{\mathbb{O}}_{\mathrm{S}}e^{-\frac{i}{\hbar}\hat{H}_0 t} \ , \\
&|\Psi_{\mathrm{I}}(t')\rangle = \hat{U}(t',-\infty)|\Phi_0\rangle \ .
\end{aligned}$$

We consider a situation where, at the time t' in the point $\mathbf{x}'$, a particle is added to the system, the state propagates and, later on, at the time t, in the position $\mathbf{x}$ the particle is eliminated. This process can be described as

$$\begin{aligned}
&\langle\Psi(t)|\hat{\psi}_{\mathrm{I}}(\mathbf{x},t)\hat{U}(t,t')\hat{\psi}_{\mathrm{I}}^+(\mathbf{x}',t')|\Psi(t')\rangle = \\
&\langle\Phi_0|\hat{U}(\infty,t)\left[\hat{U}(t,0)\hat{\psi}_{\mathrm{H}}(\mathbf{x},t)\hat{U}(0,t)\right]\hat{U}(t,t') \\
&\left[\hat{U}(t',0)\hat{\psi}_{\mathrm{H}}^+(\mathbf{x}',t')\hat{U}(0,t')\right]\hat{U}(t',-\infty)|\Phi_0\rangle =
\end{aligned}$$

$$\left\{\langle\Phi_0|\hat{U}(\infty,t)\hat{U}(t,0)\right\}\hat{\psi}_{\mathrm{H}}(\mathbf{x},t)\left[\hat{U}(0,t)\hat{U}(t,t')\hat{U}(t',0)\right]$$
$$\hat{\psi}^{+}_{\mathrm{H}}(\mathbf{x}',t')\left\{\hat{U}(0,t')\hat{U}(t',-\infty)|\Phi_0\rangle\right\}=$$
$$\langle\Psi_0|\hat{\psi}_{\mathrm{H}}(\mathbf{x},t)\hat{\psi}^{+}_{\mathrm{H}}(\mathbf{x}',t')|\Psi_0\rangle.$$

The last expression is that of the one-body Green's function for $t > t'$. The Green's function is the probability amplitude that a particle created in the point $\mathbf{x}'$ at the time t' is destroyed in the point $\mathbf{x}$ at the time t.

9.3 Two-Body Green's Function

In this section, we simplify the writing, by assuming $\langle\Psi_0|\Psi_0\rangle = 1$. The two-body Green's function is defined as

$$\begin{aligned}&(-i)^2 G(\mathbf{x}_1,t_1,\mathbf{x}_2,t_2,\mathbf{x}_3,t_3,\mathbf{x}_4,t_4)\\ \equiv\ &\langle\Psi_0|\hat{\mathbb{T}}[\hat{\psi}_{\mathrm{H}}(\mathbf{x}_1,t_1)\hat{\psi}_{\mathrm{H}}(\mathbf{x}_2,t_2)\hat{\psi}^{+}_{\mathrm{H}}(\mathbf{x}_3,t_3)\hat{\psi}^{+}_{\mathrm{H}}(\mathbf{x}_4,t_4)]|\Psi_0\rangle\ ,\end{aligned}\tag{9.58}$$

For the one-body Green's function it has been necessary to consider two cases, $t > t'$ and viceversa. For the two-body Green's function there are $4! = 24$ cases. In reality for the symmetry properties

$$G(1234) = -G(2134) = -G(1243) = G(2143)\ ,\tag{9.59}$$

there are only 6 independent cases. Out of these 6 cases only 3 have physically interesting properties.

1. $t_1, t_2 > t_3, t_4$. This case implies

$$\begin{aligned}&(-i)^2 G(\mathbf{x}_1,t_1,\mathbf{x}_2,t_2,\mathbf{x}_3,t_3,\mathbf{x}_4,t_4)\\ \equiv\ &\langle\Psi_0|\hat{\psi}_{\mathrm{H}}(\mathbf{x}_1,t_1)\hat{\psi}_{\mathrm{H}}(\mathbf{x}_2,t_2)\hat{\psi}^{+}_{\mathrm{H}}(\mathbf{x}_3,t_3)\hat{\psi}^{+}_{\mathrm{H}}(\mathbf{x}_4,t_4)|\Psi_0\rangle\ ,\end{aligned}\tag{9.60}$$

   and describes the evolution of a state to which, at the times t_3 and t_4, two particles have been added.
2. $t_1, t_2 < t_3, t_4$. This case implies

$$\begin{aligned}&(-i)^2 G(\mathbf{x}_1,t_1,\mathbf{x}_2,t_2,\mathbf{x}_3,t_3,\mathbf{x}_4,t_4)\\ \equiv\ &\langle\Psi_0|\hat{\psi}^{+}_{\mathrm{H}}(\mathbf{x}_3,t_3)\hat{\psi}^{+}_{\mathrm{H}}(\mathbf{x}_4,t_4)\hat{\psi}_{\mathrm{H}}(\mathbf{x}_1,t_1)\hat{\psi}_{\mathrm{H}}(\mathbf{x}_2,t_2)|\Psi_0\rangle\ ,\end{aligned}\tag{9.61}$$

   and describes the evolution of a state to which, at the times t_1 and t_2, two holes have been created.
3. $t_1, t_3 > t_2, t_4$. This case implies

$$(-i)^2 G(\mathbf{x}_1, t_1, \mathbf{x}_2, t_2, \mathbf{x}_3, t_3, \mathbf{x}_4, t_4)$$
$$\equiv -\langle \Psi_0 | \hat{\psi}_{\mathrm{H}}(\mathbf{x}_1, t_1) \hat{\psi}^+_{\mathrm{H}}(\mathbf{x}_3, t_3) \hat{\psi}_{\mathrm{H}}(\mathbf{x}_2, t_2) \hat{\psi}^+_{\mathrm{H}}(\mathbf{x}_4, t_4) | \Psi_0 \rangle \ , \tag{9.62}$$

and describes the time evolution of a particle-hole pair.

This latter case is that of major interest to us and it is the only case treated in this section.

Since we operate within a non-relativistic framework, the creation and destruction of a particle-hole pair occur instantaneously. Therefore, we have that

$$t_1 = t_3 = t' \ \ \mathrm{e} \ \ t_2 = t_4 = t \ . \tag{9.63}$$

For this case, we express the field operators in terms of creation and destruction operators, and obtain the expression

$$G(\mathbf{x}_1, t', \mathbf{x}_2, t, \mathbf{x}_3, t', \mathbf{x}_4, t) =$$
$$\sum_{\nu_1 \nu_2 \nu_3 \nu_4} \phi_{\nu_1}(\mathbf{x}_1) \phi^*_{\nu_3}(\mathbf{x}_3) \phi_{\nu_2}(\mathbf{x}_2) \phi^*_{\nu_4}(\mathbf{x}_4) \langle \Psi_0 | \hat{\mathbb{T}}[\hat{a}_{\nu_1}(t') \hat{a}^+_{\nu_3}(t') \hat{a}_{\nu_2}(t) \hat{a}^+_{\nu_4}(t)] | \Psi_0 \rangle$$
$$\equiv \sum_{\nu_1 \nu_2 \nu_3 \nu_4} \phi_{\nu_1}(\mathbf{x}_1) \phi_{\nu_2}(\mathbf{x}_2) \phi^*_{\nu_3}(\mathbf{x}_3) \phi^*_{\nu_4}(\mathbf{x}_4) G(\nu_1, t', \nu_2, t, \nu_3, t', \nu_4, t) \ . \tag{9.64}$$

where it is understood that all the creation and destruction operators are expressed in the Heisenberg picture. The previous equation defines a two-body Green's function depending on the quantum numbers ν characterising the single-particle states.

We define the *retarded* part of the Green's function by considering the $t' > t$ case

$$G^R(\nu_1, t', \nu_2, t, \nu_3, t', \nu_4, t) = \langle \Psi_0 | \hat{a}_{\nu_1}(t') \hat{a}^+_{\nu_3}(t') \hat{a}_{\nu_2}(t) \hat{a}^+_{\nu_4}(t) | \Psi_0 \rangle$$
$$= \langle \Psi_0 | e^{\frac{i}{\hbar}\hat{H}t'} \hat{a}_{\nu_1} e^{-\frac{i}{\hbar}\hat{H}t'} e^{\frac{i}{\hbar}\hat{H}t'} \hat{a}^+_{\nu_3} e^{-\frac{i}{\hbar}\hat{H}t'} e^{\frac{i}{\hbar}\hat{H}t} \hat{a}_{\nu_2} e^{-\frac{i}{\hbar}\hat{H}t} e^{\frac{i}{\hbar}\hat{H}t} \hat{a}^+_{\nu_4} e^{-\frac{i}{\hbar}\hat{H}t} | \Psi_0 \rangle$$
$$= \langle \Psi_0 | \hat{a}_{\nu_1} \hat{a}^+_{\nu_3} e^{-\frac{i}{\hbar}(\hat{H} - E_0)(t'-t)} \hat{a}_{\nu_2} \hat{a}^+_{\nu_4} | \Psi_0 \rangle \ , \tag{9.65}$$

where we made explicit the expression of the creation and destruction operators in the Heisenberg picture, and we used $\hat{H}|\Psi_0\rangle = E_0|\Psi_0\rangle$. In analogous manner we define the *advanced* part of the Green's function for the $t' < t$ case,

$$G^A(\nu_1, t', \nu_2, t, \nu_3, t', \nu_4, t) = \langle \Psi_0 | \hat{a}_{\nu_2} \hat{a}^+_{\nu_4} e^{+\frac{i}{\hbar}(\hat{H} - E_0)(t'-t)} \hat{a}_{\nu_1} \hat{a}^+_{\nu_3} | \Psi_0 \rangle \ . \tag{9.66}$$

We can express the two-body Green's function as

$$G(\nu_1, \nu_2, \nu_3, \nu_4, \tau) = \begin{cases} G^A(\nu_1, t', \nu_2, t, \nu_3, t', \nu_4, t) \ \text{for} \ \tau = t' - t < 0 \\ G^R(\nu_1, t', \nu_2, t, \nu_3, t', \nu_4, t) \ \text{for} \ \tau = t' - t > 0 \end{cases} . \tag{9.67}$$

9.3.1 Lehmann Representation

We define the energy dependent two-body Green's function as

$$\tilde{G}(\nu_1, \nu_2, \nu_3, \nu_4, E) = \int_{-\infty}^{\infty} d\tau\, G(\nu_1, \nu_2, \nu_3, \nu_4, \tau)\, e^{\frac{i}{\hbar}E\tau} \ , \tag{9.68}$$

where $\tau = t' - t$. We consider the retarded part

$$\tilde{G}^R(\nu_1, \nu_2, \nu_3, \nu_4, E) = \langle\Psi_0|\hat{a}_{\nu_1}\hat{a}^+_{\nu_3} \int_0^{\infty} d\tau\, e^{-\frac{i}{\hbar}(\hat{H}-E_0-E)\tau}\hat{a}_{\nu_2}\hat{a}^+_{\nu_4}|\Psi_0\rangle \ . \tag{9.69}$$

We can express the value of the time integral as

$$\begin{aligned} &\lim_{\eta\to 0}\int_0^{\infty} d\tau\, e^{\frac{i}{\hbar}(-\hat{H}+E_0+E+i\eta)\tau} \\ &= \lim_{\eta\to 0}\frac{e^{\frac{i}{\hbar}(E-\hat{H}+E_0)\tau}e^{-\eta\tau}}{\frac{i}{\hbar}(E-\hat{H}+E_0+i\eta)}\Big|_0^{\infty} = \lim_{\eta\to 0}\frac{i\hbar}{E-\hat{H}+E_0+i\eta} \ , \end{aligned} \tag{9.70}$$

therefore

$$\tilde{G}^R(\nu_1, \nu_2, \nu_3, \nu_4, E) = \hbar\langle\Psi_0|\hat{a}_{\nu_1}\hat{a}^+_{\nu_3}\frac{i}{E-\hat{H}+E_0+i\eta}\hat{a}_{\nu_2}\hat{a}^+_{\nu_4}|\Psi_0\rangle \ . \tag{9.71}$$

With an analogous calculation we obtain for the advanced part

$$\tilde{G}^A(\nu_1, \nu_2, \nu_3, \nu_4, E) = \langle\Psi_0|\hat{a}_{\nu_2}\hat{a}^+_{\nu_4}\int_{-\infty}^{0} d\tau\, e^{\frac{i}{\hbar}(\hat{H}-E_0+E)\tau}\hat{a}_{\nu_1}\hat{a}^+_{\nu_3}|\Psi_0\rangle \ , \tag{9.72}$$

since

$$\begin{aligned} &\lim_{\eta\to 0}\int_{-\infty}^{0} d\tau\, e^{\frac{i}{\hbar}(\hat{H}-E_0+E-i\eta)\tau} \\ &= \lim_{\eta\to 0}\frac{e^{\frac{i}{\hbar}(E+\hat{H}-E_0)\tau}e^{\eta\tau}}{\frac{i}{\hbar}(E+\hat{H}-E_0-i\eta)}\Big|_{-\infty}^{0} = \lim_{\eta\to 0}\frac{-i\hbar}{E+\hat{H}-E_0-i\eta} \ , \end{aligned} \tag{9.73}$$

we obtain

$$\tilde{G}^A(\nu_1, \nu_2, \nu_3, \nu_4, E) = (-1)\hbar\langle\Psi_0|\hat{a}_{\nu_2}\hat{a}^+_{\nu_4}\frac{i}{E+\hat{H}-E_0-i\eta}\hat{a}_{\nu_1}\hat{a}^+_{\nu_3}|\Psi_0\rangle \ , \tag{9.74}$$

therefore

$$
\begin{aligned}
-\frac{i}{\hbar}\tilde{G}(\nu_1,\nu_2,\nu_3,\nu_4,E) &= -\frac{i}{\hbar}(\tilde{G}^R+\tilde{G}^A) \\
&= \langle\Psi_0|\hat{a}_{\nu_1}\hat{a}^+_{\nu_3}\frac{1}{E-\hat{H}+E_0+i\eta}\hat{a}_{\nu_2}\hat{a}^+_{\nu_4}|\Psi_0\rangle \\
&- \langle\Psi_0|\hat{a}_{\nu_2}\hat{a}^+_{\nu_4}\frac{1}{E+\hat{H}-E_0-i\eta}\hat{a}_{\nu_1}\hat{a}^+_{\nu_3}|\Psi_0\rangle \ . \qquad (9.75)
\end{aligned}
$$

By inserting the completeness of the eigenfunctions of $\hat{H}$, $\sum_n |\Psi_n\rangle\langle\Psi_n| = 1$, and considering $\hat{H}|\Psi_n\rangle = E_n|\Psi_n\rangle$ we obtain the expression

$$
\begin{aligned}
-\frac{i}{\hbar}\tilde{G}(\nu_1,\nu_2,\nu_3,\nu_4,E) = \\
\sum_n\Bigg[&\frac{\langle\Psi_0|\hat{a}_{\nu_1}\hat{a}^+_{\nu_3}|\Psi_n\rangle\langle\Psi_n|\,\hat{a}_{\nu_2}\hat{a}^+_{\nu_4}|\Psi_0\rangle}{E-(E_n-E_0)+i\eta} \\
&- \frac{\langle\Psi_0|\hat{a}_{\nu_2}\hat{a}^+_{\nu_4}|\Psi_n\rangle\langle\Psi_n|\hat{a}_{\nu_1}\hat{a}^+_{\nu_3}|\Psi_0\rangle}{E+(E_n-E_0)-i\eta}\Bigg] \ . \qquad (9.76)
\end{aligned}
$$

In this expression, the states $|\Psi_n\rangle$ have the same number of particles of the ground state. The energy values related to the poles, $E = E_n - E_0$, represent the excitation energies of the A particle system.

9.4 Linear Response

Let us consider a situation in which a many-body system is subjected to an external perturbation. We can express the total Hamiltonian of the perturbed system as the sum of the Hamiltonian $\hat{H}$ that describes the system in the absence of the perturbation, plus a term $\hat{H}^{\rm ext}(t)$ that accounts for the external perturbation

$$
\hat{H}^{\rm tot} = \hat{H} + \hat{H}^{\rm ext}(t) = \hat{H} + \hat{B}F(t) \ , \qquad (9.77)
$$

where the operator $\hat{B}$ identifies the action of the external perturbation on the system. For example the perturbation can be generated by photons, electrons, hadrons, etc. The function $F(t)$ describes how the perturbation evolves in the time, and it is defined such that $\hat{F}(t) = 0$ for $t < t_0 = 0$. This means that the perturbation is switched on after a specific time, t_0 which we define as zero time.

We assume that the reaction times of the many-body system are much shorter than the timescales required for the perturbation to be switched on and off. Therefore, when the perturbation is fully applied, the Hamiltonian is given by $\hat{H}^{\rm tot} = \hat{H} + \hat{B}$. In this scenario, we can treat $\hat{B}$ as a perturbative term within the total time-dependent

Hamiltonian. Consequently, we can analyse the system's equation of motion in the interaction picture (see Sect. 6.2)

$$i\hbar\frac{\partial}{\partial t}|\Psi_{\rm I}(t)\rangle = \hat{B}_{\rm I}(t)|\Psi_{\rm I}(t)\rangle \ , \tag{9.78}$$

where

$$\hat{B}_{\rm I}(t) = e^{\frac{i}{\hbar}\hat{H}t}\hat{B}e^{-\frac{i}{\hbar}\hat{H}t} \quad \text{and} \quad |\Psi_{\rm I}(t)\rangle = e^{\frac{i}{\hbar}\hat{H}t}|\Psi(t)\rangle \ . \tag{9.79}$$

In this section, we use the convention that states and operators without subscripts are expressed in the Schrödinger picture. We integrate Eq. (9.78)

$$\begin{aligned}
i\hbar\int_{-\infty}^{t} dt'\,\frac{\partial}{\partial t'}|\Psi_{\rm I}(t')\rangle &= \int_{-\infty}^{t} dt'\,\hat{B}_{\rm I}(t')|\Psi_{\rm I}(t')\rangle \\
i\hbar\Big[|\Psi_{\rm I}(t)\rangle - |\Psi_{\rm I}(-\infty)\rangle\Big] &= \int_{-\infty}^{t} dt'\,\hat{B}_{\rm I}(t')|\Psi_{\rm I}(t')\rangle \\
|\Psi_{\rm I}(t)\rangle &= |\Psi_{\rm I}(-\infty)\rangle - \frac{i}{\hbar}\int_{-\infty}^{t} dt'\,\hat{B}_{\rm I}(t')|\Psi_{\rm I}(t')\rangle \ . \quad (9.80)
\end{aligned}$$

Since the perturbation is off when $t = -\infty$ we have that $|\Psi_{\rm I}(-\infty)\rangle = |\Psi_0\rangle$. We can express the above equation in perturbative terms by iterating the presence of $|\Psi_{\rm I}(-\infty)\rangle$

$$|\Psi_{\rm I}(t)\rangle = |\Psi_0\rangle - \frac{i}{\hbar}\int_{-\infty}^{t} dt'\,\hat{B}_{\rm I}(t')|\Psi_0\rangle \ + \ \cdots \ . \tag{9.81}$$

We call $\hat{D}$ the operator which describes how the system reacts to the external perturbation induced by the operator $\hat{B}$. The expectation value of this operator is given by

$$\begin{aligned}
&\langle\Psi_{\rm I}(t)|\hat{D}_{\rm I}(t)|\Psi_{\rm I}(t)\rangle \\
&= \left\{\langle\Psi_0| + \frac{i}{\hbar}\int_{-\infty}^{t} dt'\,\langle\Psi_0|\hat{B}_{\rm I}(t') \ + \ \cdots\right\}\hat{D}_{\rm I}(t) \\
&\quad \left\{|\Psi_0\rangle - \frac{i}{\hbar}\int_{-\infty}^{t} dt'\,\hat{B}_{\rm I}(t')|\Psi_0\rangle \ + \ \cdots\right\} \\
&= \langle\Psi_0|\hat{D}_{\rm I}(t)|\Psi_0\rangle + \frac{i}{\hbar}\int_{-\infty}^{t} dt'\langle\Psi_0|[\hat{B}_{\rm I}(t'), \hat{D}_{\rm I}(t)]|\Psi_0\rangle \ + \cdots \ . \quad (9.82)
\end{aligned}$$

We define the response function as

$$R(t'-t) = \begin{cases} 0 & \text{per } \ t' < t \\ \frac{i}{\hbar}\langle\Psi_0|[\hat{B}_{\rm I}(t'), \hat{D}_{\rm I}(t)]|\Psi_0\rangle & \text{per } \ t' > t \end{cases} \ . \tag{9.83}$$

This definition implies causality: the system cannot respond before the perturbation is switched on.

By making explicit the time dependence of $\hat{B}_I(t')$ and $\hat{D}_I(t)$,

$$\hat{B}_I(t') = e^{\frac{i}{\hbar}\hat{H}t'}\hat{B}e^{-\frac{i}{\hbar}\hat{H}t'} \quad ; \quad \hat{D}_I(t) = e^{\frac{i}{\hbar}\hat{H}t}\hat{D}e^{-\frac{i}{\hbar}\hat{H}t} \ , \tag{9.84}$$

we can express the response as

$$\begin{aligned} R(t'-t) = & \Big[\frac{i}{\hbar}\langle\Psi_0|\hat{B}e^{\frac{i}{\hbar}(\hat{H}-E_0)(t-t')}\hat{D}|\Psi_0\rangle \\ & -\frac{i}{\hbar}\langle\Psi_0|\hat{D}e^{-\frac{i}{\hbar}(\hat{H}-E_0)(t-t')}\hat{B}|\Psi_0\rangle\Big]\Theta(t'-t) \ , \end{aligned} \tag{9.85}$$

and, since it depends only on the time difference $\tau = t - t'$, by using the definition of Fourier transform, we obtain

$$\begin{aligned} R(E) = & \int_{-\infty}^{\infty} d\tau\, R(\tau)\, e^{\frac{i}{\hbar}E\tau}\Theta(-\tau) \\ = & \frac{i}{\hbar}\langle\Psi_0|\hat{B}\int_{-\infty}^{0} d\tau e^{\frac{i}{\hbar}(\hat{H}-E_0+E)\tau}\hat{D}|\Psi_0\rangle \\ & -\frac{i}{\hbar}\langle\Psi_0|\hat{D}\int_{-\infty}^{\infty} d\tau e^{-\frac{i}{\hbar}(\hat{H}-E_0-E)\tau}\hat{B}|\Psi_0\rangle \\ = & \lim_{\eta\to 0}\Big[-\langle\Psi_0|\hat{B}(\hat{H}-E_0+E+i\eta)^{-1}\hat{D}|\Psi_0\rangle \\ - & \langle\Psi_0|\hat{D}(\hat{H}-E_0-E-i\eta)^{-1}\hat{B}|\Psi_0\rangle\Big] \ . \end{aligned} \tag{9.86}$$

We insert the completeness $\sum_n |\Psi_n\rangle\langle\Psi_n| = 1$ and obtain

$$R(E) = \sum_n \left[\frac{\langle\Psi_0|\hat{D}|\Psi_n\rangle\langle\Psi_n|\hat{B}|\Psi_0\rangle}{E-(E_n-E_0)+i\eta} - \frac{\langle\Psi_0|\hat{B}|\Psi_n\rangle\langle\Psi_n|\hat{D}|\Psi_0\rangle}{E+(E_n-E_0)+i\eta}\right] \ . \tag{9.87}$$

The poles of $R(E)$ correspond to the excitation energies of the system. For each positive pole there is a negative pole, equal in absolute value to the positive one.

We consider the Dirac expression

$$\frac{1}{x'-x\pm i\eta} = \mathcal{P}\frac{1}{x'-x} \mp i\pi\delta(x-x') \ , \tag{9.88}$$

where $\mathcal{P}$ indicates the principal part, therefore

$$\delta(x-x') = -\frac{1}{\pi}\Im\left(\frac{1}{x'-x\pm i\eta}\right) \ , \tag{9.89}$$

with the symbol $\Im$ indicating the imaginary part.

We assume $\hat{D} = \hat{B}$, as it usually happens, and consider only positive energies. The transition probability from the ground state to an excited state is given by

$$S(E) = -\frac{1}{\pi}\Im\big(R(E)\big) = \sum_n |\langle\Psi_0|\hat{B}|\Psi_n\rangle|^2 \delta\big(E - (E_n - E_0)\big) \ . \tag{9.90}$$

This is the traditional expression obtained by applying the time-dependent perturbation theory [1]. Assuming that $\hat{B}$ is a one-body operator

$$\hat{B} = \sum_{\nu_1\nu_2} B_{\nu_1\nu_2}\hat{a}_{\nu_1}\hat{a}^+_{\nu_2} \ \text{ and } \ B_{\nu_1\nu_2} = \int d^3r\, \phi^*_{\nu_1}(\mathbf{r})\,\hat{B}(\mathbf{r})\,\phi_{\nu_2}(\mathbf{r}) \ , \tag{9.91}$$

we obtain

$$\begin{aligned} R(E) = \sum_{\nu_1\nu_2}\sum_{\nu_3\nu_4}\sum_n \Bigg[& B_{\nu_1\nu_2}B^*_{\nu_3\nu_4}\frac{\langle\Psi_0|\hat{a}_{\nu_1}\hat{a}^+_{\nu_2}|\Psi_n\rangle\langle\Psi_n|\hat{a}_{\nu_3}\hat{a}^+_{\nu_4}|\Psi_0\rangle}{E - (E_n - E_0) + i\eta} \\ & - B_{\nu_3\nu_4}B^*_{\nu_1\nu_2}\frac{\langle\Psi_0|\hat{a}_{\nu_3}\hat{a}^+_{\nu_4}|\Psi_n\rangle\langle\Psi_n|\hat{a}_{\nu_1}\hat{a}^+_{\nu_2}|\Psi_0\rangle}{E + (E_n - E_0) + i\eta}\Bigg] \ , \end{aligned} \tag{9.92}$$

Since $\hat{B}$ is hermitian, $B_{\nu_1\nu_2} = B^*_{\nu_2\nu_1}$ and the indexes ν are dummy, we can write

$$\begin{aligned} R(E) &= \sum_{\nu_1\nu_2}\sum_{\nu_3\nu_4} B_{\nu_1\nu_2}B^*_{\nu_3\nu_4} \\ & \quad \sum_n \Bigg[\frac{\langle\Psi_0|\hat{a}_{\nu_1}\hat{a}^+_{\nu_2}|\Psi_n\rangle\langle\Psi_n|\hat{a}_{\nu_3}\hat{a}^+_{\nu_4}|\Psi_0\rangle}{E - (E_n - E_0) + i\eta} - \frac{\langle\Psi_0|\hat{a}_{\nu_3}\hat{a}^+_{\nu_4}|\Psi_n\rangle\langle\Psi_n|\hat{a}_{\nu_1}\hat{a}^+_{\nu_2}|\Psi_0\rangle}{E + (E_n - E_0) + i\eta}\Bigg] \\ &= \sum_{\nu_1\nu_2}\sum_{\nu_3\nu_4} B_{\nu_1\nu_2}B^*_{\nu_3\nu_4}(-i)\tilde{G}(\nu_1, \nu_3, \nu_2, \nu_4, E) \ , \end{aligned} \tag{9.93}$$

where, in the last step, we considered the expression (9.76) of the two-body Green's function in Lehmann representation. The transition probability is given by

$$S(E) = -\frac{1}{\pi}\Im\big(R(E)\big) = \sum_{\nu_1\nu_2}\sum_{\nu_3\nu_4} B_{\nu_1\nu_2}B^*_{\nu_3\nu_4}\frac{\Im}{\pi}\Big(i\hbar\tilde{G}(\nu_1, \nu_3, \nu_2, \nu_4, E)\Big) \ . \tag{9.94}$$

9.5 Equation of Motion

In this section we obtain a set of equations describing the time evolution of the Green's functions. Let us consider the total hamiltonian as sum of one-body term $\hat{H}_0$ and a two-body term $\hat{H}_1$.

$$\hat{H} = \hat{H}_0 + \hat{H}_1 = \int d^3x\, \hat{\psi}^+(\mathbf{x})\hat{h}(\mathbf{x})\hat{\psi}(\mathbf{x})$$
$$+ \frac{1}{2}\int d^3x\, d^3y\, \hat{\psi}^+(\mathbf{x})\hat{\psi}^+(\mathbf{y})\hat{V}(\mathbf{x},\mathbf{y})\hat{\psi}(\mathbf{y})\hat{\psi}(\mathbf{x}) \ , \tag{9.95}$$

where $\hat{h}(\mathbf{x})$, in addition to the kinetic energy term, can also contain a MF, one-body, potential term. By using the techniques adopted to obtain (9.20) and (9.23), we have

$$[\hat{\psi}(\mathbf{r}), \hat{H}_0] = \hat{h}(\mathbf{r})\hat{\psi}(\mathbf{r}), \tag{9.96}$$
$$\left[\hat{\psi}(\mathbf{r}), \hat{H}_1\right] = \int d^3x\, \hat{\psi}^+(\mathbf{x})\hat{V}(\mathbf{r},\mathbf{x})\hat{\psi}(\mathbf{x})\hat{\psi}(\mathbf{r}), \tag{9.97}$$
$$\left[\hat{\psi}^+(\mathbf{r}), \hat{H}_0\right] = -\hat{\psi}^+(\mathbf{r})\hat{h}(\mathbf{r}), \tag{9.98}$$
$$\left[\hat{\psi}^+(\mathbf{r}), \hat{H}_1\right] = -\int d^3x\, \hat{\psi}^+(\mathbf{r})\hat{\psi}^+(\mathbf{x})\hat{V}(\mathbf{r},\mathbf{x})\hat{\psi}(\mathbf{x})\,. \tag{9.99}$$

From the definition (9.5) of one-body Green's function we obtain

$$\begin{aligned} i\frac{\partial}{\partial t} G(\mathbf{x},t,\mathbf{x}',t') &= \langle\Psi_0|\hat{\psi}_{\mathrm{H}}(\mathbf{x},t)\hat{\psi}^+_{\mathrm{H}}(\mathbf{x}',t')|\Psi_0\rangle \frac{\partial\,\theta(t-t')}{\partial t} \\ &- \langle\Psi_0|\hat{\psi}^+_{\mathrm{H}}(\mathbf{x}',t')\hat{\psi}_{\mathrm{H}}(\mathbf{x},t)|\Psi_0\rangle \frac{\partial\,\theta(t'-t)}{\partial t} \\ &+ \langle\Psi_0|\frac{\partial\hat{\psi}_{\mathrm{H}}(\mathbf{x},t)}{\partial t}\hat{\psi}^+_{\mathrm{H}}(\mathbf{x}',t')|\Psi_0\rangle\theta(t-t') \\ &- \langle\Psi_0|\hat{\psi}^+_{\mathrm{H}}(\mathbf{x}',t')\frac{\partial\hat{\psi}_{\mathrm{H}}(\mathbf{x},t)}{\partial t}|\Psi_0\rangle\theta(t'-t) \ . \end{aligned}$$

By considering the equations of motion in the Heisenberg picture,

$$i\hbar\frac{\partial}{\partial t}\hat{\psi}_H = [\hat{\psi}_H, \hat{H}] \ , \tag{9.100}$$

we can write

$$\begin{aligned} i\frac{\partial}{\partial t} G(\mathbf{x},t,\mathbf{x}',t') &= \langle\Psi_0|\hat{\psi}_{\mathrm{H}}(\mathbf{x},t)\hat{\psi}^+_{\mathrm{H}}(\mathbf{x}',t')|\Psi_0\rangle\delta(t-t') \\ &- \langle\Psi_0|\hat{\psi}^+_{\mathrm{H}}(\mathbf{x}',t')\hat{\psi}_{\mathrm{H}}(\mathbf{x},t)|\Psi_0\rangle\left(-\delta(t-t')\right) \\ &+ \langle\Psi_0|[\hat{\psi}_{\mathrm{H}}(\mathbf{x},t), \hat{H}]\hat{\psi}^+_{\mathrm{H}}(\mathbf{x}',t')|\Psi_0\rangle\theta(t-t')(i\hbar)^{-1} \\ &- \langle\Psi_0|\hat{\psi}^+_{\mathrm{H}}(\mathbf{x}',t')[\hat{\psi}_{\mathrm{H}}(\mathbf{x},t), \hat{H}]|\Psi_0\rangle\theta(t'-t)(i\hbar)^{-1} \ . \end{aligned}$$

The first two terms can be written as anti-commutator between $\hat{\psi}$ and $\hat{\psi}^+$

$$\begin{aligned} &\langle\Psi_0|\hat{\psi}_{\mathrm{H}}(\mathbf{x},t)\hat{\psi}^+_{\mathrm{H}}(\mathbf{x}',t') + \hat{\psi}^+_{\mathrm{H}}(\mathbf{x}',t')\hat{\psi}_{\mathrm{H}}(\mathbf{x},t)|\Psi_0\rangle \\ &= \langle\Psi_0|\left\{\hat{\psi}_{\mathrm{H}}(\mathbf{x},t), \hat{\psi}^+_{\mathrm{H}}(\mathbf{x}',t')\right\}|\Psi_0\rangle = \delta(\mathbf{x}-\mathbf{x}')\,. \end{aligned} \tag{9.101}$$

By considering the commutators between hamiltonian and field operators written above we have

$$
\begin{aligned}
& i\frac{\partial}{\partial t} G(\mathbf{x}, t, \mathbf{x}', t') = \\
& \delta(\mathbf{x}-\mathbf{x}')\delta(t-t') + \hat{h}(\mathbf{x})\langle\Psi_0|\hat{\mathbb{T}}[\hat{\psi}_{\mathrm{H}}(\mathbf{x}, t)\hat{\psi}^{+}_{\mathrm{H}}(\mathbf{x}', t')]|\Psi_0\rangle(i\hbar)^{-1} \\
+ & \langle\Psi_0|\hat{\mathbb{T}}\left[[\hat{\psi}_{\mathrm{H}}(\mathbf{x}, t), H_1]\hat{\psi}^{+}_{\mathrm{H}}(\mathbf{x}', t')\right]|\Psi_0\rangle(i\hbar)^{-1} \ ,
\end{aligned}
\tag{9.102}
$$

where, as usual, $\hat{\mathbb{T}}$ indicates the time-ordering operator.

The second term is the definition of one-body Green's function, therefore we can write

$$
\begin{aligned}
& \left[i\hbar\frac{\partial}{\partial t} - \hat{h}(\mathbf{x})\right] G(\mathbf{x}, t, \mathbf{x}', t') \\
= \ & \hbar\delta(\mathbf{x}-\mathbf{x}')\delta(t-t') - i\langle\Psi_0|\hat{\mathbb{T}}\left[[\hat{\psi}_{\mathrm{H}}(\mathbf{x}, t), \hat{H}_1]\hat{\psi}^{+}_{\mathrm{H}}(\mathbf{x}', t')\right]|\Psi_0\rangle \ .
\end{aligned}
\tag{9.103}
$$

By considering Eq. (9.97) the last term can be written as

$$
\begin{aligned}
& \langle\Psi_0|\hat{\mathbb{T}}\left[\int d^3y\, \hat{\psi}^{+}_{\mathrm{H}}(\mathbf{y}, t)\hat{V}(\mathbf{x}, \mathbf{y})\hat{\psi}_{\mathrm{H}}(\mathbf{y}, t)\hat{\psi}_{\mathrm{H}}(\mathbf{x}, t)\hat{\psi}^{+}_{\mathrm{H}}(\mathbf{x}', t')\right]|\Psi_0\rangle \\
= & \int d^3y\, \hat{V}(\mathbf{x}, \mathbf{y})\langle\Psi_0|\hat{\mathbb{T}}\left[\hat{\psi}_{\mathrm{H}}(\mathbf{y}, t)\hat{\psi}_{\mathrm{H}}(\mathbf{x}, t)\hat{\psi}^{+}_{\mathrm{H}}(\mathbf{y}, t)\hat{\psi}^{+}_{\mathrm{H}}(\mathbf{x}', t')\right]|\Psi_0\rangle \\
= & \int d^3y\, \hat{V}(\mathbf{x}, \mathbf{y})(i)^2 G(\mathbf{y}, t, \mathbf{x}, t, \mathbf{y}, t, \mathbf{x}', t') \ ,
\end{aligned}
\tag{9.104}
$$

where we used the definition of the two-body Green's function. The ambiguity in the choice of the time to be associated to $\mathbf{y}$ and $\mathbf{x}$ is clarified by considering that the three field operators, depending on these variables, are part of a single block defined at the time t. The equation of motion can be written as

$$
\begin{aligned}
& \left[i\hbar\frac{\partial}{\partial t} - \hat{h}(\mathbf{x})\right] G(\mathbf{x}, t, \mathbf{x}', t') \\
= \ & \hbar\delta(\mathbf{x}-\mathbf{x}')\delta(t-t') + i\int d^3y\, \hat{V}(\mathbf{x}, \mathbf{y})G(\mathbf{y}, t, \mathbf{x}, t, \mathbf{y}, t, \mathbf{x}', t') \ .
\end{aligned}
\tag{9.105}
$$

For a system of non-interacting particles $\hat{V}(\mathbf{x}, \mathbf{y}) = 0$ therefore we can write

$$
\left[i\hbar\frac{\partial}{\partial t} - \hat{h}(\mathbf{x})\right] G^0(\mathbf{x}, t, \mathbf{x}', t') = \hbar\delta(\mathbf{x}-\mathbf{x}')\delta(t-t') \ .
\tag{9.106}
$$

This expression shows that, for a system of non interacting particles, the Green's function defined in this chapter coincide with the resolvent of the Schrödinger equation, as it has been used in Sect. 8.4.

The equation of motion (9.105) can be expressed in integral form as

$$G(\mathbf{x}, t, \mathbf{x}', t') = G^0(\mathbf{x}, t, \mathbf{x}', t') + \frac{i}{\hbar} \int d^3y\, d^3z\, G^0(\mathbf{x}, t, \mathbf{y}, t)\, \hat{V}(\mathbf{y}, \mathbf{z})\, G(\mathbf{z}, t, \mathbf{y}, t, \mathbf{z}, t, \mathbf{x}', t') \ . \tag{9.107}$$

The equation of motion of the one-body Green's function contains the two-body Green's function in both its differential (9.105) or integral (9.107) expressions. In general, it is possible to define a n-body Green's function as [2]

$$(i)^n G_n(\mathbf{x}_1, t1, \cdots, \mathbf{x}_n, t_n; \mathbf{x}'_1, t'_1, \cdots, \mathbf{x}'_n, t'_n) \equiv \frac{\langle \Psi_0 | \hat{\mathbb{T}} \left[\hat{\psi}_{\mathrm{H}}(\mathbf{x}_1, t_1) \cdots \hat{\psi}_{\mathrm{H}}(\mathbf{x}_n, t_n) \hat{\psi}^+_{\mathrm{H}}(\mathbf{x}'_1, t'_1) \cdots \hat{\psi}^+_{\mathrm{H}}(\mathbf{x}'_n, t'_n) \right] | \Psi_0 \rangle}{\langle \Psi_0 | \Psi_0 \rangle} \ . \tag{9.108}$$

The equation of motion for a n - body Green's function requires the information on the $n + 1$ - body Green's function.

$$\begin{aligned} &\left[i\hbar \frac{\partial}{\partial t_1} - h(\mathbf{x}_1) \right] G_n(\mathbf{x}_1, t_1, \cdots, \mathbf{x}_n, t_n; \mathbf{x}'_1, t'_1, \cdots, \mathbf{x}'_n, t'_n) \\ &= \hbar \sum_{i=1}^{n} \delta(\mathbf{x}_1 - \mathbf{x}_i) \delta(t_1 - t_i) (-1)^{n-i} \\ &\quad G_{n-1}(\mathbf{x}_1, t_1, \cdots, \mathbf{x}_{n-1}, t_{n-1}; \mathbf{x}'_1, t'_1, \cdots, \mathbf{x}'_{n-1}, t'_{n-1}) \\ &+ i \int d^3y\, \hat{V}(\mathbf{x}, \mathbf{y}) G_{n+1}(\mathbf{x}_1, t_1, \cdots, \mathbf{x}_{n+1}, t_{n+1}; \mathbf{x}'_1, t'_1, \cdots, \mathbf{x}'_{n+1}, t'_{n+1}) \ . \end{aligned} \tag{9.109}$$

This is a system of coupled equations that connect Green's functions with different numbers of particles. The problem is formulated through a set of equations that become increasingly complex. If it were possible to break the hierarchy by making an assumption about the n-body Green's function, then one could compute the Green's functions with fewer particles. However, from a practical standpoint, this is a very complicated process, and typically, a perturbative approach, such as the one that will presented in Chap. 10, is preferred.

References

1. A. Messiah, *Quantum Mechanics* (North Holland, Amsterdam, 1961)
2. E.K.U. Gross, E. Runge, O. Heinonen, *Many-particle Theory* (Adam Hilger, Bristol, 1991)

Chapter 10
Perturbative Description of the Green's Function

10.1 Introduction

The perturbation theory developed in Chap. 6 can be applied to the calculation of Green's functions. We will first examine the case of the one-body Green's function, and then, at the end of the chapter, we will address the two-body Green's function.

We recall some relations presented in Sect. 9.2.3.

$$\hat{U}(t,0) = e^{-\frac{i}{\hbar}\hat{H}_1 t} \;\; ; \;\; \hat{U}(0,t) = e^{\frac{i}{\hbar}\hat{H}_1 t} \tag{10.1}$$

$$\hat{\mathbb{O}}_{\rm H}(t) = e^{\frac{i}{\hbar}\hat{H}_1 t}\hat{\mathbb{O}}_{\rm I}(t)e^{-\frac{i}{\hbar}\hat{H}_1 t} = \hat{U}(0,t)\hat{\mathbb{O}}_{\rm I}\hat{U}(t,0) \tag{10.2}$$

$$|\Psi_{\rm H}\rangle = |\Psi_{\rm I}(0)\rangle = |\Psi_{\rm S}(0)\rangle = \hat{U}_\epsilon(0,-\infty)|\Phi_0\rangle = |\Psi_0\rangle \;. \tag{10.3}$$

where $\hat{\mathbb{O}}$ is a generic operator, and the meaning of the other symbols is the same as in Chap. 6.

We express the expectation value of an operator $\hat{\mathbb{O}}$ in the Heisenberg picture by utilizing the technique of adiabatically switching on the interaction.

$$\frac{\langle\Psi_0|\hat{\mathbb{O}}_{\rm H}(t)|\Psi_0\rangle}{\langle\Psi_0|\Psi_0\rangle} = \lim_{\epsilon\to 0}\frac{\langle\Phi_0|\hat{U}_\epsilon(+\infty,0)\left[\hat{U}_\epsilon(0,t)\hat{\mathbb{O}}_{\rm I}(t)\hat{U}_\epsilon(t,0)\right]\hat{U}_\epsilon(0,-\infty)|\Phi_0\rangle}{\langle\Phi_0|\hat{U}_\epsilon(\infty,-\infty)|\Phi_0\rangle}$$
$$= \lim_{\epsilon\to 0}\frac{\langle\Phi_0|\hat{U}_\epsilon(+\infty,t)\hat{\mathbb{O}}_{\rm I}(t)\hat{U}_\epsilon(t,-\infty)|\Phi_0\rangle}{\langle\Phi_0|\hat{U}_\epsilon(+\infty,-\infty)|\Phi_0\rangle} \;.$$

© The Author(s), under exclusive license to Springer Nature Switzerland AG 2026

G. Co', *Concepts in Quantum Many-Body Physics*, UNITEXT for Physics,
https://doi.org/10.1007/978-3-032-08920-5_10

Since

$$\hat{U}_\epsilon(+\infty, t)\, \hat{\mathbb{O}}_\mathrm{I}(t)\, \hat{U}_\epsilon(t, -\infty) =$$
$$\sum_{n=0}^{\infty}(-i)^n \frac{1}{n!} \int_t^{\infty} dt_1 \cdots \int_t^{\infty} dt_n\, e^{-\epsilon(|t_1|+\cdots|t_n|)}\, \hat{\mathbb{T}}\left[\hat{H}_{\mathrm{I},1}(t_1)\cdots \hat{H}_{\mathrm{I},1}(t_n)\right] \hat{\mathbb{O}}_\mathrm{I}(t)$$
$$\sum_{m=0}^{\infty}(-i)^m \frac{1}{m!} \int_{-\infty}^{t} dt'_1 \cdots \int_{-\infty}^{t} dt_m\, e^{-\epsilon(|t'_1|+\cdots|t'_m|)}\, \hat{\mathbb{T}}\left[\hat{H}_{\mathrm{I},1}(t'_1)\cdots \hat{H}_{\mathrm{I},1}(t'_m)\right] , \quad (10.4)$$

we can rewrite the one-body Green's function as

$$iG(x, y) = \sum_{\mu=0}^{\infty}(-i)^\mu \frac{1}{\mu!}$$
$$\int_{-\infty}^{\infty} dt_1 \cdots \int_{-\infty}^{\infty} dt_\mu \frac{\langle\Phi_0|\hat{\mathbb{T}}\left[\hat{H}_{\mathrm{I},1}(t_1)\cdots \hat{H}_{\mathrm{I},1}(t_\mu)\hat{\psi}_I(x)\hat{\psi}^+_\mathrm{I}(y)\right]|\Phi_0\rangle}{\langle\Phi_0|\hat{U}(\infty, -\infty)|\Phi_0\rangle} , \quad (10.5)$$

where we used the symbol $x \equiv (\mathbf{x}, t_x)$, and analogously for y. In the above expression we have already carried out the limit for $\epsilon \to 0$. Furthermore, we have considered that $\hat{H}_1$ is hermitian, therefore the expression (10.4) is valid for both $\hat{\mathbb{O}}_\mathrm{I}(t)$ and for $\hat{\mathbb{O}}^+_\mathrm{I}(t)$.

We consider an instantaneous interaction

$$\hat{\mathcal{W}}(x, x') \equiv \hat{V}(\mathbf{x}, \mathbf{x}')\delta(t - t') , \quad (10.6)$$

therefore we can express the perturbative term of the hamiltonian as

$$\hat{H}_{\mathrm{I},1}(t, t') = \frac{1}{2}\int d^3x \int d^3x' \hat{\psi}^+_\mathrm{I}(x)\hat{\psi}^+_\mathrm{I}(x')\hat{\mathcal{W}}(x, x')\hat{\psi}_\mathrm{I}(x')\hat{\psi}_\mathrm{I}(x)$$
$$= \frac{1}{2}\int d^3x \int d^3x' \hat{\psi}^+_\mathrm{I}(x)\hat{\psi}^+_\mathrm{I}(x')\hat{V}(\mathbf{x}, \mathbf{x}')\delta(t - t')\hat{\psi}_\mathrm{I}(x')\hat{\psi}_\mathrm{I}(x) . \quad (10.7)$$

By explicitly writing out the first term of the sum, we can express the numerator of Eq. (10.5) as follows:

$$iG^{\mathrm{num}}(x, y) = \langle\Phi_0|\hat{\mathbb{T}}\left[\hat{\psi}_\mathrm{I}(x)\hat{\psi}^+_\mathrm{I}(y)\right]|\Phi_0\rangle$$
$$+\ (-i)\int_{-\infty}^{\infty} dt_1 \langle\Phi_0|\hat{\mathbb{T}}\left[\hat{H}_{\mathrm{I},1}(t_1)\hat{\psi}_\mathrm{I}(x)\hat{\psi}^+_\mathrm{I}(y)\right]|\Phi_0\rangle + \cdots ,$$

The first term of this expression is the one-body Green's function of a system of non-interacting particles, which we call G^0. The presence of $\delta(t_1 - t'_1)$ allows us to insert an integral on t'_1. Therefore, we can write

$$
\begin{aligned}
iG^{\text{num}}(x,y) = \;& iG^0(x,y) + (-i)\int_{-\infty}^{\infty} dt_1 \langle\Phi_0|\hat{\mathbb{T}}\Big[\frac{1}{2}\int d^3x_1 \int d^3x_1' \hat{\psi}_{\text{I}}^+(x_1)\hat{\psi}_{\text{I}}^+(x_1') \\
& \hat{V}(\mathbf{x}_1,\mathbf{x}_1')\delta(t_1-t_1')\hat{\psi}_{\text{I}}(x_1')\hat{\psi}_{\text{I}}(x_1)\hat{\psi}_{\text{I}}(x)\hat{\psi}_{\text{I}}^+(y)\Big]|\Phi_0\rangle + \cdots \\
= \;& iG^0(x,y) \\
& + (-i)\frac{1}{2}\int d^4x_1 \int d^4x_1' \, \hat{V}(\mathbf{x}_1,\mathbf{x}_1')\delta(t_1-t_1') \\
& \langle\Phi_0|\hat{\mathbb{T}}\left[\hat{\psi}_{\text{I}}^+(x_1)\hat{\psi}_{\text{I}}^+(x_1')\hat{\psi}_{\text{I}}(x_1')\hat{\psi}_{\text{I}}(x_1)\hat{\psi}_{\text{I}}(x)\hat{\psi}_{\text{I}}^+(y)\right]|\Phi_0\rangle + \cdots \; .
\end{aligned}
\tag{10.8}
$$

The evaluation of the above equation is carried out using the techniques presented in Chap. 6. To apply Wick's theorem, it is necessary to define the contraction between two field operators. This is relatively straightforward, as the field operators can be expressed in terms of creation and annihilation operators. We have demonstrated that the anti-commutation properties of these operators are preserved in each picture. In the specific case of the interaction picture, we have

$$
\hat{a}_{\text{I},k}(t) = \hat{a}_k e^{-i\omega_k t} \;\; ; \;\; \hat{a}_{\text{I},k}^+(t) = \hat{a}_k^+ e^{i\omega_k t} \; . \tag{10.9}
$$

The field operators are expressed in terms of creation and destruction operators as

$$
\hat{\psi}_{\text{I}}^+(x) = \sum_k \phi_k^*(\mathbf{x})\hat{a}_{\text{I},k}^+ \;\; ; \;\; \hat{\psi}_{\text{I}}(x) = \sum_k \phi_k(\mathbf{x})\hat{a}_{\text{I},k} \; . \tag{10.10}
$$

where the $\phi_k(\mathbf{x})$ is a single-particle wave function. We can express the contraction as

$$
\begin{aligned}
\overline{\hat{\psi}_{\text{I}}^+(x)\hat{\psi}_{\text{I}}(y)} &= \sum_k\sum_{k'} \phi_k^*(\mathbf{x})\phi_{k'}(\mathbf{y})\overline{\hat{a}_{\text{I},k}^+(t_x)\hat{a}_{\text{I},k'}(t_y)} \\
&= \sum_k\sum_{k'} \phi_k^*(\mathbf{x})e^{-i\omega_k t_x}\phi_{k'}(\mathbf{y})e^{i\omega_{k'} t_y}\overline{\hat{a}_k^+\hat{a}_{k'}} \\
&= \sum_k\sum_{k'} \phi_k^*(\mathbf{x})e^{-i\omega_k t_x}\phi_{k'}(\mathbf{y})e^{i\omega_{k'} t_y}\left\{\hat{\mathbb{T}}[\hat{a}_k^+\hat{a}_{k'}] - \hat{\mathbb{N}}[\hat{a}_k^+\hat{a}_{k'}]\right\} \\
&= \hat{\mathbb{T}}[\hat{\psi}_{\text{I}}^+(x)\hat{\psi}_{\text{I}}(y)] - \hat{\mathbb{N}}[\hat{\psi}_{\text{I}}^+(x)\hat{\psi}_{\text{I}}(y)] \; ,
\end{aligned}
$$

therefore

$$
\begin{aligned}
\overline{\hat{\psi}_{\text{I}}(x)\hat{\psi}_{\text{I}}^+(y)} &= \langle\Phi_0|\hat{\psi}_{\text{I}}(x)\hat{\psi}_{\text{I}}^+(y)|\Phi_0\rangle \\
&= \langle\Phi_0|\hat{\mathbb{T}}[\hat{\psi}_{\text{I}}(x)\hat{\psi}_{\text{I}}^+(y)]|\Phi_0\rangle - \langle\Phi_0|\hat{\mathbb{N}}[\hat{\psi}_{\text{I}}(x)\hat{\psi}_{\text{I}}^+(y)]|\Phi_0\rangle \\
&= \langle\Phi_0|\hat{\mathbb{T}}[\hat{\psi}_{\text{I}}(x)\hat{\psi}_{\text{I}}^+(y)]|\Phi_0\rangle = iG^0(x,y) \; .
\end{aligned}
\tag{10.11}
$$

where we used the fact that the expectation value of a normal ordered product of operators between unperturbed ground states is zero, by definition.

The calculation of Eq. (10.8) necessitates the use of Wick's theorem, which involves calculating contractions that, as we have shown above, can be expressed in terms of unperturbed Green's functions. This implies that the Green's function of an interacting system can be represented as a perturbative expansion in terms of the Green's functions of non-interacting particles.

10.2 Goldstone-Feynman Diagrams

We consider the perturbative expansion (10.8) of the numerator of the one-body Green's function truncated at the first order

$$
\begin{aligned}
&iG^{(1),\mathrm{num}}(x, y) = iG^0(x, y) + \\
&(-i)\frac{1}{2}\int d^4x_1 \int d^4x_1'\, \hat{\mathbb{W}}(x_1, x_1') \\
&\langle \Phi_0 | \hat{\mathbb{T}}\left[\hat{\psi}_{\mathrm{I}}^+(x_1)\hat{\psi}_{\mathrm{I}}^+(x_1')\hat{\psi}_{\mathrm{I}}(x_1')\hat{\psi}_{\mathrm{I}}(x_1)\hat{\psi}_{\mathrm{I}}(x)\hat{\psi}_{\mathrm{I}}^+(y)\right] |\Phi_0\rangle \ .
\end{aligned}
$$

Let us apply the Wick's theorem to evaluate the expectation value of the second term:

$$
\begin{aligned}
&\langle \Phi_0 | \hat{\mathbb{T}}\left[\hat{\psi}_{\mathrm{I}}^+(x_1)\hat{\psi}_{\mathrm{I}}^+(x_1')\hat{\psi}_{\mathrm{I}}(x_1')\hat{\psi}_{\mathrm{I}}(x_1)\hat{\psi}_{\mathrm{I}}(x)\hat{\psi}_{\mathrm{I}}^+(y)\right] |\Phi_0\rangle \\
= \; &\langle \Phi_0 | \hat{\psi}_{\mathrm{I}}^+(x_1)\hat{\psi}_{\mathrm{I}}^+(x_1')\hat{\psi}_{\mathrm{I}}(x_1')\hat{\psi}_{\mathrm{I}}(x_1)\hat{\psi}_{\mathrm{I}}(x)\hat{\psi}_{\mathrm{I}}^+(y) |\Phi_0\rangle \\
+ \; &\langle \Phi_0 | \hat{\psi}_{\mathrm{I}}^+(x_1)\hat{\psi}_{\mathrm{I}}^+(x_1')\hat{\psi}_{\mathrm{I}}(x_1')\hat{\psi}_{\mathrm{I}}(x_1)\hat{\psi}_{\mathrm{I}}(x)\hat{\psi}_{\mathrm{I}}^+(y) |\Phi_0\rangle \\
+ \; &\langle \Phi_0 | \hat{\psi}_{\mathrm{I}}^+(x_1)\hat{\psi}_{\mathrm{I}}^+(x_1')\hat{\psi}_{\mathrm{I}}(x_1')\hat{\psi}_{\mathrm{I}}(x_1)\hat{\psi}_{\mathrm{I}}(x)\hat{\psi}_{\mathrm{I}}^+(y) |\Phi_0\rangle \\
+ \; &\langle \Phi_0 | \hat{\psi}_{\mathrm{I}}^+(x_1)\hat{\psi}_{\mathrm{I}}^+(x_1')\hat{\psi}_{\mathrm{I}}(x_1')\hat{\psi}_{\mathrm{I}}(x_1)\hat{\psi}_{\mathrm{I}}(x)\hat{\psi}_{\mathrm{I}}^+(y) |\Phi_0\rangle \\
+ \; &\langle \Phi_0 | \hat{\psi}_{\mathrm{I}}^+(x_1)\hat{\psi}_{\mathrm{I}}^+(x_1')\hat{\psi}_{\mathrm{I}}(x_1')\hat{\psi}_{\mathrm{I}}(x_1)\hat{\psi}_{\mathrm{I}}(x)\hat{\psi}_{\mathrm{I}}^+(y) |\Phi_0\rangle \\
+ \; &\langle \Phi_0 | \hat{\psi}_{\mathrm{I}}^+(x_1)\hat{\psi}_{\mathrm{I}}^+(x_1')\hat{\psi}_{\mathrm{I}}(x_1')\hat{\psi}_{\mathrm{I}}(x_1)\hat{\psi}_{\mathrm{I}}(x)\hat{\psi}_{\mathrm{I}}^+(y) |\Phi_0\rangle \ .
\end{aligned}
\tag{10.12}
$$

By expressing the contractions in terms of unperturbed Green's function, the above equation can be written as

$$iG^{(1),\mathrm{num}}(x,y) = iG^0(x,y) + (-i)\frac{1}{2}\int d^4x_1 \int d^4x_1' \, \hat{\mathcal{W}}(x_1,x_1')$$

$$\begin{aligned}
\Big\{ & [iG^0(x,y)][-iG^0(x_1,x_1)][-iG^0(x_1',x_1')]_{\mathrm{A}} \\
- & [iG^0(x,y)][-iG^0(x_1,x_1')][-iG^0(x_1',x_1)]_{\mathrm{B}} \\
+ & [-iG^0(x,x_1)][-iG^0(x_1,x_1')][iG^0(x_1',y)]_{\mathrm{C}} \\
- & [-iG^0(x,x_1)][-iG^0(x_1',x_1')][iG^0(x_1,y)]_{\mathrm{D}} \\
+ & [-iG^0(x,x_1')][-iG^0(x_1',x_1)][iG^0(x_1,y)]_{\mathrm{E}} \\
- & [-iG^0(x,x_1')][-iG^0(x_1,x_1)][iG^0(x_1',y)]_{\mathrm{F}} \Big\} .
\end{aligned} \tag{10.13}$$

where the subscripts A, $\cdots$, F, refer to the diagrams of Fig. 10.1.

Also in this case, a graphical representation is useful. The unperturbed one-body Green's function $G^0(x,y)$ is depicted as an oriented line extending from y to x, since $t_x > t_y$. The interaction is represented by a dashed line. The various terms of the perturbative expansion (10.13) are illustrated by the diagrams shown in Fig. 10.1.

The A and B terms in Fig. 10.1 are sub-units composed of unlinked diagrams. The numerator of the Green's function can be described as shown in Fig. 10.2, specifically

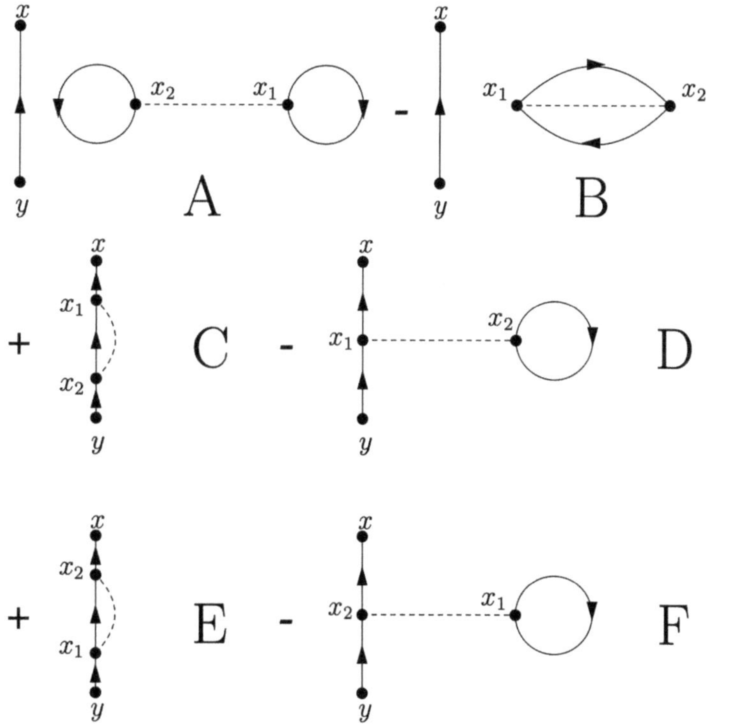

Fig. 10.1 Goldstone diagrams representing the terms of Eq. 10.13

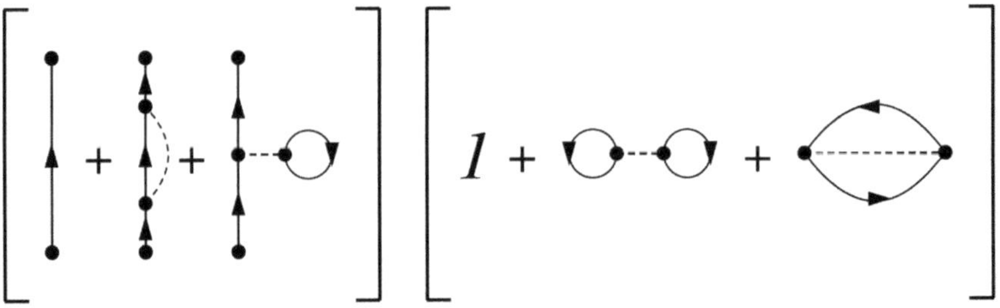

Fig. 10.2 Factorisation of Eq. 10.13

as the product of two blocks of different types of diagrams. The first block consists of all the linked diagrams that include the points x and y. These diagrams are represented in the left bracket of the figure and are multiplied by those in the right bracket, where the points x and y do not appear.

This latter set includes all the terms in the denominator that do not contain the points x and y. Similarly to what occurs in Eq. (6.56), the denominator cancels the unlinked diagrams in the numerator. Consequently, the Green's function can be expressed solely in terms of linked diagrams that include the x and y coordinates. Diagrams that are topologically equivalent, such as C and E, as well as D and F in Fig. 10.1, contribute identically.

The one-body Green's function in configuration space depends on two coordinates. For translationally invariant systems, it is convenient to use the Fourier transform of this function, defined in momentum-energy space, as it depends on a single four-dimensional variable

$$G(x, y) = \frac{1}{(2\pi)^4} \int d^4k\, e^{ik\cdot(x-y)} \tilde{G}(\mathbf{k}, \omega) \ , \tag{10.14}$$

$$\tilde{G}(k) = \int d^4(x - y)\, e^{-ik\cdot(x-y)} G(x, y) \ . \tag{10.15}$$

where we defined $d^4k \equiv d^3k\, d\omega$ and $k \cdot x = \omega t \ - \mathbf{k} \cdot \mathbf{x}$. We obtain an analogous expression for G^0.

For instantaneous interactions depending only on the difference between the coordinates of the two interacting particles, we obtain

$$\begin{aligned} \hat{\mathcal{W}}(x, y) &= \frac{1}{(2\pi)^4} \int d^4k\, e^{-ik\cdot(x-y)} \hat{\tilde{\mathcal{W}}}(k) \\ &= \frac{1}{(2\pi)^3} \int d^3k\, e^{-i\mathbf{k}\cdot(\mathbf{x}-\mathbf{y})} \hat{\tilde{V}}(\mathbf{k}) \delta(t_x - t_y) \ . \end{aligned} \tag{10.16}$$

By substituting these definitions into the expression for the Green's function, we can define the Fourier transform of expressions that can be interpreted as Goldstone-Feynman diagrams. A notable feature of this representation in momentum space is that momentum is conserved at each vertex.

10.3 Dyson's Equation and Self-Energy

The expression (10.5) indicates that the Green's function of the interacting system comprises the unperturbed Green's function along with all the linked diagrams that can be inserted between the coordinates x and y. The mathematical representation of this observation is

$$G(x, y) = G^0(x, y) + \int d^4x_1 \int d^4x_1'\, G^0(x, x_1)\underline{\hat{\Sigma}}(x_1, x_1')G^0(x_1', y) \ , \quad (10.17)$$

which defines the quantity $\underline{\hat{\Sigma}}$ called **self-energy**.

The graphical representation of this expression is shown in Fig. 10.3. The thicker oriented line represents the Green's function G of the interacting system, while the thinner lines indicate the Green's function G^0 of the non-interacting system. The blob represents the self-energy $\underline{\hat{\Sigma}}$.

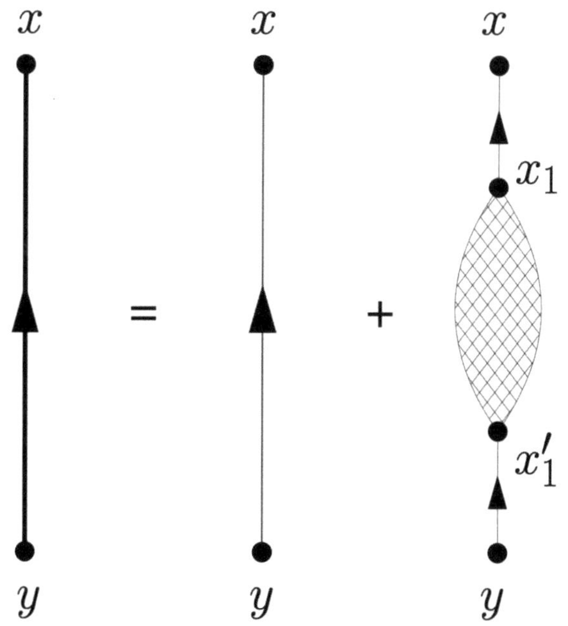

Fig. 10.3 Graphical representation of the self-energy

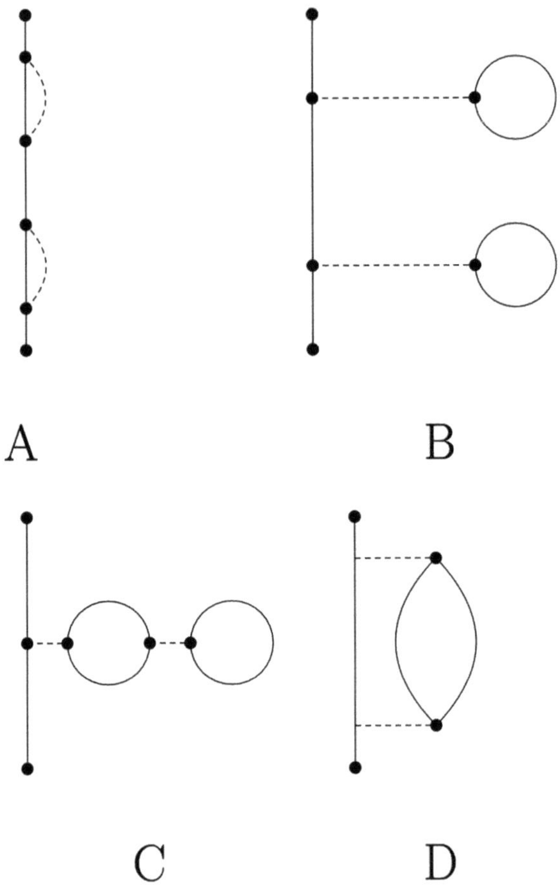

Fig. 10.4 The proper insertions of the interaction lines are those of the C and D diagrams

From a computational perspective, it is more convenient to use the proper self-energy. In perturbative expansions, there are diagrams that can be separated by cutting a line representing G^0. For instance, diagrams A and B in Fig. 10.4 can be separated in this manner. However, there are also diagrams, such as C and D in the figure, that cannot be divided into independent parts.

The proper **self-energy** is defined as the sum of all diagrams that cannot be separated. Clearly, the self-energy can be obtained by summing all possible insertions of proper self-energy. In Fig. 10.5, the blob with two sets of dashed lines represents the full self-energy, while the blobs with a single set of dashed lines denote the proper self-energies. The mathematical expression of this concept is given by

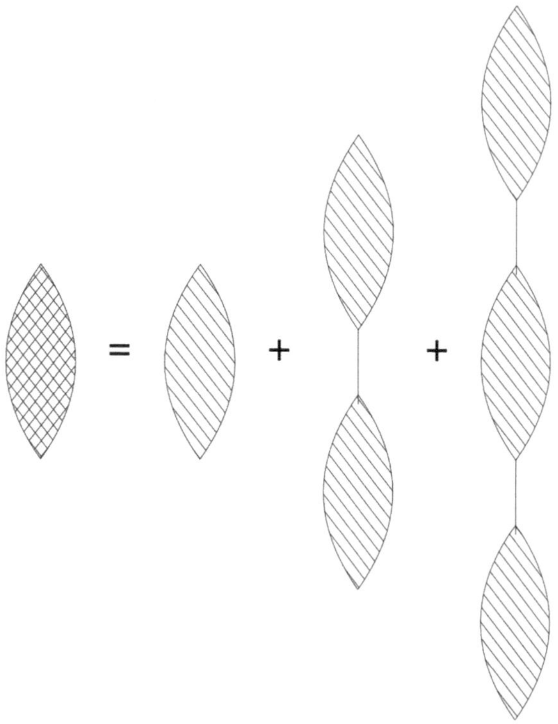

Fig. 10.5 The self-energy can be rewritten as a sum of terms of proper self-energy insertions

$$
\begin{aligned}
&\underline{\hat{\Sigma}}(x_1, x_1') = \hat{\Sigma}(x_1, x_1') + \int d^4x_2 \int d^4x_2' \, \hat{\Sigma}(x_1, x_2) G^0(x_2, x_2') \hat{\Sigma}(x_2', x_1') \\
&+ \int d^4x_2 \int d^4x_2' \int d^4x_3 \int d^4x_3' \, \hat{\Sigma}(x_1, x_2) \\
&G^0(x_2, x_2') \hat{\Sigma}(x_2', x_3) G^0(x_3, x_3') \hat{\Sigma}(x_3', x_1') + \cdots \quad .
\end{aligned}
$$

By inserting this definition of proper self-energy in Eq. (10.17) we obtain

$$
\begin{aligned}
&G(x, y) = G^0(x, y) + \int d^4x_1 \int d^4x_1' \, G^0(x, x_1) \hat{\Sigma}(x_1, x_1') G^0(x_1', y) \\
&+ \int d^4x_1 \int d^4x_1' \int d^4x_2 \int d^4x_2' \, G^0(x, x_1) \hat{\Sigma}(x_1, x_1') G^0(x_1', x_2) \\
&\hat{\Sigma}(x_2, x_2') G^0(x_2', y) + \cdots \quad . \qquad (10.18)
\end{aligned}
$$

This expression can be written as

$$
G(x, y) = G^0(x, y) + \int d^4x_1 \int d^4x_1' \, G^0(x, x_1) \hat{\Sigma}(x_1, x_1') G(x_1', y) \quad . \qquad (10.19)
$$

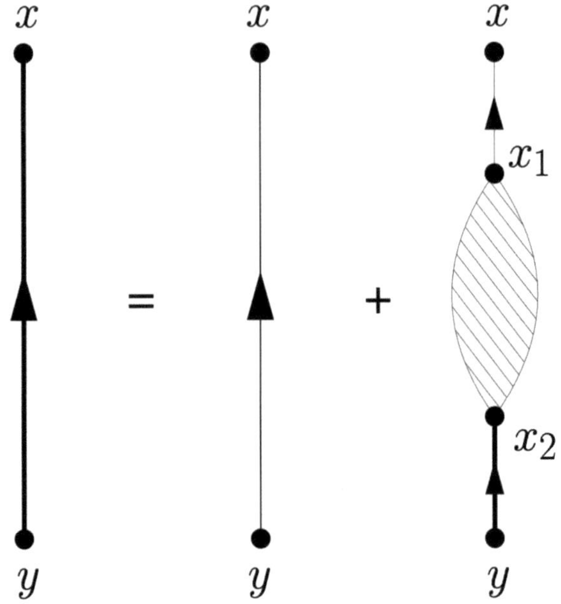

Fig. 10.6 Graphical representation of the Dyson's equation

and it is known in the literature as **Dyson's equation**. Its graphical representation is illustrated in Fig. 10.6.

In the right part of this equation, it is possible to perform a perturbative expansion of both the Green's function G and the self-energy $\hat{\Sigma}$. This is particularly evident for the Green's function. For the self-energy, it is useful to consider the diagram C in Fig. 10.4. In this diagram, the first order in the interaction includes a single loop, while the complete diagram corresponds to the second order.

Dyson's equation takes on a simpler form when applied to an infinite and homogeneous system of fermions. In this case, we can define

$$G(x, y) = \frac{1}{(2\pi)^4} \int d^4k e^{-ik\cdot(x-y)} \tilde{G}(k), \tag{10.20}$$

and

$$\hat{\Sigma}(x, y) = \frac{1}{(2\pi)^4} \int d^4k e^{-ik\cdot(x-y)} \tilde{\hat{\Sigma}}(k) \ . \tag{10.21}$$

By inserting these expressions in the Dyson's equation (10.19), and by using the conservation of the four-momentum k, we obtain

$$G(x, y) = \frac{1}{(2\pi)^4} \int d^4k e^{-ik\cdot(x-y)} \left[\tilde{G}^0(k) + \tilde{G}^0(k)\tilde{\hat{\Sigma}}(k)\tilde{G}(k) \right] , \tag{10.22}$$

therefore, the expression of the Dyson's equation in momentum space is

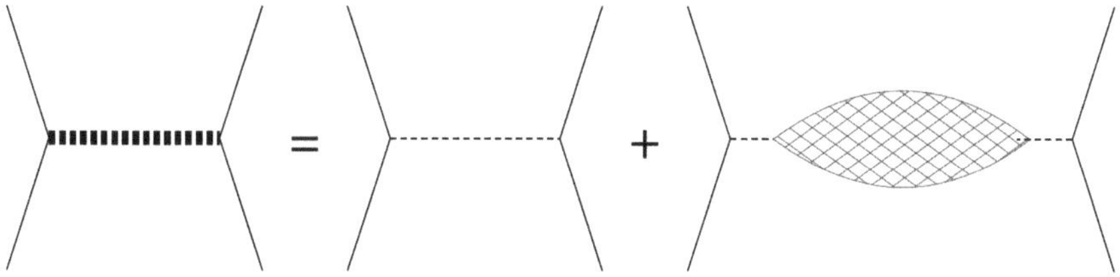

Fig. 10.7 Representation of the polarisation operator

$$\tilde{G}(k) = \left[\tilde{G}^0(k) + \tilde{G}^0(k)\tilde{\hat{\Sigma}}(k)\tilde{G}(k)\right] , \tag{10.23}$$

from which

$$\tilde{G}(k) = \frac{\tilde{G}^0(k)}{1 - \tilde{G}^0(k)\tilde{\hat{\Sigma}}(k)} = \frac{1}{[\tilde{G}^0(k)]^{-1} - \tilde{\hat{\Sigma}}(k)} . \tag{10.24}$$

Since

$$\tilde{G}^0(\mathbf{k}, \omega) = \left[\frac{\Theta(k - k_F)}{\omega - \omega_k + i\eta} + \frac{\Theta(k_F - k)}{\omega - \omega_k - i\eta}\right] , \tag{10.25}$$

we obtain

$$[\tilde{G}^0(k)]^{-1} = \omega - \omega_k , \tag{10.26}$$

and then

$$\tilde{G}(k) = \frac{1}{\omega - \omega_k - \tilde{\hat{\Sigma}}(k)} . \tag{10.27}$$

We can approach the interaction in a manner similar to how we derived Dyson's equation. In Fig. 10.7, we provide a graphical representation illustrating how to consider the interaction between two generic particle or hole lines by defining the *polarisation operator* $\underline{\hat{\Pi}}$

$$\hat{\mathcal{W}}(x, y) = \hat{\mathcal{W}}_0(x, y) + \int d^4x_1 \int d^4x_1' \, \hat{\mathcal{W}}_0(x, x_1)\underline{\hat{\Pi}}(x_1, x_1')\hat{\mathcal{W}}_0(x_1', y) . \tag{10.28}$$

The equation above indicates that the interaction between two particles in the medium consists of the bare interaction, represented in vacuum as $\hat{\mathcal{W}}_0$, plus all the terms associated with medium polarisation, which we describe in terms of virtual particle-hole excitations. The polarisation operator $\underline{\hat{\Pi}}$ encompasses all of these excitations.

In Fig. 10.8, we present two of the possible diagrams included in the polarisation operator $\underline{\hat{\Pi}}$. Diagram A is classified as a proper diagram because it cannot be separated by cutting an interaction line. In contrast, Diagram B contains three interaction lines, but it is possible to cut the second line, allowing it to be separated into proper diagrams.

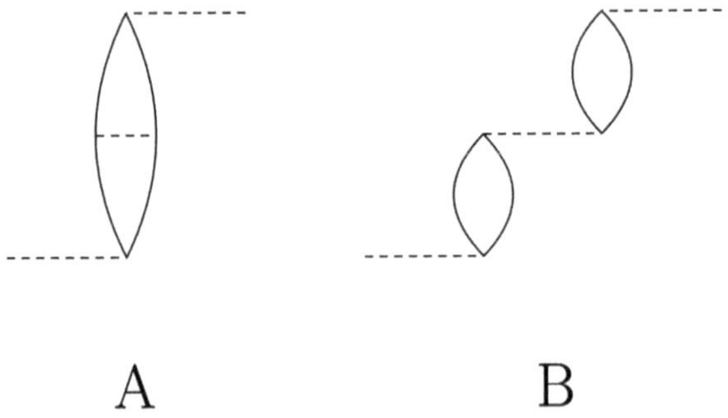

Fig. 10.8 Two diagrams included in the polarisation operator $\underline{\hat{\Pi}}$

We can describe the interaction in terms of proper diagrams by defining a proper polarisation operator $\hat{\Pi}$ as one that includes only proper diagrams

$$\begin{aligned}\underline{\hat{\Pi}}(x_1, x_1') &= \hat{\Pi}(x_1, x_1') + \int d^4x_2 \int d^4x_2' \, \hat{\Pi}(x_1, x_2)\hat{\mathcal{W}}_0(x_2, x_2')\hat{\Pi}(x_2', x_1') \\ &+ \int d^4x_2 \int d^4x_2' \int d^4x_3 \int d^4x_3' \, \hat{\Pi}(x_1, x_2)\hat{\mathcal{W}}_0(x_2, x_2')\hat{\Pi}(x_2', x_3) \\ &\quad \hat{\mathcal{W}}_0(x_3, x_3')\hat{\Pi}(x_3', x_1') + \cdots \\ &= \hat{\Pi}(x_1, x_1') + \int d^4x_2 \int d^4x_2' \, \hat{\Pi}(x_1, x_2)\hat{\mathcal{W}}_0(x_2, x_2')\underline{\hat{\Pi}}(x_2', x_1') \ . \end{aligned} \tag{10.29}$$

Therefore

$$\begin{aligned}\hat{\mathcal{W}}(x, y) &= \hat{\mathcal{W}}_0(x, y) + \int d^4x_1 \int d^4x_1' \, \hat{\mathcal{W}}_0(x, x_1)\hat{\Pi}(x_1, x_1')\hat{\mathcal{W}}_0(x_1', y) \\ &+ \int d^4x_1 \int d^4x_1' \int d^4x_2 \int d^4x_2' \, \hat{\mathcal{W}}_0(x, x_1)\hat{\Pi}(x_1, x_2) \\ &\quad \hat{\mathcal{W}}_0(x_2, x_2')\hat{\Pi}(x_2', x_1')\hat{\mathcal{W}}_0(x_1', y) + \cdots \\ &= \hat{\mathcal{W}}_0(x, y) + \int d^4x_1 \int d^4x_1' \, \hat{\mathcal{W}}_0(x, x_1)\hat{\Pi}(x_1, x_1')\hat{\mathcal{W}}(x_1', y) \ . \end{aligned} \tag{10.30}$$

which is the analogous of the Dyson's equation for the interaction.

In this case, for translationally invariant systems that are infinite and homogeneous, it is convenient to define the Fourier transforms of $\hat{\mathcal{W}}$ and $\hat{\Pi}$, which we denote by simply indicating their dependence on the four-vector k:

$$\hat{\mathcal{W}}(k) = \hat{\mathcal{W}}_0(k) + \hat{\mathcal{W}}_0(k)\underline{\hat{\Pi}}(k)\hat{\mathcal{W}}_0(k) \tag{10.31}$$

$$\hat{\mathcal{W}}(k) = \hat{\mathcal{W}}_0(k) + \hat{\mathcal{W}}_0(k)\hat{\Pi}(k)\hat{\mathcal{W}}(k) \ . \tag{10.32}$$

From the second equation, above, we obtain an expression for $\hat{\mathcal{W}}(k)$

$$\hat{\mathcal{W}}(k) = \frac{\hat{\mathcal{W}}_0(k)}{1 - \hat{\mathcal{W}}_0(k)\hat{\Pi}(k)} \equiv \frac{\hat{\mathcal{W}}_0(k)}{\hat{\mathcal{K}}(k)} \quad , \tag{10.33}$$

where the last expression defines the dielectric function $\hat{\mathcal{K}}$ which modifies the bare interaction $\hat{\mathcal{W}}_0$ for the presence of the medium.

10.4 Hartree–Fock

The simplest solution to Dyson's equation (10.19) consists in considering only the insertion of a single interaction line. This approximation is illustrated diagrammatically in Fig. 10.9. A slightly more refined approximation involves considering all possible self-energy insertions in the Green's function lines of the two terms shown in Fig. 10.9. This procedure is illustrated in Fig. 10.10. The results of this infinite number of self-energy insertions is illustrated in Fig. 10.11.

Let us consider the Dyson's equation (10.19)

$$G(x, y) = G^0(x, y) + \int d^4x_1 \int d^4x_1' G^0(x, x_1)\hat{\Sigma}(x_1, x_1')G(x_1', y) \ .$$

In the approximation presented in Fig. 10.9 the self-energy can be expressed as

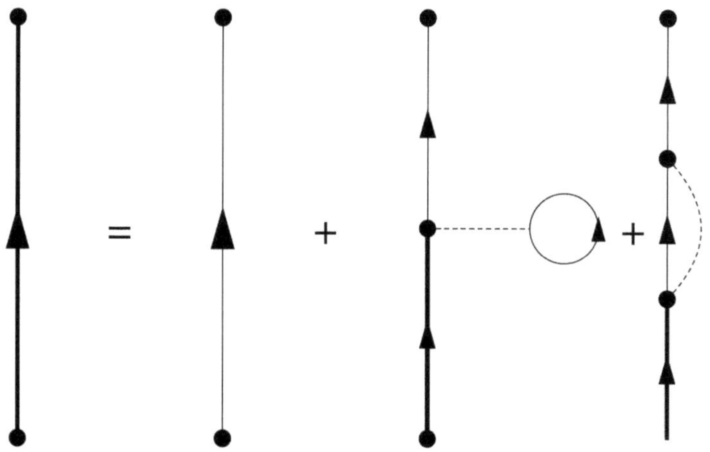

Fig. 10.9 First order of the perturbation expansion of the Green's function described in terms of the Dyson's equation

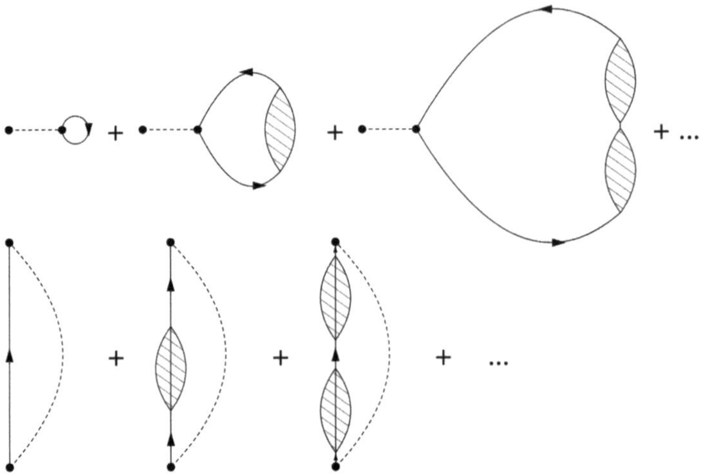

Fig. 10.10 Self-energy insertions in the diagrams of Fig. 10.9

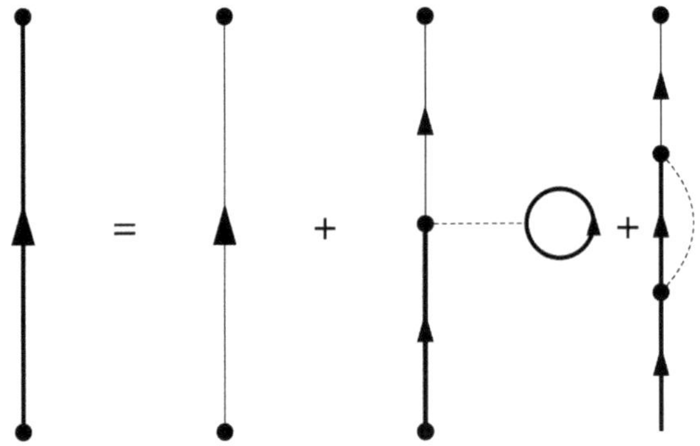

Fig. 10.11 Representation of the HF approximation in the Dyson's equation

$$\hbar\hat{\Sigma}^{(I)}(x_1, x_1') = -i\Big[\delta^4(x_1 - x_1')$$
$$\int d^4x_2 \hat{\mathcal{W}}(x_1, x_2)G^0(x_2, x_2) - \hat{\mathcal{W}}(x_1, x_1')G^0(x_1, x_1')\Big] \,. \tag{10.34}$$

and, in the approximation of Fig. 10.11 as

$$\hbar\hat{\Sigma}^{(II)}(x_1, x_1') = -i\Big[\delta^4(x_1 - x_1')$$
$$\int d^4x_2 \hat{\mathcal{W}}(x_1, x_2)G(x_2, x_2) - \hat{\mathcal{W}}(x_1, x_1')G(x_1, x_1')\Big] \,. \tag{10.35}$$

This approximation does not include diagrams such as those presented in Fig. 10.12. Below, we demonstrate how using approximation 10.35 in the Dyson equation leads to the traditional Hartree-Fock equations in coordinate space.

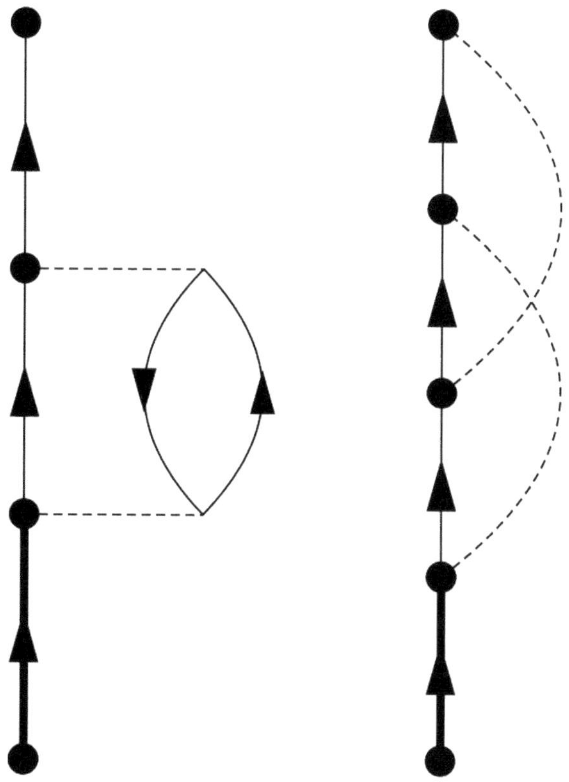

Fig. 10.12 Example of diagrams not contained in the HF approximation

If the hamiltonian $\hat{H}$ is time independent, the interaction $\hat{\mathscr{W}}$ can be written as:

$$\hat{\mathscr{W}}(x_1, x_1') = \hat{V}(\mathbf{x}_1, \mathbf{x}_1')\delta(t_1 - t_1') \ .$$

We define the Fourier transforms about the time variable as

$$G(\mathbf{x}t, \mathbf{x}'t') = \frac{1}{2\pi}\int d\omega e^{-i\omega(t-t')}\tilde{G}(\mathbf{x}, \mathbf{x}', \omega) \ , \tag{10.36}$$

$$G^0(\mathbf{x}t, \mathbf{x}'t') = \frac{1}{2\pi}\int d\omega e^{-i\omega(t-t')}\tilde{G}^0(\mathbf{x}, \mathbf{x}', \omega) \ , \tag{10.37}$$

$$\hat{\Sigma}(\mathbf{x}t, \mathbf{x}'t') = \hat{\Sigma}(\mathbf{x}, \mathbf{x}')\delta(t - t') = \frac{1}{2\pi}\int d\omega e^{-i\omega(t-t')}\hat{\Sigma}(\mathbf{x}, \mathbf{x}') \ . \tag{10.38}$$

The Dyson's equation in mixed representation is

$$\begin{aligned} \tilde{G}(\mathbf{x}, \mathbf{y}, \omega) &= \tilde{G}^0(\mathbf{x}, \mathbf{y}, \omega) \\ &+ \int d^3x_1 \int d^3x_1' \, \tilde{G}^0(\mathbf{x}, \mathbf{x}_1, \omega)\hat{\Sigma}(\mathbf{x}_1, \mathbf{x}_1')\tilde{G}(\mathbf{x}_1', \mathbf{y}, \omega) \ . \end{aligned} \tag{10.39}$$

In the approximation (II) the self-energy becomes

$$\hbar\hat{\Sigma}^{(II)}(x_1, x_1') = \hbar\hat{\Sigma}^{(II)}(\mathbf{x}_1, \mathbf{x}_1')\delta(t_1 - t_1') = -i\delta(t_1 - t_1')$$
$$\times\Big[\delta^3(\mathbf{x}_1 - \mathbf{x}_1')\int d^4x_2\hat{V}(\mathbf{x}_1 - \mathbf{x}_2)\frac{1}{2\pi}\int d\omega e^{-i\omega(t_2 - t_2^+)}\tilde{G}(\mathbf{x}_2, \mathbf{x}_2, \omega)$$
$$-\hat{V}(\mathbf{x}_1 - \mathbf{x}_1')\frac{1}{2\pi}\int d\omega e^{-i\omega(t_1 - t_1^+)}\tilde{G}(\mathbf{x}_1, \mathbf{x}_1', \omega)\Big] . \tag{10.40}$$

In the above equation $t^+ > t$. We shall use this property to properly define the step function. We indicate as $\phi_j^0(\mathbf{r})$ the single particle wave functions of the one-body hamiltonian $\hat{H}_0$

$$\hat{H}_0 = \sum_j h_j \; ; \; \hat{h}_j\phi_j^0(\mathbf{r}) = \left[-\frac{\hbar^2\nabla^2}{2m} + \hat{\mathcal{W}}(\mathbf{r})\right]\phi_j^0(\mathbf{r}) = \epsilon_j\phi_j^0(\mathbf{r}) \; , \tag{10.41}$$

where $\hat{\mathcal{W}}(\mathbf{r})$ is the MF potential. We use the basis formed by these single-particle states to express the field operators in the interaction picture

$$\hat{\psi}_{\rm I}(\mathbf{x}, t) = \sum_k \hat{a}_{{\rm I},k}(t)\phi_k^0(\mathbf{x}) \; ; \; \hat{\psi}_{\rm I}^+(\mathbf{x}, t) = \sum_k \hat{a}_{{\rm I},k}^+(t)(\phi_k^0(\mathbf{x}))^* \; . \tag{10.42}$$

Therefore

$$iG^0(\mathbf{x}t, \mathbf{x}'t') = \langle\Phi_0|\hat{\mathbb{T}}\left[\hat{\psi}_{\rm I}(\mathbf{x}, t)\hat{\psi}_{\rm I}^+(\mathbf{x}', t')\right]|\Phi_0\rangle$$
$$= \langle\Phi_0|\sum_k \hat{a}_k e^{-i\omega_k t}\phi_k^0(\mathbf{x})\sum_{k'}\hat{a}_{k'}^+e^{i\omega_{k'}t'}(\phi_{k'}^0(\mathbf{x}'))^*|\Phi_0\rangle\Theta(t - t')\Theta(\omega_k - \epsilon_{\rm F})$$
$$-\langle\Phi_0|\sum_{k'}\hat{a}_{k'}^+e^{i\omega_{k'}t'}(\phi_{k'}^0(\mathbf{x}'))^*\sum_k \hat{a}_k e^{-i\omega_k t}\phi_k^0(\mathbf{x})|\Phi_0\rangle\Theta(t' - t)\Theta(\epsilon_{\rm F} - \omega_k)$$
$$= \sum_k \phi_k^0(\mathbf{x})(\phi_k^0(\mathbf{x}'))^* e^{-i\omega_k(t-t')}$$
$$\left[\Theta(t - t')\Theta(\omega_k - \epsilon_{\rm F}) - \Theta(t' - t)\Theta(\epsilon_{\rm F} - \omega_k)\right] , \tag{10.43}$$

where we defined $\omega_k = \epsilon_k/\hbar$, and used

$$\langle\Phi_0|\hat{a}_k\hat{a}_{k'}^+|\Phi_0\rangle = \delta_{k'k}\Theta(\omega_k - \epsilon_{\rm F}) \; ; \; \langle\Phi_0|\hat{a}_{k'}^+\hat{a}_k|\Phi_0\rangle = \delta_{k'k}\Theta(\epsilon_{\rm F} - \omega_k) \; . \tag{10.44}$$

By considering the integral representation of $\Theta(t - t')$, see Eq. (9.45), we obtain

$$\tilde{G}^0(\mathbf{x}, \mathbf{x}', \omega) = \sum_k \phi_k^0(\mathbf{x})(\phi_k^0(\mathbf{x}'))^*\left[\frac{\Theta(\epsilon_k - \epsilon_{\rm F})}{\omega - \omega_k + i\eta} + \frac{\Theta(\epsilon_{\rm F} - \epsilon_k)}{\omega - \omega_k - i\eta}\right] . \tag{10.45}$$

With the ϕ^0 obtained from Eq. (10.41), we can calculate the G^0 which can be used to activate an iterative procedure implying the Dyson's equation (10.39) and the self-energy equation (10.40).

It possible to search for G an expression analogous to that of Eq. (10.45) where the ϕ^0 are substituted by generic ϕ eigenstates of a new hamiltonian and related to the previous single-particle wave functions by a unitary transformation (we discuss this point in more detail in Sect. 15.2.2). The Green's function has the following expression

$$\tilde{G}(\mathbf{x}, \mathbf{x}', \omega) = \sum_k \phi_k(\mathbf{x})\phi_k^*(\mathbf{x}')\left[\frac{\Theta(\epsilon_k - \epsilon_{\rm F})}{\omega - \omega_k + i\eta} + \frac{\Theta(\epsilon_k - \epsilon_{\rm F})}{\omega - \omega_k - i\eta}\right] . \tag{10.46}$$

By inserting this expression in Eq. (10.40) we encounter integrals of the type

$$\frac{1}{2\pi i}\int d\omega \frac{e^{i\omega|t-t^+|}}{\omega-\omega_k+i\eta}=0 \;;\; \frac{1}{2\pi i}\int d\omega \frac{e^{i\omega|t-t^+|}}{\omega-\omega_k-i\eta}=\Theta(-|t-t^+|) \;, \quad (10.47)$$

therefore, we obtain the expressions

$$\hat{\Sigma}^{(II)}(x_1,x_1')=\hat{\Sigma}^{(II)}(\mathbf{x}_1,\mathbf{x}_1')\delta(t_1-t_1'), \quad (10.48)$$

with

$$\begin{aligned}
\hbar\hat{\Sigma}^{(II)}(\mathbf{x}_1,\mathbf{x}_1') &= -i\Big[\delta^3(\mathbf{x}_1-\mathbf{x}_1')\int d^3x_2\hat{V}(\mathbf{x}_1-\mathbf{x}_2) \\
& i\sum_k \phi_k(\mathbf{x}_2)\phi_k^*(\mathbf{x}_2)\Theta(\epsilon_{\rm F}-\epsilon_k)-i\hat{V}(\mathbf{x}_1-\mathbf{x}_1')\sum_k \phi_k(\mathbf{x}_1)\phi_k^*(\mathbf{x}_1')\Theta(\epsilon_{\rm F}-\epsilon_k)\Big] \\
&= \delta^3(\mathbf{x}_1-\mathbf{x}_1')\int d^3x_2\hat{V}(\mathbf{x}_1-\mathbf{x}_2)\rho(\mathbf{x}_2) \\
& -\hat{V}(\mathbf{x}_1-\mathbf{x}_1')\sum_k \phi_k(\mathbf{x}_1)\phi_k^*(\mathbf{x}_1')\Theta(\epsilon_{\rm F}-\epsilon_k) \;, \quad (10.49)
\end{aligned}$$

where we used the expression of the density in the MF model

$$\rho(\mathbf{x})=\sum_k^{\epsilon_{\rm F}}\phi_k(\mathbf{x})\phi_k^*(\mathbf{x}) \;. \quad (10.50)$$

We define the operator

$$\hat{\mathcal{L}}\equiv\hbar\omega-\hat{H}_0=\hbar\omega+\frac{\hbar^2\nabla^2}{2m}-\hat{\mathcal{W}}(\mathbf{x}) \;, \quad (10.51)$$

and we apply it to the unperturbed Green's function

$$\begin{aligned}
&\hat{\mathcal{L}}\,\tilde{G}^0(\mathbf{x}_1,\mathbf{x}_1',\omega) \\
&= \sum_k(\hbar\omega-\epsilon_k^0)\phi_k^0(\mathbf{x}_1)\phi_k^{0,*}(\mathbf{x}_1')\left[\frac{\Theta(\epsilon_k^0-\epsilon_{\rm F}^0)}{\omega-\omega_k+i\eta}+\frac{\Theta(\epsilon_{\rm F}^0-\epsilon_k^0)}{\omega-\omega_k-i\eta}\right] \\
&= \hbar\sum_k\phi_k^0(\mathbf{x}_1)\phi_k^{0,*}(\mathbf{x}_1')\left[\Theta(\epsilon_k^0-\epsilon_{\rm F}^0)+\Theta(\epsilon_{\rm F}^0-\epsilon_k^0)\right]=\hbar\delta^3(\mathbf{x}_1-\mathbf{x}_1') \;, \quad (10.52)
\end{aligned}$$

where we used the completeness of the ϕ^0, and $\omega_k=\epsilon_k^0/\hbar$. The previous expression indicates that the action of the operator $\hat{\mathcal{L}}$ is analogous to the multiplication by $(\tilde{G}^0)^{-1}$. Let us apply this operator to the Dyson's equation (10.39)

$$\begin{aligned}
&\hat{\mathcal{L}}\,\tilde{G}(\mathbf{x}_1,\mathbf{x}_1',\omega) \\
&= \hat{\mathcal{L}}\,\tilde{G}^0(\mathbf{x}_1,\mathbf{x}_1',\omega)+\int d^3x_2\int d^3x_2'\,\hat{\mathcal{L}}\,\tilde{G}^0(\mathbf{x}_1,\mathbf{x}_2,\omega)\hat{\Sigma}(\mathbf{x}_2,\mathbf{x}_2')\,\tilde{G}(\mathbf{x}_2',\mathbf{x}_1',\omega) \\
&= \hbar\delta^3(\mathbf{x}_1-\mathbf{x}_1')+\hbar\int d^3x_2'\hat{\Sigma}(\mathbf{x}_1,\mathbf{x}_2')\,\tilde{G}(\mathbf{x}_2',\mathbf{x}_1',\omega) \;. \quad (10.53)
\end{aligned}$$

By making explicit the expression of $\hat{\mathcal{L}}$ we obtain

$$\left[\hbar\omega + \frac{\hbar^2\nabla_1^2}{2m} - \hat{\mathcal{W}}(\mathbf{x}_1)\right]\sum_k \phi_k(\mathbf{x}_1)\phi_k^*(\mathbf{x}_1')\left[\frac{\Theta(\epsilon_k - \epsilon_F)}{\omega - \omega_k + i\eta} + \frac{\Theta(\epsilon_F - \epsilon_k)}{\omega - \omega_k - i\eta}\right]$$
$$= \hbar\delta^3(\mathbf{x}_1 - \mathbf{x}_1')$$
$$+ \int d^3x_2 \hbar\hat{\Sigma}(\mathbf{x}_1, \mathbf{x}_2)\sum_k \phi_k(\mathbf{x}_2)\phi_k^*(\mathbf{x}_1')\left[\frac{\Theta(\epsilon_k - \epsilon_F)}{\omega - \omega_k + i\eta} + \frac{\Theta(\epsilon_F - \epsilon_k)}{\omega - \omega_k - i\eta}\right] . \quad (10.54)$$

By multiplying by $\phi_j(\mathbf{x}_1')$ and integrating on $\mathbf{x}_1'$, for the orthonormality properties of the ϕ_k the sum disappears and we obtain the equation

$$\left[\hbar\omega + \frac{\hbar^2\nabla_1^2}{2m} - \hat{\mathcal{W}}(\mathbf{x}_1)\right]\phi_j(\mathbf{x}_1)\left[\frac{\Theta(\epsilon_j - \epsilon_F)}{\omega - \omega_j + i\eta} + \frac{\Theta(\epsilon_F - \epsilon_j)}{\omega - \omega_j - i\eta}\right] = \hbar\phi_j(\mathbf{x}_1)$$
$$+ \int d^3x_2\, \hbar\hat{\Sigma}(\mathbf{x}_1, \mathbf{x}_2)\phi_j(\mathbf{x}_2)\left[\frac{\Theta(\epsilon_j - \epsilon_F)}{\omega - \omega_j + i\eta} + \frac{\Theta(\epsilon_F - \epsilon_j)}{\omega - \omega_j - i\eta}\right] . \quad (10.55)$$

By multiplying by $\omega - \omega_j \equiv \omega - \epsilon_j/\hbar$ we obtain

$$\left[-\frac{\hbar^2\nabla_1^2}{2m} + \hat{\mathcal{W}}(\mathbf{x}_1)\right]\phi_j(\mathbf{x}_1) + \int d^3x_2'\, \hbar\hat{\Sigma}(\mathbf{x}_1, \mathbf{x}_2')\phi_j(\mathbf{x}_2) = \epsilon_j\phi_j(\mathbf{x}_1) , \quad (10.56)$$

We insert the expression (10.49) of the self-energy and obtain

$$\left[-\frac{\hbar^2\nabla_1^2}{2m} + \hat{\mathcal{W}}(\mathbf{x}_1)\right]\phi_j(\mathbf{x}_1) + \int d^3x_2\, \hat{V}(\mathbf{x}_1 - \mathbf{x}_2)\rho(\mathbf{x}_2)\phi_j(\mathbf{x}_1)$$
$$- \int d^3x_2\, \hat{V}(\mathbf{x}_1 - \mathbf{x}_2)\sum_k \phi_k(\mathbf{x}_1)\phi_k^*(\mathbf{x}_2)\phi_j(\mathbf{x}_2)\Theta(\epsilon_F - \epsilon_k)$$
$$= \epsilon_j\phi_j(\mathbf{x}_1) . \quad (10.57)$$

which is the traditional form of expressing the HF equations, (see Sect. 15.2.2).

10.5 Bethe-Salpeter Equation

In the calculation of the Green's function using the Dyson equation, as illustrated in Fig. 10.13, each individual term in the perturbative sum remains finite only if the potential is also finite. Generally, the presence of a strongly repulsive core in the bare interaction between two particles causes each term in the sum to be significantly larger than the value that needs to be evaluated. This issue is similar to the one discussed in Chap. 8 regarding the calculation of the system energy, which was addressed using Brueckner's theory.

Each thin line in Fig. 10.13 represents the Green's function G^0, which can be expressed in terms of single-particle wave functions ϕ_k, as indicated by Eqs. (10.45) and (10.46). This implies that each intersection between a dashed line and a continuous line corresponds to the product $\hat{V}\phi_k$, which combines the microscopic interaction

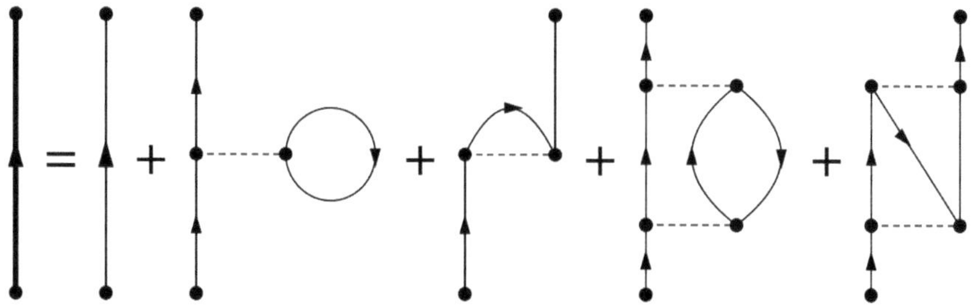

Fig. 10.13 Some diagrams included in the Dyson equation. The thick line indicates the full Green's function G, the thin lines G^0 and the dashed lines the interaction $\hat{V}$

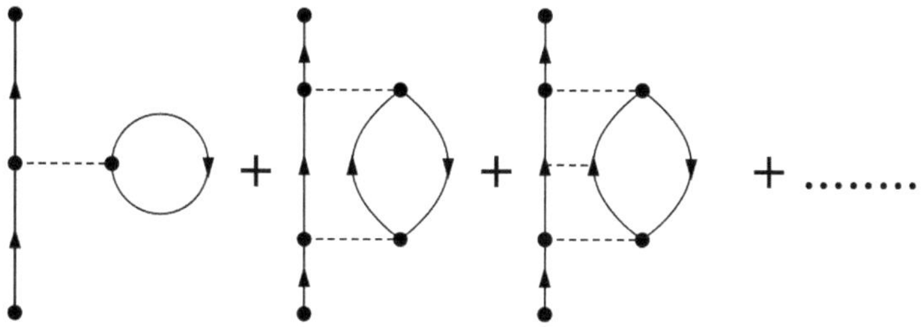

Fig. 10.14 Ladder diagrams approximation

with the unperturbed wave function. As discussed in Chap. 8, this product becomes infinite for potentials that diverge. By employing a strategy similar to that presented in Chap. 8, we construct an effective interaction by summing an infinite set of diagrams, ensuring that the product $\hat{V}^{\text{eff}}\phi_k$ remains finite.

We consider all the interactions connecting the particle lines. This approximation, known as ladder diagrams, is illustrated for the first three terms in Fig. 10.14. It is important to note that the ladder approximation sums all the infinite interaction terms associated with the particle line. This restriction is necessary because, during the interaction process, the fermion modifies its state. Since all the hole states are occupied, using the particle state helps avoid issues related to the Pauli exclusion principle.

We consider the interaction term between two particles, as represented by the diagrams in the upper part of Fig. 10.15, and denote it as $\hat{\Sigma}_{\text{L}}$, where the subscript L indicates the ladder approximation. The expression for the interaction can be written as

$$\hat{\Sigma}_{\text{L}}(x_1, x_1') = \hat{\mathcal{W}}(x_1, x_1')$$
$$+ \hat{\mathcal{W}}(x_1, x_1') \int d^4x_2 \int d^4x_2' \, G^0(x_1, x_2)\hat{\mathcal{W}}(x_2, x_2')G^0(x_2', x_1')$$
$$+ \hat{\mathcal{W}}(x_1, x_1') \int d^4x_2 \int d^4x_2' \int d^4x_3 \int d^4x_3'$$

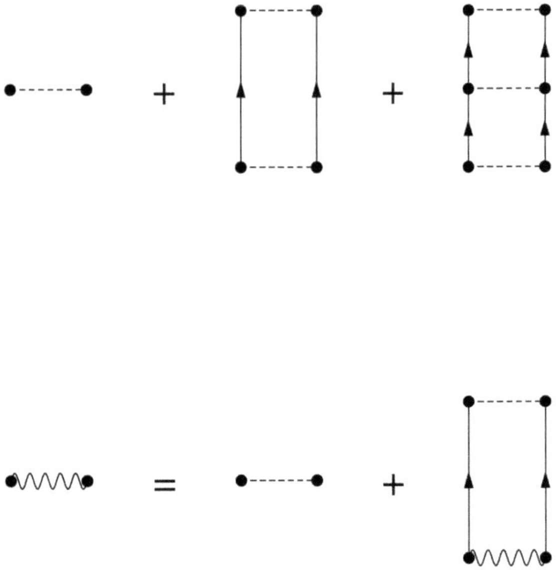

Fig. 10.15 Graphical representation of Eq. (10.59)

$$
\begin{aligned}
& G^0(x_1, x_2)\hat{\mathcal{W}}(x_2, x_2')G^0(x_2, x_3)\hat{\mathcal{W}}(x_3, x_3')G^0(x_3', x_2')G^0(x_2', x_1') + \cdots \\
= \; & \hat{\mathcal{W}}(x_1, x_1')\Big\{1 + \int d^4x_2 \int d^4x_2' \, G^0(x_1, x_2)\hat{\mathcal{W}}(x_2, x_2')\Big[1 + \\
& \int d^4x_3 \int d^4x_3' G^0(x_2, x_3)\hat{\mathcal{W}}(x_3, x_3')\Big(1 + \cdots\Big)G^0(x_3', x_2')\Big] \\
& G^0(x_2', x_1')\Big\} \; . \qquad (10.58)
\end{aligned}
$$

The above expression iterates the same integration kernel. By considering the expansion to all orders, we obtain the following expression

$$
\begin{aligned}
\hat{\Sigma}_{\mathrm{L}}(x_1, x_1') = \; & \hat{\mathcal{W}}(x_1, x_1')\Big[1 + \\
& \int d^4x_2 \int d^4x_2' \, G^0(x_1, x_2)\hat{\Sigma}_{\mathrm{L}}(x_2, x_2')G^0(x_2', x_1')\Big] \; . \qquad (10.59)
\end{aligned}
$$

The graphical representation of this expression is provided by the diagrams in the lower part of Fig. 10.15. Equation (10.59) is known as the **Bethe-Salpeter equation**, although it is actually the ladder approximation of the Bethe-Salpeter equation.

This equation also includes products between the microscopic bare potential $\hat{\mathcal{W}}$ and the non-interacting Green's function G^0. However, these terms are multiplied by $\hat{\Sigma}_{\mathrm{L}}$, which ensures that the integrand remains finite.

The above expression of the Bethe-Salpeter equation is written in terms of G^0, but it can also utilise a G obtained from a Hartree-Fock calculation.

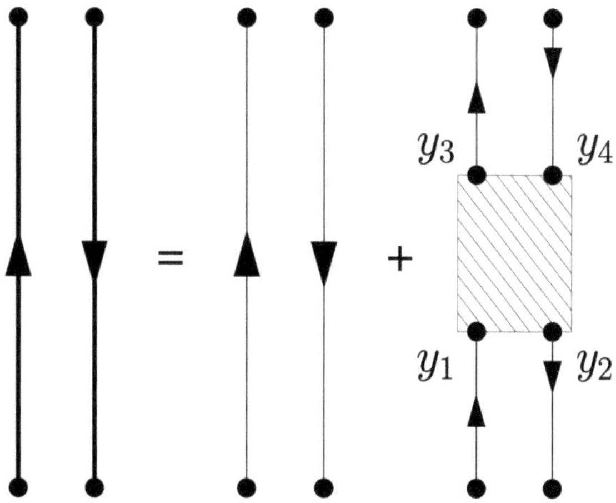

Fig. 10.16 Graphical representation of Eq. (10.60). The dashed part represent the kernel $\hat{\mathcal{K}}$

10.6 Random Phase Approximation

It is possible to formulate an equation analogous to Dyson's equation (10.19) for the two-body Green's function. Here, we focus solely on the two-body Green's functions that describe the time evolution of a particle-hole pair. In this context, the equation analogous to Dyson's equation is:

$$\begin{aligned} G(x_1, x_2, x_3, x_4) &= G^0(x_1, x_2, x_3, x_4) \\ + &\int d^4y_1\, d^4y_2\, d^4y_3\, d^4y_4 \\ &G^0(x_1, x_2, y_1, y_2)\underline{\hat{\mathcal{K}}}(y_1, y_2, y_3, y_4)G^0(y_3, y_4, x_3, x_4)\ , \end{aligned} \tag{10.60}$$

where the term $\underline{\hat{\mathcal{K}}}$ contains all the linked diagrams which can be inserted between particle and hole lines. Figure 10.16 offers a graphical representation of the Dyson equation (10.60), where the two-body Green's function is depicted by a pair of oriented lines, and the kernel is represented by the shaded box (Fig. 10.17).

In this case as well, it is possible to define proper, irreducible diagrams. In Fig. 10.18, diagram A is reducible because it is possible to cut the external particle and hole lines, resulting in two diagrams that are already present in the perturbative expansion of the two-body Green's function. In contrast, this is not possible for diagram B, which is referred to as irreducible or proper. By considering only the insertion of proper diagrams, Eq. (10.60) can be formally rewritten as

$$G(x_1, x_2, x_3, x_4) = G^0(x_1, x_2, x_3, x_4) +$$
$$\int d^4y_1\, d^4y_2\, d^4y_3\, d^4y_4\, G^0(x_1, x_2, y_1, y_2)$$
$$\hat{\mathcal{K}}(y_1, y_2, y_3, y_4) G(y_3, y_4, x_3, x_4) \ , \tag{10.61}$$

where the kernel $\hat{\mathcal{K}}$ contains the insertion of all the irreducible diagrams.

The graphical representation of Eq. (10.61) is displayed in Fig. 10.18, where the doubly shaded box represents the kernel containing only proper insertions of the interaction. The Random Phase Approximation (RPA) involves substituting the operator $\hat{\mathcal{K}}$ in the previous equation with the interaction $\hat{\mathcal{W}}$, which depends solely on two coordinates.

$$\hat{\mathcal{K}}^{\mathrm{RPA}}(y_1, y_2, y_3, y_4) =$$
$$\hat{\mathcal{W}}(y_1, y_4)\,[\delta(y_1 - y_2)\delta(y_3 - y_4) - \delta(y_1 - y_3)\delta(y_2 - y_4)] \ , \tag{10.62}$$

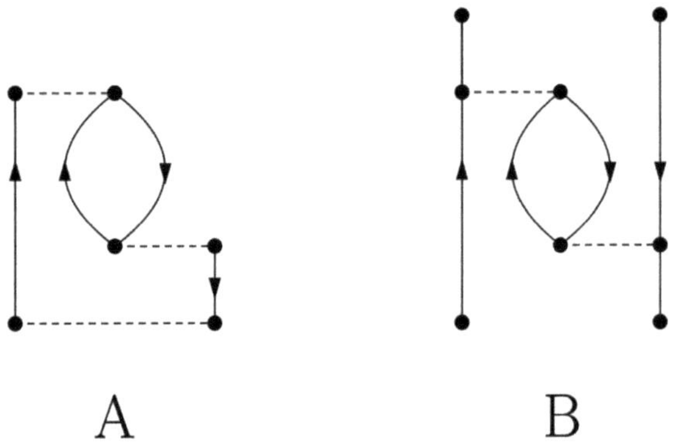

Fig. 10.17 The A diagram is reducible while the B diagram is irreducible (proper)

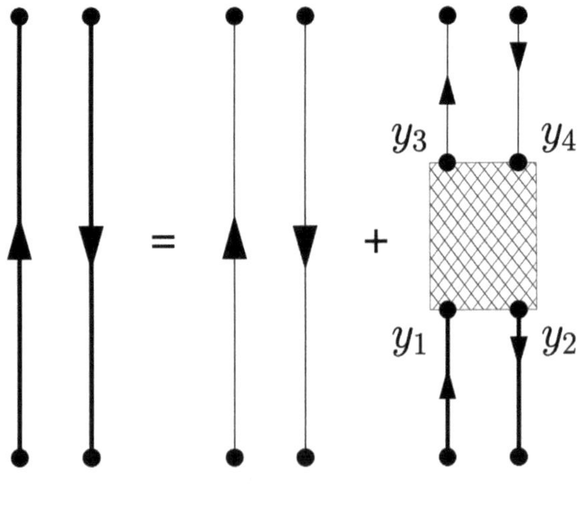

Fig. 10.18 Graphical representation of Eq. (10.61). The doubly dashed part represent the kernel $\mathcal{K}$

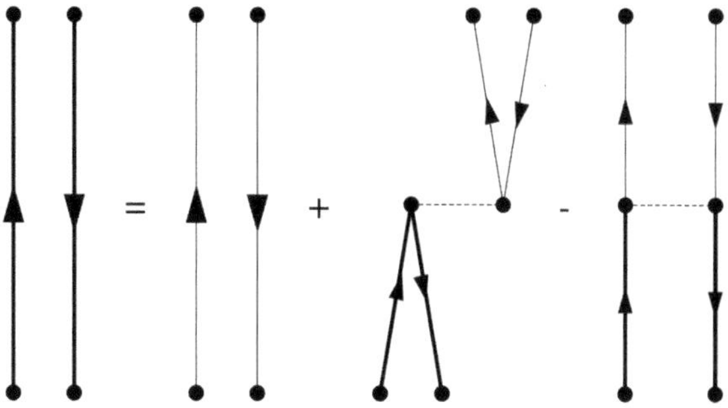

Fig. 10.19 Graphical representation of the RPA equation (10.63)

therefore

$$
\begin{aligned}
G^{\mathrm{RPA}}(x_1, x_2, x_3, x_4) &= G^0(x_1, x_2, x_3, x_4) \\
+ \int d^4y_1\, d^4y_2\, & G^0(x_1, x_2, y_1, y_1)\hat{\mathcal{W}}(y_1, y_2)G^{\mathrm{RPA}}(y_2, y_2, x_3, x_4) \\
\int d^4y_1\, d^4y_2\, & G^0(x_1, x_2, y_1, y_2)\hat{\mathcal{W}}(y_1, y_2)G^{\mathrm{RPA}}(y_1, y_2, x_3, x_4) \ , \qquad (10.63)
\end{aligned}
$$

where we separated the direct and the exchange terms. The graphical representation of the above equation is given Figs. 10.19 and 10.20.

Let us consider the expression of the two-body Green's function in Lehmann representation given by Eq. (9.76). In a translational invariant system, since $|\Psi_n\rangle$ and $\langle\Psi_n|$ indicate the same excited state the terms

$$
\langle\Psi_0|a_{k_1}a^+_{k_2}|\Psi_n\rangle\langle\Psi_n|a_{k_3}a^+_{k_4}|\Psi_0\rangle \ ,
$$

implies

$$
\delta_{k_2,k_3}\theta(k_2 - k_F)\delta_{k_4,k_1}\theta(k_4 - k_F) \ .
$$

We define the Fourier transforms of the two-body Green's function, and also of the interaction kernel, depending on two momenta

$$
\begin{aligned}
& G(x_1, x_2, x_3, x_4) = \\
& \quad (2\pi)^{-8} \int d^4k_1\, d^4k_2 e^{-ik_1(x_1-x_4)} e^{ik_2(x_2-x_3)} \tilde{G}(k_1, k_2) \ , \qquad (10.64) \\
& \hat{\mathcal{K}}(x_1, x_2, x_3, x_4) = \\
& \quad (2\pi)^{-8} \int d^4k_1\, d^4k_2 e^{-ik_1(x_1-x_4)} e^{ik_2(x_2-x_3)} \tilde{\hat{\mathcal{K}}}(k_1, k_2) \ . \qquad (10.65)
\end{aligned}
$$

By substituting these definitions into Eq. (10.61) and applying the RPA assumption (10.63) to the kernel, we derive an expression for Dyson's equation in the context

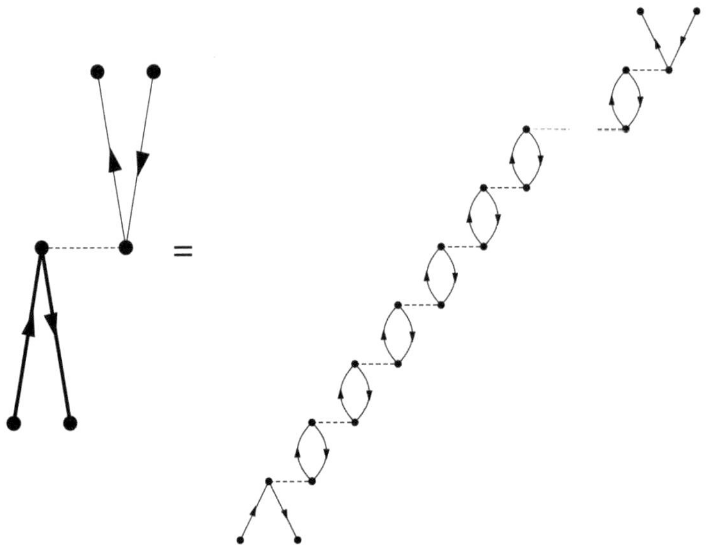

Fig. 10.20 Graphical representation of the ring diagrams approximation of the RPA

of the RPA in momentum space. A lengthy calculation, although straightforward, reveals that by considering only the direct term from Eq. (10.63), the equation simplifies to a purely algebraic form

$$\tilde{G}^{\mathrm{RPA,D}}(k_1, k_2) = \tilde{G}^0(k_1, k_2) + \tilde{G}^0(k_1, k_2)\hat{\tilde{\mathcal{U}}}(k_1 - k_2 = q)\tilde{G}^{\mathrm{RPA,D}}(k_1, k_2) \ . \tag{10.66}$$

or

$$\tilde{G}^{\mathrm{RPA,D}}(k_1, k_1 + q) = \frac{\tilde{G}^0(k_1, k_1 + q)}{1 - \tilde{G}^0(k_1, k_1 + q)\hat{\tilde{\mathcal{W}}}(q)} . \tag{10.67}$$

This expression, known as the ring approximation of the RPA, is commonly used to calculate the linear response of infinite fermion systems to an external probe. The exchange term cannot be factored into a simple algebraic expression.

Part III
Theories Inspired to Statistical Mechanics

Chapter 11
Correlated Basis Function Theory

11.1 Introduction

In the previous chapters, we emphasised that a naive application of perturbation theory should be avoided in the description of many-body systems due to the strongly repulsive core of the bare particle-particle interaction.

In Part II, using language and terminology inspired by Quantum Field Theory, we demonstrated how this issue can be addressed. The interaction between particles in a medium is modified compared to that in a vacuum. Brueckner's theory, discussed in Chap. 8, serves as a typical example of how to derive this type of effective interaction, which is defined by Eq. (8.2).

A different approach to the many-body problem arises from Statistical Mechanics. In this case, the focus shifts from the interaction itself to the wave function that describes the relative motion of the interacting pair of particles. This situation is illustrated in Fig. 8.1, where ϕ represents the unperturbed wave function of the two interacting particles. The overlap of this wave function with the potential results in large matrix elements that complicate the application of perturbation theory.

By multiplying ϕ by a function known as correlation function, which prevents the two particles from coming too close to each other, it becomes possible to circumvent the issue of the strongly repulsive core. The correlation function is the fundamental quantity upon which the theories presented in this Part III are based. The language and terminology used in these theories differ significantly from those employed in Part II.

In this chapter, we present the **Correlated Basis Function** (CBF) theory. The starting point of this theory is the search for solutions of the many-body Schrödinger equation by using the variational principle (see Appendix A). The equation to be solved is

$$\delta E[\Psi_{\rm T}] = \delta \frac{\langle \Psi_{\rm T} | \hat{H} | \Psi_{\rm T} \rangle}{\langle \Psi_{\rm T} | \Psi_{\rm T} \rangle} = 0 \ , \tag{11.1}$$

© The Author(s), under exclusive license to Springer Nature Switzerland AG 2026

G. Co', *Concepts in Quantum Many-Body Physics*, UNITEXT for Physics,
https://doi.org/10.1007/978-3-032-08920-5_11

where $\hat{H}$ is the hamiltonian describing the system of interacting particles, and $|\Psi_T\rangle$ is a many-body trial wave function. In the specific case of the CBF theory, the search for the minimum is carried out by considering only wave functions which can be expressed as

$$|\Psi_T(1, ..., A)\rangle = F(1, ..., A)\,|\Phi(1, ..., A)\rangle \ , \tag{11.2}$$

where $|\Phi\rangle$ is the state describing the system of non-interacting particles.

The second assumption of the theory, usually called Jastrow's anstatz [1], is that the many-body correlation function F can be written as a product of two-body correlation functions

$$F(1, ..., A) = \prod_{j>i=1}^{A} f(r_{ij}) \ , \tag{11.3}$$

where $f(r_{ij})$ is a function of the distance between the i and j particles of the system.

These assumptions are the same as those adopted in the Variational Monte Carlo (VMC) approach presented in Sect. 4.3. The key difference between these two theories lies in the methods used to calculate the expectation value of the Hamiltonian in Eq. (11.1). In the case of VMC, the expectation value is obtained by using Monte Carlo multidimensional integration techniques, as discussed in Chap. 4. In contrast, the CBF approach employs a cluster expansion proposed by Joseph Mayer and Maria Goeppert-Mayer to describe liquids within the framework of classical Statistical Mechanics [2]. The particles, correlated by the function f, generate clusters. The CBF theory, after conducting a topological analysis of the various clusters, successfully formulates a set of integral equations that sum all the clusters of a certain type in closed form. This set of equations is known as the Hypernetted Chain for systems composed of classical particles or bosons [3, 4]. The extension of this theory to fermionic many-body systems, which is far from trivial, was formulated in the early 1970s by Fantoni and Rosati [5].

We present first the HNC equations whose formulation is relatively simpler and it will be used to clarify the essential points of the theory. After that, we shall discuss the extension to fermionic system which has to account for the Pauli exclusion principle.

11.2 The Hypernetted Chain (HNC) Equations

The cluster expansion techniques and the Hypernetted Chain (HNC) equations were initially developed to address classical particles [2–4, 6]. Extending these techniques to bosonic systems is straightforward, so we can immediately consider a system of A identical bosons, assuming zero spin for simplicity, contained within a volume $\mathcal{V}$. The thermodynamic limit is achieved by allowing both A and $\mathcal{V}$ to approach infinity while keeping the particle density $\rho = A/\mathcal{V}$. finite. The system is homogeneous and exhibits translational invariance, meaning that the particle density remains constant. The wave function describing this system can be expressed as:

$$\Phi(x_1, x_2, \ldots, x_A) = \hat{\mathcal{S}}\Big(\phi_1(x_1)\cdots\phi_A(x_A)\Big) , \tag{11.4}$$

where $\hat{\mathcal{S}}$ is an operator which symmetrizes the total wave function for the exchange of two particles. In the previous equation, $\phi_i(x_i)$ indicates the single particle wave functions generated by a translational invariant MF model. We called x_i the generalized coordinates of the i-th particle, that is, in addition to the space coordinates, also spin, eventually isospin and other quantum numbers which characterize the particles considered.

In this description of the system ground state, all the bosons occupy the same single-particle state, which corresponds to the lowest energy level. In analogy to the case presented in Sect. 2.3, the single-particle wave functions are eigenstates of the momentum $\mathbf{p} = \hbar\mathbf{k}$, and they can be expressed as:

$$\phi_j(x_j) = \frac{1}{\sqrt{\mathcal{V}}}\, e^{i\mathbf{k}_j\cdot\mathbf{r}_j} . \tag{11.5}$$

In this case, the generalised coordinate x corresponds to $\mathbf{r}$. The uncorrelated density of the system can be obtained using the Eqs. (11.4) and (11.5),

$$\begin{aligned}
\rho_0(x) &= \langle\Phi|\sum_{i=1}^{A}\delta(x - x_i)|\Phi\rangle \\
&= \sum_{i=1}^{A}\int dx_1\cdots d_A\phi^*(x_1)\cdots\phi^*(x_a)\delta(x - x_i)\phi(x_1)\cdots\phi(x_a) \\
&= A\phi^*(x)\phi(x) = \frac{A}{\mathcal{V}} = \rho .
\end{aligned} \tag{11.6}$$

which is a constant, as expected.

As we have already mentioned, the goal is to solve the Schrödinger equation using the variational principle with the trial wave function given in (11.2), where the correlation function is defined by the expression in (11.3).

In the calculation of the energy functional (11.1) it is very convenient the use of a quantity called two-body distribution function (TBDF) defined as:

$$\begin{aligned}
g(x_1, x_2) &= \frac{A(A-1)\displaystyle\int dx_3\ldots dx_A\Psi_{\mathrm{T}}^*(x_1, x_2, \ldots, x_A)\Psi_{\mathrm{T}}(x_1, x_2, \ldots, x_A)}{\rho^2\displaystyle\int dx_1 dx_2\ldots dx_A\Psi_{\mathrm{T}}^*(x_1, x_2, \ldots, x_A)\Psi_{\mathrm{T}}(x_1, x_2, \ldots, x_A)} \\
&= \frac{A(A-1)\displaystyle\int dx_3\ldots dx_A F^2(x_1, x_2, \ldots, x_A)|\Phi(x_1, x_2, \ldots, x_A)|^2}{\rho^2\displaystyle\int dx_1 dx_2\ldots dx_A F^2(x_1, x_2, \ldots, x_A)|\Phi(x_1, x_2, \ldots, x_A)|^2} ,
\end{aligned} \tag{11.7}$$

where we utilised the fact that F is a real scalar function. The expectation value of every two-body operator is given by integrating the product of the operator and the TBDF on the two coordinates x_1 e x_2:

$$\langle\hat{\mathbb{O}}\rangle = \rho^2 \int dx_1 dx_2\, g(x_1, x_2)\, \hat{\mathbb{O}}(x_1, x_2) \ . \tag{11.8}$$

The many-body effects are described by the TBDF by means of the correlation functions, and they are independent of the operator $\hat{\mathbb{O}}(x_1, x_2)$.

For the bosonic system under consideration, we have that

$$|\Phi|^2 = \prod_{i=1}^{A} \phi^*(x_i)\phi(x_i) = \frac{1}{\mathcal{V}^A} = \frac{\rho^A}{A^A} \ . \tag{11.9}$$

By considering the equations (11.4), (11.5), and (11.6), along with the expressions (11.2) and (11.3), the numerator, $\mathcal{N}$, and the denominator, $\mathcal{D}$, of Eq. (11.7) can be expressed, respectively, as

$$\begin{aligned}\mathcal{N} &= \frac{A(A-1)}{\rho^2} \int dx_3 \ldots dx_A \prod_{i<j} f^2(r_{ij}) |\Phi(x_1, x_2, \ldots, x_A)|^2 \\ &= \frac{A(A-1)}{\rho^2} \frac{\rho^A}{A^A} \int dx_3 dx_4 ... dx_A \prod_{i<j} f^2(r_{ij}) \ ,\end{aligned} \tag{11.10}$$

and

$$\mathcal{D} = \frac{\rho^A}{A^A} \int dx_1 dx_2 ... dx_A \prod_{i<j} f^2(r_{ij}) \ . \tag{11.11}$$

We define a new function $h(r_{ij})$ as

$$f^2(r_{ij}) = 1 + h(r_{ij}) \ , \tag{11.12}$$

and, by inserting it in the Jastrow's ansatz of the many-body correlation function, we obtain,

$$\begin{aligned}\prod_{i<j} f^2(r_{ij}) &= \prod_{i<j} \Big(1 + h(r_{ij})\Big) \\ &= \Big[\Big(1 + h(r_{12})\Big)\Big(1 + h(r_{13})\Big) \cdots \Big(1 + h(r_{23})\Big) \cdots\Big] \ . \end{aligned} \tag{11.13}$$

The cluster expansion consists in grouping the terms containing the same number of h-functions

$$\prod_{i<j} f^2(r_{ij}) = 1 + \sum_{i<j} h(r_{ij}) + \sum_{i<j,k<l} h(r_{ij})h(r_{kl}) + \cdots \ . \tag{11.14}$$

Each term of this sum forms a cluster of particles.

Let us consider the numerator $\mathcal{N}$, defined in Eq. (11.10). In the product of the correlation functions it is convenient to factorize the $f^2(r_{12})$ term outside the integral. By using the expansion (11.14) we can express $\mathcal{N}$ as:

$$\begin{aligned}
\mathcal{N} = & \frac{A(A-1)}{\rho^2} \frac{\rho^A}{A^A} f^2(r_{12}) \int dx_3 dx_4 \cdots dx_A \\
& \Big[\underbrace{1}_{\mathbf{A}} + \underbrace{\sum_{2<i} h(r_{1i})}_{\mathbf{B}} + \underbrace{\sum_{2<i} h(r_{2i})}_{\mathbf{C}} + \underbrace{\sum_{2<i<j} h(r_{ij})}_{\mathbf{D}} \\
& + \underbrace{\sum_{2<i} h(r_{1i})h(r_{2i})}_{\mathbf{E}} + \underbrace{\sum_{2<i<j} h(r_{1i})h(r_{ij})}_{\mathbf{F}} + \underbrace{\sum_{2<i<j} h(r_{2i})h(r_{ij})}_{\mathbf{G}} \\
& + \underbrace{\sum_{2<i<j} h(r_{1i})h(r_{2j})}_{\mathbf{H}} + \cdots \Big] ,
\end{aligned} \tag{11.15}$$

where we have explicitly written the terms up to two h-functions.

We calculate here the contribution of each term of Eq. (11.15) by considering that

$$\int dx = \mathcal{V} = \frac{A}{\rho} .$$

The contribution of the term **A** is:

$$\int dx_3 dx_4 \cdots dx_A = \mathcal{V}^{A-2} \ . \tag{11.16}$$

The contribution of the term **B** is:

$$\begin{aligned}
& \int dx_3 dx_4 \cdots dx_A \sum_{2<i} h(r_{1i}) \\
= \ & \mathcal{V}^{A-3} \sum_{2<i} \int dx_i h(r_{1i}) = \mathcal{V}^{A-3}(A-2) \int dx_i h(r_{1i}) \ ,
\end{aligned} \tag{11.17}$$

where the $A-2$ term arises from the summation on i and it is related to the fact that all the integrals on the generic coordinates are equal. Since the system is homogeneous, we have that $h(r_{1,i}) = h(r_2, i)$, therefore the contribution of the term **C** is the same as that of **B**.

For the term **D** we have:

$$\int dx_3 dx_4 \cdots dx_A \sum_{2<i} \sum_{i<j} h(r_{ij}) = \mathcal{V}^{A-4}(A-2)(A-3) \int dx_i dx_j h(r_{ij}) \ , \quad (11.18)$$

the $A-2$ term arises from the sum on i, while the term $A-3$ from the sum on j.

The expression of the term **E** is:

$$\int dx_3 dx_4 \cdots dx_A \sum_{2<i} h(r_{1i}) h(r_{2i}) = \mathcal{V}^{A-3}(A-2) \int dx_i h(r_{1i}) h(r_{2i}) \ . \quad (11.19)$$

For the **F** term we have:

$$\int dx_3 dx_4 \cdots dx_A \sum_{2<i<j} h(r_{1i}) h(r_{ij})$$
$$= \mathcal{V}^{A-4}(A-2)(A-3) \int dx_i dx_j h(r_{1i}) h(r_{ij}) \ . \quad (11.20)$$

Also in this case, since the system is homogeneous and has translational invariance, the contribution of the **G** terms is identical to that of the **F** term.

The contribution of the **H** term is

$$\int dx_3 dx_4 \cdots dx_A \sum_{2<i<j} h(r_{1i}) h(r_{2j})$$
$$= \mathcal{V}^{A-4}(A-2)(A-3) \int dx_i h(r_{1i}) \int dx_j h(r_{1j}) \ . \quad (11.21)$$

A convenient method for studying the various terms of the cluster expansion involves using the graphical representation introduced by Mayer and Goeppert-Mayer [2]. In this representation, the h-functions are depicted as dashed lines connecting black or white points. The coordinates x over which integrations are performed are represented by black dots, referred to as **internal points**. When there is no integration on the coordinates, they are indicated by white dots and are called **external points**. In our calculations, we identify these external points with the labels 1 and 2.

The various terms of Eq. (11.15) are illustrated by the diagrams in Fig. 11.1. The uncorrelated term is represented by diagram **A**. In this equation, a correlation function $f^2(r_{12})$ is included, which is understood in the graphical representation. Diagrams **B** and **C** correspond to the respective terms of Eq. (11.15), where the h-function connects an external point. It is clear that they yield identical contributions. Similarly, diagrams **F** and **G** also produce identical contributions. Furthermore, it is possible to interchange the indexes i and j in diagrams **D, F, G, H**, which would result in double counting. Therefore, the contribution from each of these terms should be divided by a factor of 2.

In general, a diagram with n internal points has to be multiplied by a factor

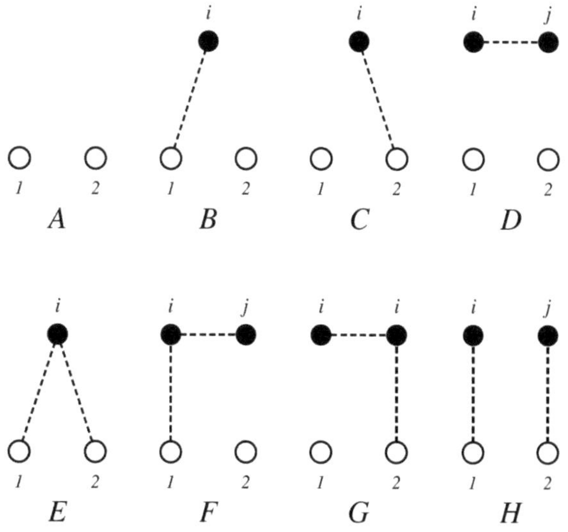

Fig. 11.1 Graphical representation of the terms contributing to Eq. (11.15). The dashed lines indicate the h-functions, the black dots indicate the internal points over which integrations are carried out, and the white dots the external points we have labelled as 1 and 2

$$\frac{1}{n!}\prod_{k=1}^{n}(A-1-k) \ .$$

By considering the results of Eqs. (11.16)–(11.21), and the symmetry properties emerging from the analysis of the diagrams of Fig. 11.1, we obtain the expression

$$\begin{aligned}
\mathcal{N} = f^2(r_{12})\frac{(A-1)}{A}\Bigg[&1 + 2(A-2)\frac{\rho}{A}\int dx_i h(r_{1i}) \\
&+ \frac{(A-2)(A-3)}{2}\frac{\rho^2}{A^2}\int dx_i dx_j h(r_{ij}) \\
&+ (A-2)\frac{\rho}{A}\int dx_i h(r_{1i})h(r_{2i}) \\
&+ 2\frac{(A-2)(A-3)}{2}\frac{\rho^2}{A^2}\int dx_i dx_j h(r_{1i})h(r_{ij}) \\
&+ \frac{(A-2)(A-3)}{2}\frac{\rho^2}{A^2}\int dx_i h(r_{1i})\int dx_j h(r_{2j}) + \ldots\Bigg] \ . \qquad (11.22)
\end{aligned}$$

The same strategy can be adopted in the study of the denominator $\mathcal{D}$, Eq. (11.11), which can be written as:

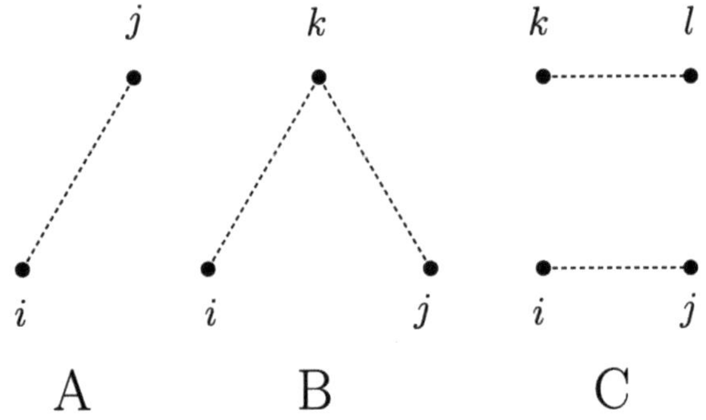

Fig. 11.2 Graphical representation of some terms contributing to Eq. (11.23)

$$\mathcal{D} = \frac{\rho^A}{A^A} \int dx_1 dx_2 ... dx_A \Big[1 + \underbrace{\sum_{i<j} h(r_{ij})}_{\mathbf{A}} + \underbrace{\sum_{i<j<k} h(r_{ik}) h(r_{jk})}_{\mathbf{B}} + \underbrace{\sum_{i<j<k<l} h(r_{ij}) h(r_{kl})}_{\mathbf{C}} + \ldots \Big] . \tag{11.23}$$

The first term is

$$\frac{\rho^A}{A^A} \int dx_1 dx_2 ... dx_A = \frac{\rho^A}{A^A} \mathcal{V}^A = 1 \ , \tag{11.24}$$

and it is omitted in the graphical representation of Fig. 11.2.

The denominator does not have external points, for this reasons all the dots in Fig. 11.2 are black, indicating an integration. In the diagram **A** single h-function connects the i and j points. The diagrams **B** and **C** have 2 h-functions but in **B** there are 3 points which all connected, while in **C** there are 4 points forming 2 unliked clusters.

The contribution of the term **A** of Eq. (11.23) is given by

$$\frac{\rho^A}{A^A} \int dx_1 dx_2 ... dx_A \sum_{i<j} h(r_{ij}) = \frac{\rho^2}{A^2} \frac{A(A-1)}{2} \int dx_i dx_j h(r_{ij}) \ ,$$

where the factor $1/2$ is inserted to avoid the double counting due to the possibility of exchanging the indexes i and j.

The contribution of the term **B** of Eq. (11.23) is:

$$\frac{\rho^A}{A^A}\int dx_1 dx_2 ... dx_A \sum_{i<j<k} h(r_{ik})h(r_{jk})$$
$$= \frac{1}{3!}\frac{A(A-1)(A-2)}{A^3}\rho^3 \int dx_i dx_j dx_k h(r_{ik})h(r_{jk}) \; , \tag{11.25}$$

where the factor $A(A-1)(A-2)$ accounts for the condition that the indexes i, j and k are constrained by the requirement $i < j < k$. The factor 3! arises from the symmetry achieved by interchanging the three indexes.

The contribution of **C** term of Eq. (11.23) can be expressed as

$$\frac{\rho^A}{A^A}\int dx_1 dx_2 ... dx_A \sum_{i<j<k<l} h(r_{ij})h(r_{kl})$$
$$= \frac{1}{4\cdot 3}A(A-1)\frac{\rho^2}{A^2}\int dx_i dx_j h(r_{ij})$$
$$\frac{(A-2)(A-3)}{2}\frac{\rho^2}{A^2}\int dx_k dx_l h(r_{kl}) \; . \tag{11.26}$$

The global contribution of this equation is given by the product of two separated integrals, a first one on x_i and x_j and a second one on x_k and x_l.

By using the above expressions we can write the denominator $\mathcal{D}$ as:

$$\mathcal{D} = 1 + \frac{\rho^2}{A^2}\frac{A(A-1)}{2}\int dx_i dx_j h(r_{ij})$$
$$+\frac{1}{3!}\frac{A(A-1)(A-2)}{A^3}\rho^3 \int dx_i dx_j dx_k h(r_{ik})h(r_{jk})$$
$$+\frac{\rho^2}{A^2}\int dx_i dx_j h(r_{ij}) \; \frac{(A-2)(A-3)}{2} \; \frac{\rho^2}{A^2}\int dx_k dx_l h(r_{kl}) \tag{11.27}$$
$$+\cdots \; .$$

To effectively manage the complex set of clusters, it is beneficial to make the most of the graphical representation of the various terms involved in the calculation of the TBDF. Therefore, we will perform a topological classification of the different diagrams.

(a) *Linked and unliked diagrams*

From a graphical perspective, the unlinked diagrams are easily identifiable as they consist of separate components. For instance, the diagrams labeled **D** and **H** in Fig. 11.1, as well as the diagram **C** in Fig. 11.2, exemplify this characteristic.

Mathematically, Eqs. (11.18, 11.21) and (11.28) illustrate that the contributions from these types of diagrams are derived by multiplying the integrals that correspond to the individual components of the diagrams.

In contrast, linked diagrams cannot be represented as products of independent parts. This is evident in diagrams **B, C, E, F, G** shown in Fig. 11.1.

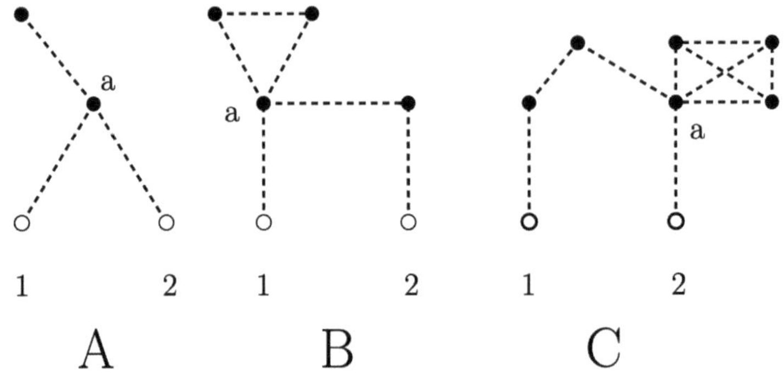

Fig. 11.3 Examples of reducible diagrams in the numerator of the TBDF, Eq. (11.22)

Additionally, the components of the unlinked diagrams in the numerator that do not include the two external points are also found within the set of diagrams in the denominator. For example, the correlated pair of indexes i and j in the **D** diagram from Fig. 11.1 is identical to that in the **A** diagram from Fig. 11.2.

(b) *Reducible diagrams*

The linked diagrams in Fig. 11.3 exhibit the property of being reducible. From a graphical perspective, these types of diagrams are characterised by the presence of at least one internal point that connects the portion of the diagram containing external points to a section that includes only internal points. Due to the translational invariance of the system, the contributions from these two sections can be factorized. In bosonic systems, both unlinked diagrams and reducible ones can be expressed as products of a term containing external points multiplied by one or more terms that contain only internal points.

The factorized portion of these diagrams that does not include external points can be simplified, up to order $1/A$, using the diagrams from the denominator. A formal proof of this property can be found in Ref. [5]. Below, we outline how this property arises when only the linear terms in h are taken into account.

Let us consider the TBDF $g(r_{12})$. From the expressions (11.22) of the numerator and (11.28) of the denominator we consider only the terms with a single h-line.

$$g(r_{12}) =$$

$$f^2(r_{12})\frac{(A-1)}{A}\frac{\left[1+2(A-2)\frac{\rho}{A}\int dx_i h(r_{1i})+\frac{(A-2)(A-3)}{2}\frac{\rho^2}{A^2}\int dx_i dx_j h(r_{ij})\right]}{1+\frac{\rho^2}{A^2}\frac{A(A-1)}{2}\int dx_i dx_j h(r_{ij})}\ .$$

We make a power expansion of the denominator and we consider only the terms with a single h line

$$g(r_{12}) = f^2(r_{12})\frac{(A-1)}{A}\left(1 + 2(A-2)\frac{\rho}{A}\int dx_i h(r_{1i}) + \frac{(A-2)(A-3)}{2}\frac{\rho^2}{A^2}\int dx_i dx_j h(r_{ij}) - \frac{\rho^2}{A^2}\frac{A(A-1)}{2}\int dx_i dx_j h(r_{ij})\right) . \quad (11.28)$$

Since we are dealing with a homogeneous system having translational invariance we have

$$\frac{\rho}{A}\int dx_i h(r_{1i}) = \frac{\rho}{A}\int d^3r\, h(r) , \quad (11.29)$$

and

$$\frac{\rho^2}{A^2}\int dx_i dx_j h(r_{ij}) = \frac{\rho}{A}\int d^3r\, h(r) . \quad (11.30)$$

where we have defined $r = |\mathbf{r}|$. By inserting Eqs. (11.28,11.28) in (11.28), after some algebra, we obtain

$$g(r_{12}) = f^2(r_{12})\left(1 - \frac{1}{A}\right)\left(1 - \frac{\rho}{A}\int dx_i h(r_{1i})\right). \quad (11.31)$$

In the limit $A \to \infty$ the linear terms in h vanishes and we have $g(r_{12}) = f^2(r_{12})$. In effect these linear terms are either unlinked as the **D** diagram of Fig. 11.1, or reducible as the **B** and **C** diagrams of the same figure.

In summary, we can state that in the expression (11.7) of the TBDF, the diagrams in the denominator effectively cancel the factorised contributions from the unlinked and reducible diagrams present in the numerator, up to terms of order $1/A$. For this reason, in a system containing an infinite number of particles, the TBDF can be expressed as sum of linked and irreducible diagrams, all containing the external points 1 and 2:

$$g(r_{12}) = f^2(r_{12}) \sum_{\text{all orders}} Y_{\text{irr}}(r_{12}), \quad (11.32)$$

where we have indicated with Y_{irr} the sum on all the linked irreducible diagrams. In the above equation we considered that, because of the translational invariance of the system, the TBDF depends only on the relative distance of the two external points r_{12}.

(c) *Simple and composite diagrams*

The irreducible diagrams can be further classified into *simple* and *composite*, which we denote as $S(r_{12})$ and $C(r_{12})$, respectively. Composite diagrams are formed by components that are connected solely through the two external points, as illustrated in Fig. 11.4. These composite diagrams can be expressed in terms of simple diagrams. Since there is no integration over the external points, the contribution of a composite

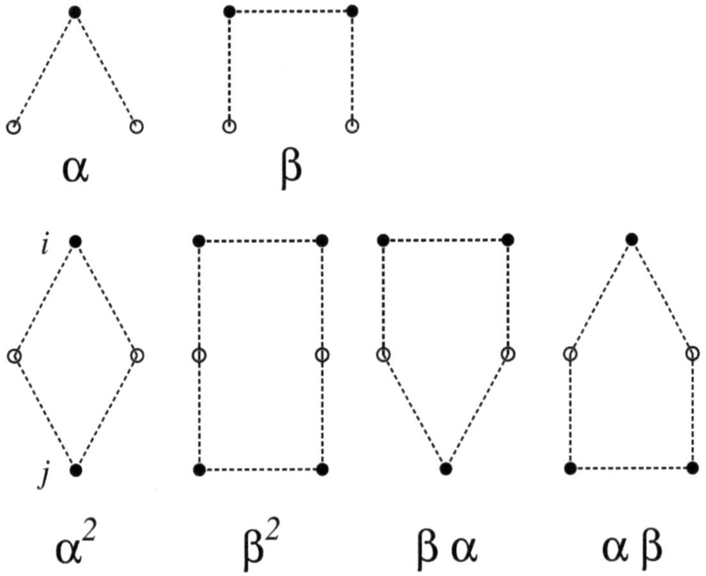

Fig. 11.4 Example of formation of composite diagrams starting from simple diagrams

diagram is given by the product of the simple diagrams connected to the external points.

Let us denote $S^2(r_{12})$ as the contribution from the sum of all composite diagrams formed by the product of only two simple diagrams, similar to the composite diagrams shown in Fig. 11.4. Since exchanging all the particles of one sub-diagram with those of the other results in the same composite diagram, we need to multiply $S^2(r_{12})$ by a factor of $1/2$ to avoid double counting. Below, we will clarify how this $1/2$ factor arises by examining the diagrams in Fig. 11.4.

Let us suppose to build composite diagrams with the two $\boldsymbol{\alpha}$ and $\boldsymbol{\beta}$ diagrams of Fig. 11.4. The contribution of the diagrams obtained by iterating twice $\boldsymbol{\alpha}$ and $\boldsymbol{\beta}$ is given by $(\boldsymbol{\alpha}+\boldsymbol{\beta})^2 = \boldsymbol{\alpha}^2+\boldsymbol{\beta}^2+2\boldsymbol{\alpha}\boldsymbol{\beta}$. The diagrams $\boldsymbol{\alpha}^2$ and $\boldsymbol{\beta}^2$ have symmetry 1/2. In the evaluation of the numerator (11.22) it necessary to insert this factor since by interchanging the i and j indexes one obtains the same contribution. On the other hand, the sums of Eq. (11.22) are restricted by the condition $j > i$ and this implies that only one of the two topologically identical terms appears. The same idea is valid for the $\boldsymbol{\beta}^2$ diagram. For example the expression for $\boldsymbol{\alpha}^2$ is

$$\sum_{i>j}\int dx_i h(r_{1i})h(r_{i2})\int dx_j h(r_{1j})h(r_{j2})$$
$$=\frac{1}{2}\sum_{i,j}\int dx_i h(r_{1i})h(r_{i2})\int dx_j h(r_{1j})h(r_{j2}) \;.$$

The diagrams $\boldsymbol{\alpha\beta}$ and $\boldsymbol{\beta\alpha}$ are topologically identical and they have to appear only once, therefore a factor $1/2$ must multiply $(\boldsymbol{\alpha} + \boldsymbol{\beta})^2$.

The same procedure can be used to build composite diagrams with n simple diagrams. The symmetry properties used to obtain the 1/2 factor for $S^2(r_{12})$ are valid for each $S^n(r_{12})$ and require the presence of a $1/n!$ factor to avoid multiple counting. The total sum of the composite diagrams can be expressed in terms of simple diagrams as:

$$C(r_{12}) = \frac{S^2(r_{12})}{2!} + \frac{S^3(r_{12})}{3!} + \frac{S^4(r_{12})}{4!} + \dots . \tag{11.33}$$

By using this result, the TBDF, Eq. (11.32) can be rewritten as:

$$\begin{aligned} g(r_{12}) &= f^2(r_{12})\,[1 + S(r_{12}) + C(r_{12})] \\ &= f^2(r_{12}) \left[1 + S(r_{12}) + \frac{S^2(r_{12})}{2!} + \frac{S^3(r_{12})}{3!} + \dots \right] \\ &= f^2(r_{12}) \exp[S(r_{12})] \, , \end{aligned} \tag{11.34}$$

where the last equality appears since the system has an infinite number of particles.

The above equation expresses the TBDF only in terms of simple diagrams which can be classified as *nodal* or *elementary*, sometime called *bridge*, diagrams.

(d) *Nodal and elementary (bridge) diagrams*

In a nodal diagram, there is at least one point through which all paths connecting one external point to the others must pass. These special internal points are referred to as *nodes*. Diagrams that do not contain nodes are called elementary. Examples of both nodal and elementary diagrams are presented in Fig. 11.5.

The calculation of the TBDF can be summarised as follows: there is no contribution from unlinked and reducible diagrams, while the contribution from composite diagrams can be expressed as an infinite sum of simple diagrams, as indicated by Eq. (11.34). Let $N(r_{12})$ represent the contribution of all the nodal diagrams, and $E(r_{12})$ that of the elementary diagrams. The TBDF can then be expressed as:

$$\begin{aligned} g(r_{12}) &= f^2(r_{12}) \exp[N(r_{12}) + E(r_{12})] \\ &= [1 + h(r_{12})]\,[1 + N(r_{12}) + E(r_{12}) + \dots] \\ &= 1 + N(r_{12}) + X(r_{12}) \, . \end{aligned} \tag{11.35}$$

This equation defines the diagrams contained in $X(r_{12})$, which are normally called non-nodal since they do not have nodes.

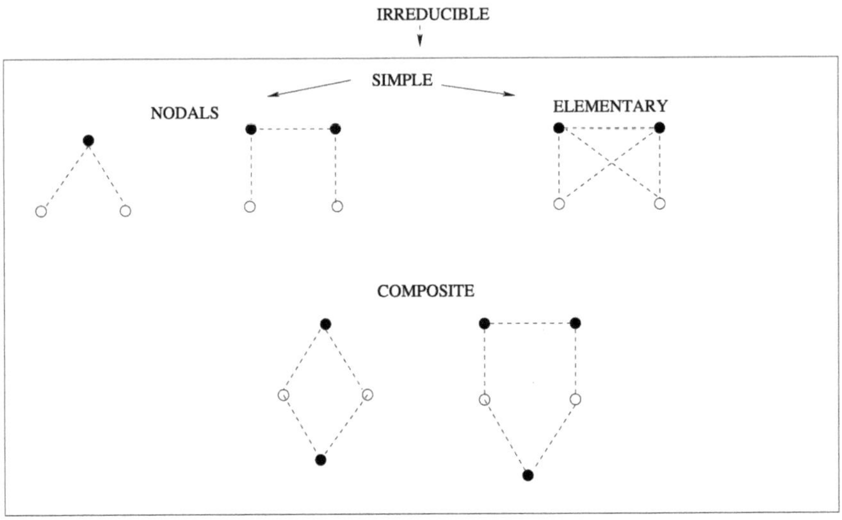

Fig. 11.5 Examples of irreducible diagrams, classified as composite and simple. The latter ones are classified as nodal and elementary

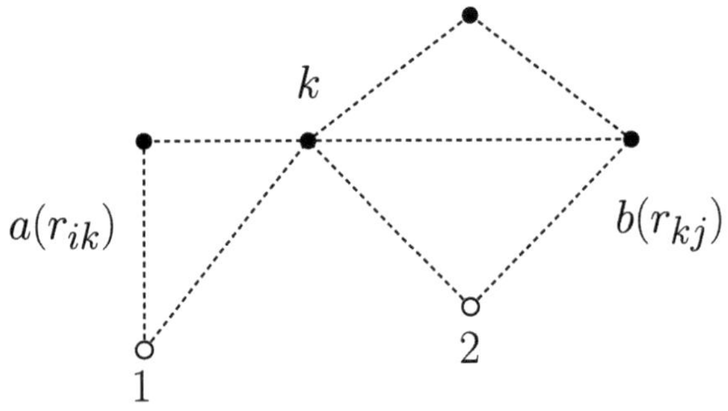

Fig. 11.6 Example of nodal diagram. The contribution of the part of the diagram to the left of the node k is called $a(r_{1k})$ and that to the right as $b(r_{k2})$

$$
\begin{aligned}
X(r_{12}) &= g(r_{12}) - 1 - N(r_{12}) \\
&= h(r_{12})[1 + N(r_{12})] + [1 + h(r_{12})](E(r_{12}) + C(r_{12})]. \qquad (11.36)
\end{aligned}
$$

A nodal diagram can be viewed as composed of interconnected parts at the nodal point. Consequently, each nodal diagram can be derived by integrating two or more functions that represent the contributions of the various components within the diagram. For example, the nodal diagram shown in Fig. 11.6 togliere la virgola has points 1 and 2 as external points and point k as a node. The contributions from the two parts of the diagram are denoted as $a(r_{1k})$ and $b(r_{k2})$. The total contribution to the TBDF (11.35) for this diagram is:

$$
\int dx_k \, a(r_{ik}) b(r_{kj}) \rho(x_k), \qquad (11.37)
$$

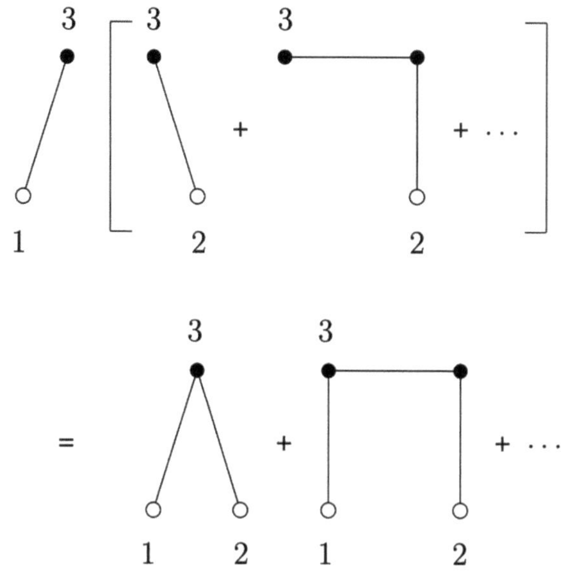

Fig. 11.7 Graphical representation of Eq. (11.38). The full lines indicate the X function defined in Eq. (11.36)

Here, the density $\rho(x_k)$ is associated with the integration point x_k to ensure proper normalisation. In the case we are considering, due to the homogeneity and translational invariance of the bosonic system, the density remains constant. As a result, it can be factored out of the integral.

The global contribution $N(r_{12})$ of all the nodal diagrams between the points 1 and 2 can be obtained as folding product in the node x_k of all the irreducible diagrams which can be built between 1 and k and between k and 2:

$$N(r_{12}) = \int dx_k \, X(r_{1k})\rho(x_k)[N(r_{k2}) + X(r_{k2})] \, . \tag{11.38}$$

Each nodal diagram contains at least one node, and every path connecting the external points 1 and 2 must pass through all the nodes in the diagram. The equation above indicates that the section of the diagram between point 1 and the first node k, which is a non-nodal diagram, must be folded at x_k in one of two ways:
i) with a non-nodal diagram, resulting in a diagram with a single node, or
ii) with a nodal diagram, which produces a diagram with multiple nodes. Folding $N(r_{ik})$ with $N(r_{kj})$ would create diagrams that are already included in $N(r_{ij})$ and, therefore, this option is not considered.

We show in Fig. 11.7 a representation of Eq. (11.38). In this figure the $X(r_{ij})$ function defined in Eq. (11.36) is indicated with a full line joining the points i and j.

At the top of the figure, a diagram featuring the external point 1 must be combined with the diagrams in the square brackets, all of which include the external point 2. The folding point is the internal point 3. This folding results in the diagrams shown in the lower part of the figure.

The first diagram in the lower section contains one nodal point. This is derived from combining the complete set of non-nodal functions $X(r_{13})$ and $X(r_{32})$. The second diagram features two nodal points and is obtained by combining the function $X(r_{13})$ with a nodal diagram contained within $N(r_{32})$.

The set of Eqs. (11.35), (11.36), and (11.38) is referred to as the **HyperNetted Chain** (HNC) equations. Equation (11.35) expresses the TBDF in terms of simple diagrams by summing the composite diagrams in a closed form. Meanwhile, Equation (11.38) calculates the contributions from all the nodal diagrams. Unfortunately, there is no closed expression available for calculating the elementary diagrams; they must be included individually. Calculations of the TBDF that do not account for the contributions of the elementary diagrams are termed HNC/0. When the contribution of the simplest elementary diagram, as shown in Fig. 11.5, is included, the calculation of the TBDF is referred to as HNC/4, since this diagram involves four particles.

The HNC equations are solved using an iterative procedure. The input required for the calculation is the correlation function $f(r_{12})$. Starting with the ansatz $N(r_{12}) = E(r_{12}) = 0$, which leads to $X(r_{12}) = f^2(r_{12}) - 1$, we generate a new set of nodal diagrams using Eq. (11.38). With this new set of nodal diagrams, we calculate $g(r_{12})$ using Eq. (11.35). This calculation allows us to evaluate the non-nodal $X(r_{12})$ diagrams through Eq. (11.36), which are then reused in Eq. (11.38) to compute another set of nodal diagrams. This iterative process continues until convergence is achieved.

11.2.1 The Calculation of the Energy

The solution of the HNC provides the TBDF which is used to obtain the energy value (11.1) with the hamiltonian

$$\hat{H} = -\sum_{i=1}^{A} \frac{\hbar^2}{2m} \nabla_i^2 + \sum_{i<j=1}^{A} \hat{V}_{ij} \,. \tag{11.39}$$

The expectation value of the energy is calculated by using the trial wave function

$$\Psi_{\rm T}(1,\ldots,A) = F(1,\ldots,A)\Phi(1,\ldots,A) = \mathcal{S}\Big(\prod_{i<j} f_{ij}\Big)\Phi(1,\ldots,A) \,. \tag{11.40}$$

The calculation of the contribution from the two-body potential term V_{ij} is straightforward, as indicated by Eq. (11.8). However, the evaluation of the kinetic energy term is more complex. The commonly adopted procedure divides this contribution into three parts

$$\langle T \rangle \equiv T_\phi + T_F - T_{c.m.} \; . \tag{11.41}$$

The first part is a term where the derivatives act on the MF wave function Φ

$$T_\phi \equiv -\frac{\hbar^2}{4m}\left(\langle \Phi^* F^2 \sum_{i=1}^{A} \nabla_i^2 \Phi \rangle - \langle \sum_{i=1}^{A} (\nabla_i \Phi^*) \cdot F^2 \nabla_i \Phi \rangle \right) , \tag{11.42}$$

the second part is a term where the derivatives act on the correlation function

$$T_F \equiv -\frac{\hbar^2}{4m} \langle \Phi^* \Big[F\Big(\sum_{i=1}^{A} \nabla_i^2 F \Big) - \sum_{i=1}^{A} (\nabla_i F)^2 \Big] \Phi \rangle \; , \tag{11.43}$$

and the third part is a term giving the center of mass contribution:

$$T_{c.m.} = -\frac{\hbar^2}{2mA} \langle \Psi_{\rm T}^* \Big(\sum_{i=1}^{A} \nabla_i \Big)^2 \Psi_{\rm T} \rangle \; . \tag{11.44}$$

In the previous equations the symbol $\langle\rangle$ indicates

$$\langle X \rangle = \frac{\int dx_1, \ldots, dx_A X(x_1, \ldots, x_A)}{\langle \Psi_{\rm T} | \Psi_{\rm T} \rangle} \; . \tag{11.45}$$

The term T_F is handled in analogous manner of the two-body term of the interaction $\hat{V}_{ij}$ and inserted in Eq. (11.8). The evaluation of the other terms is more complicated, but it does not present special problems.

11.3 The Fermi Hypernetted Chain (FHNC) Equations

In describing a fermionic system, it is essential to take into account the Pauli exclusion principle. The extension of the HNC theory to these systems is referred to as the Fermi HyperNetted Chain (FHNC). In this context, the MF wave function Φ (as shown in Eq. 11.2) is a Slater determinant of single-particle wave functions ϕ_a.

The expression of the TBDF (11.7) contains the squared module of the uncorrelated wave function. The $|\Phi|^2$ can be expressed as:

$$|\Phi(1, 2, \ldots, A)|^2 = \begin{vmatrix} \rho_0(x_1, x_1) & \rho_0(x_1, x_2) & \ldots & \rho_0(x_1, x_A) \\ \rho_0(x_2, x_1) & \rho_0(x_2, x_2) & \ldots & \rho_0(x_2, x_A) \\ \vdots & \vdots & \ddots & \vdots \\ \rho_0(x_A, x_1) & \rho_0(x_A, x_2) & \ldots & \rho_0(x_A, x_A) \end{vmatrix} , \tag{11.46}$$

were the elements of the above determinant have been defined as:

$$\rho_0(x_i, x_j) = \sum_a \phi_a^*(x_i)\phi_a(x_j) \; . \tag{11.47}$$

The sum is performed over all occupied single-particle states, which, for the ground state, include all states below the Fermi surface. Eq. (11.47) defines the uncorrelated One-Body Density Matrix (OBDM), which serves as the fundamental component in the calculation of the TBDF in fermionic systems.

A fundamental property of the OBDM, due to the orthonormality of the single-particle wave functions, is:

$$\int dx_j \rho_0(x_i, x_j)\rho_0(x_j, x_k) = \rho_0(x_i, x_k) \; , \tag{11.48}$$

where with the integral sign we indicate the integration on the space coordinate and also the sum on the third components of spin, and eventually, of isospin.

The sub-determinants are defined as:

$$\Delta_p(1, ..., p) = \begin{vmatrix} \rho_0(x_1, x_1) & \rho_0(x_1, x_2) & \dots & \rho_0(x_1, x_p) \\ \rho_0(x_2, x_1) & \rho_0(x_2, x_2) & \dots & \rho_0(x_2, x_p) \\ \vdots & \vdots & \ddots & \vdots \\ \rho_0(x_p, x_1) & \rho_0(x_p, x_2) & \dots & \rho_0(x_p, x_p) \end{vmatrix} \quad p \leq A \; . \tag{11.49}$$

For the property (11.48) of the OBDM, the sub-determinants satisfy the following relation

$$\int dx_{p+1}\Delta_{p+1}(1, ..., p+1) = (A-p)\Delta_p(1, ..., p) \; , \tag{11.50}$$

and, by iterating it we have that

$$\int dx_{p+1}...dx_A \Delta_A(1, ..., A) = (A-p)!\, \Delta_p(1, ..., p) \; . \tag{11.51}$$

This implies that:

$$\Delta_p = 0 \; , \qquad \text{if} \qquad p > A \; . \tag{11.52}$$

As example, we show these properties in the Δ_3 case.

$$\Delta_3(1,2,3) = \begin{vmatrix} \rho_0(x_1,x_1) & \rho_0(x_1,x_2) & \rho_0(x_1,x_3) \\ \rho_0(x_2,x_1) & \rho_0(x_2,x_2) & \rho_0(x_2,x_3) \\ \rho_0(x_3,x_1) & \rho_0(x_3,x_2) & \rho_0(x_3,x_3) \end{vmatrix}$$
$$= \rho(x_1,x_1)\begin{vmatrix} \rho_0(x_2,x_2) & \rho_0(x_2,x_3) \\ \rho_0(x_3,x_2) & \rho_0(x_3,x_3) \end{vmatrix}$$
$$- \rho(x_1,x_2)\begin{vmatrix} \rho_0(x_2,x_1) & \rho_0(x_2,x_3) \\ \rho_0(x_3,x_1) & \rho_0(x_3,x_3) \end{vmatrix}$$
$$+ \rho(x_1,x_3)\begin{vmatrix} \rho_0(x_2,x_1) & \rho_0(x_2,x_2) \\ \rho_0(x_3,x_1) & \rho_0(x_3,x_2) \end{vmatrix} .$$

For the normalisation of the OBDM and the property (11.48) we obtain

$$\begin{aligned} \int d^3r_3\, \Delta_3(1,2,3) &= \rho(x_1,x_1)\rho(x_2,x_2)A - \rho(x_1,x_1)\rho(x_1,x_2) \\ &- \rho(x_1,x_2)\rho(x_2,x_1)A + \rho(x_1,x_2)\rho(x_2,x_1) \\ &+ \rho(x_2,x_1)\rho(x_1,x_2) - \rho(x_1,x_1)\rho(x_2,x_2) \\ &= (A-1)\rho(x_1,x_1)\rho(x_2,x_2) - (A-1)\rho(x_2,x_1)\rho(x_1,x_2) \\ &= (A-1)\begin{vmatrix} \rho_0(x_1,x_1) & \rho_0(x_1,x_2) \\ \rho_0(x_2,x_1) & \rho_0(x_2,x_2) \end{vmatrix}. \end{aligned}$$

The properties of the OBDM and of the sub-determinants mentioned above depend solely on the orthonormality of the single-particle wave functions, rather than their specific expression. Therefore, these properties are applicable to both infinite and finite fermion systems.

We write the TBDF (11.7) as

$$g(x_1,x_2) = \frac{\mathcal{N}}{\mathcal{D}}. \tag{11.53}$$

By utilizing the trial wave function (11.2) in conjunction with the Jastrow ansatz (11.3), along with the definition of the h-function (11.12) and the definition of the sub-determinant (11.49), both the numerator, $\mathcal{N}$, and the denominator, $\mathcal{D}$, can be expressed as sums of terms that are characterized by the number of h-functions involved.

$$\mathcal{N} = \frac{A(A-1)}{\rho^2} f^2(r_{12})$$
$$\int dx_3 \dots dx_A \left(1 + \sum_{i<j} h_{ij} + \sum_{i<j<k} h_{ij}h_{jk} + \dots \right) \Delta_A \, , \tag{11.54}$$

$$\mathcal{D} = \int dx_1 \dots dx_A \left(1 + \sum_{i<j} h_{ij} + \sum_{i<j<k} h_{ij}h_{jk} + \dots \right) \Delta_A \, . \tag{11.55}$$

In every term present in $\mathcal{N}$ and $\mathcal{D}$, it is possible to perform the integration over the x variables that are not accounted for by the h functions. In these calculations, we utilized the property (11.50) of the sub-determinants.

We express the numerator as

$$\mathcal{N} = A! \frac{f^2(r_{12})}{\rho^2}$$

$$\sum_{p=2}^{A} \frac{1}{(p-2)!} \int dx_3 \dots dx_p W^{(p)}(1, 2; \dots, p) \Delta_p(1, \dots, p), \qquad (11.56)$$

where the functions $W^{(p)}$ indicate the sum of all the products of h functions involving p points, two of them are the external point x_1 and x_2. For example

$$W^{(3)}(1, 2; i) = h_{1i} + h_{2i} + h_{1i} h_{2i} ,$$

and

$$W^{(4)}(1, 2; i, j) = h_{ij} + h_{1i} h_{1j} + h_{2i} h_{2j} + h_{1i} h_{ij} + h_{2i} h_{ij} + h_{1i} h_{2j}$$
$$+ h_{1i} h_{2j} h_{ij} + h_{1i} h_{1j} h_i + h_{1i} h_{1j} h_{2j} h_{2i} h_{ij} + \cdots ,$$

The graphical representation of each term in Eq. (11.56) must take into account not only the *dynamical correlations* h, which are included in the $W^{(p)}$ term, but also the *statistical correlations* arising from the Pauli exclusion principle, represented by the sub-determinant Δ_p. To facilitate this, we introduce a new graphical symbol for the OBDM $\rho_0(x_i, x_j)$, depicted as an oriented line connecting the points x_i and x_j. Since Δ_p is a determinant of the OBDM, the oriented lines form closed loops that do not overlap and touch the $q \leq p$ particles involved. The determinant structure of Δ_p indicates that the contribution of each loop is characterised by a sign of $(-1)^{q-1}$.

We consider here below the case of Δ_3.

$$\Delta_3(1, 2, 3) = \begin{vmatrix} \rho_0(x_1, x_1) & \rho_0(x_1, x_2) & \rho_0(x_1, x_3) \\ \rho_0(x_2, x_1) & \rho_0(x_2, x_2) & \rho_0(x_2, x_3) \\ \rho_0(x_3, x_1) & \rho_0(x_3, x_2) & \rho_0(x_3, x_3) \end{vmatrix} =$$

$$+ \rho_0(x_1, x_1) \rho_0(x_2, x_2) \rho_0(x_3, x_3) \longrightarrow \mathbf{A}$$
$$+ \rho_0(x_1, x_2) \rho_0(x_2, x_3) \rho_0(x_3, x_1) \longrightarrow \mathbf{B}$$
$$+ \rho_0(x_2, x_1) \rho_0(x_3, x_2) \rho_0(x_1, x_3) \longrightarrow \mathbf{C}$$
$$- \rho_0(x_1, x_3) \rho_0(x_2, x_2) \rho_0(x_3, x_1) \longrightarrow \mathbf{D}$$
$$- \rho_0(x_3, x_2) \rho_0(x_2, x_3) \rho_0(x_1, x_1) \longrightarrow \mathbf{E}$$
$$- \rho_0(x_2, x_1) \rho_0(x_1, x_2) \rho_0(x_3, x_3) \longrightarrow \mathbf{F}. \qquad (11.57)$$

The graphical representation of these six terms id given in Fig. 11.8.

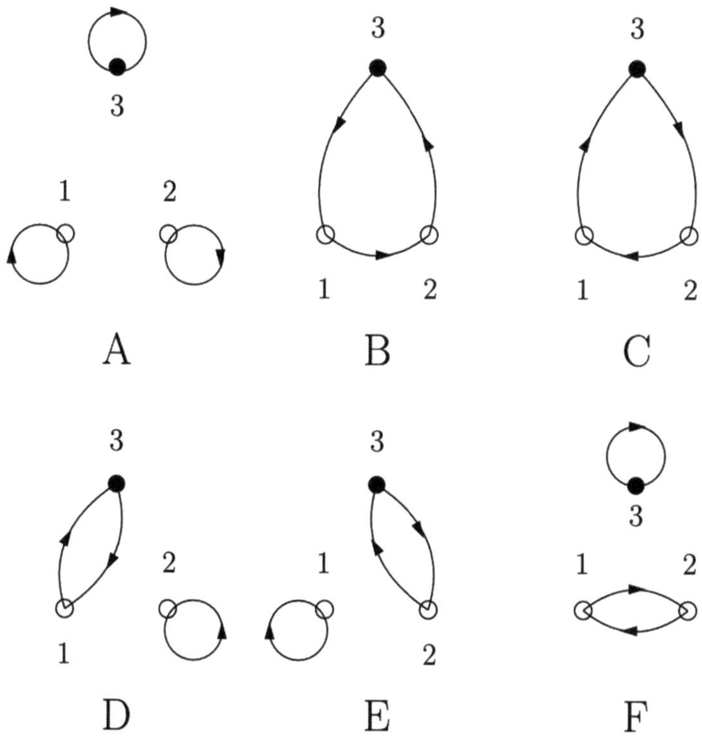

Fig. 11.8 Diagrams of Δ_3, Eq. (11.57)

We can analyse the denominator $\mathcal{D}$ in the same manner:

$$\mathcal{D} = A! \sum_{p=0}^{A} \frac{1}{p!} \int dx_1 ... dx_p Z^{(p)}(1, ..., p) \Delta_p(1, ..., p) \; . \tag{11.58}$$

where the $Z^{(p)}(1, ..., p)$ functions contain only internal points.

It is possible to formally extend the upper limits of the two sums of the numerators and the denominator to infinity by utilizing the property (11.52) of the subdeterminants.

In analogy to the bosonic case, at this stage of the calculation of $g(r_{12})$, a topological classification of the various diagrams present in $\mathcal{N}$ and $\mathcal{D}$ is necessary.

(a) *Linked and unlinked diagrams*

The diagrams illustrating the $W^{(p)}$ functions include terms that involve the external points x_1 and x_2. We refer to the portion of the diagrams where the external points are connected to a set of internal points, either through dynamical or statistical correlations, as *linked*. The sections of the diagrams that are not connected to the external points can be factored out in each integral of Eq. (11.59). Taking these considerations into account, we express the numerator $\mathcal{N}$ as follows:

$$\mathcal{N} = \frac{f^2(r_{12})}{\rho^2}\left[\Delta_2(1,2) + \sum_{p=3}^{\infty}\frac{1}{(p-2)!}\int dx_3...dx_p\,\mathcal{L}_p(1,2;...,p)\right]$$
$$\left[A!\sum_{q=1}^{\infty}\frac{1}{q!}\int dx_3\ldots dx_q\,Z^{(q)}(1,2;\ldots,q)\Delta_p(1,...,q)\right]. \quad (11.59)$$

In the expression above, $\mathcal{L}_p$ represents the sum of all linked diagrams that connect p points through statistical or dynamical correlations. The second term in the equation is identical to the expression for the denominator, as shown in Eq. (11.59). Consequently, the diagrams in the denominator cancel out the factorised unlinked portion of the diagrams in the numerator. The TBDF can be expressed by considering only the linked diagrams $\mathcal{L}_n(1, 2, i_3, ..., i_n)$

$$g(x_1, x_2) = g(r_{12})$$
$$= \frac{f^2(r_{12})}{\rho^2}\left[\Delta_2(1,2) + \sum_{p=3}^{\infty}\frac{1}{(p-2)!}\int dx_3...dx_p\,\mathcal{L}_p(1,2;...,p)\right]. \quad (11.60)$$

(b) *Reducible diagrams*
In bosonic systems, the contribution from reducible diagrams is canceled by the denominator, up to a factor of $1/A$. In contrast, in fermionic systems, this contribution is exactly zero in infinite systems. This concept is illustrated in Fig. 11.9.

In an infinite system, for any given linked diagram, such as diagram **A** in Fig. 11.9, it is always possible to find another diagram that differs only by the presence of a statistical loop connecting two internal points, like diagram **B** in the figure. Due to the properties of the sub-determinants, the contribution of diagram **B** is identical to that of diagram **A**, but with a factor of -1. This factor represents the contribution from the two-point closed loop. As indicated in the lower part of the figure, the total contribution sums to zero. A more formal proof can be found in Ref. [5].

(c) *Simple and composite diagrams*
Similar to the bosonic system, composite diagrams are constructed by connecting simple diagrams at their external points. In the fermionic case, it is essential to consider not only the dynamical correlations but also the statistical correlations. The overall contribution of the composite diagrams is determined by summing the simple diagrams, which are categorised into nodal and elementary types.

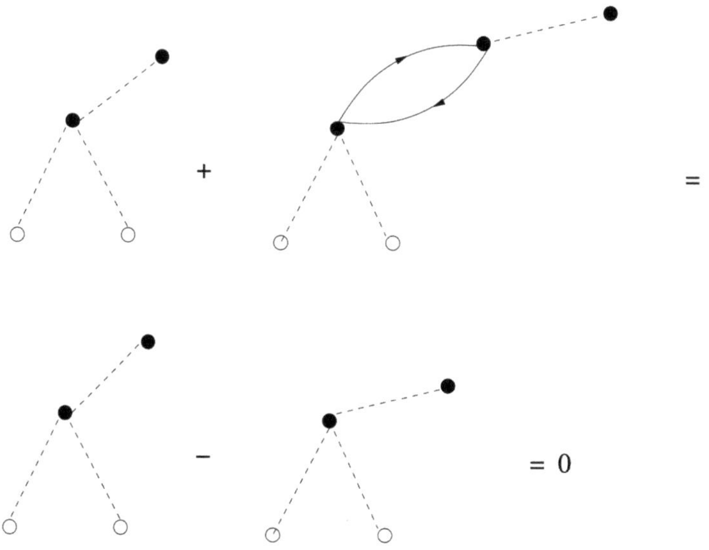

Fig. 11.9 Example of cancellation of two FHNC reducible diagrams The statistical loop in the higher part introduces a minus sign and, therefore, the total contribution of the two diagrams is zero

(d) *Nodal and elementary*
The definitions of nodal and elementary diagrams are similar to those in the bosonic case, but they take into account both dynamical and statistical correlations. The presence of statistical correlations prevents the formulation of a single integral equation for calculating all the nodal diagrams, as was done in Eq. (11.38). However, it is possible to establish a system of interconnected integral equations that describe the contributions of the nodal diagrams, categorised by the type of correlation reaching the external points [5].

The graphical examples of the various types of diagrams needed to derive the different integral equations are presented in Fig. 11.10. In diagrams **A** and **B**, the external points are reached solely through dynamical correlations. For this reason, they are labeled with the sub-indexes *dd* (dynamical-dynamical). Diagrams **C** and **D** feature dynamical correlations reaching point 1 and two statistical correlations reaching point 2. The sub-indexes for these diagrams are *de* (dynamical-exchange). Diagrams **E** and **F** are designated as *ee* (exchange-exchange) because, in both cases, the external points are accessed via statistical lines. It is also useful to define diagrams that are reached by a statistical correlation starting from point 1 and arriving at point 2, forming an open ring. These diagrams are referred to as *cc* (cyclic-cyclic) and do not directly contribute to the calculation of the TBDF.

As in the bosonic case, as seen in Eq. (11.38), the total contribution of the nodal diagrams in the present scenario can also be obtained by performing a folding integral of the various terms associated with the diagrams at the nodal point. In fermionic

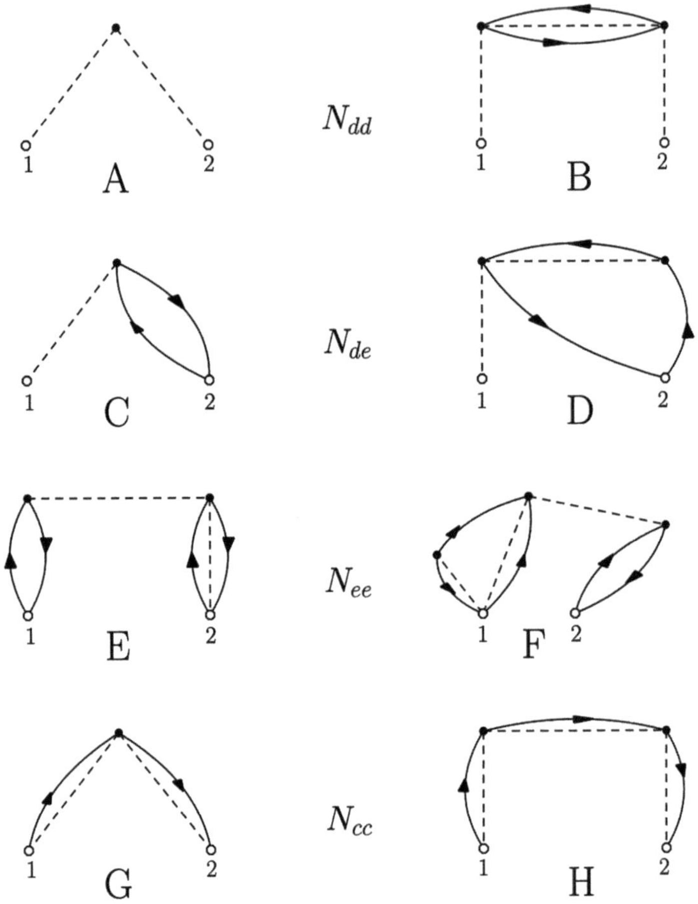

Fig. 11.10 The different types of nodal diagrams needed for the FHNC equations. The sub-indexes categorise the diagrams based on the type of correlation that reaches the external points 1 and 2

systems, the Pauli exclusion principle prevents the folding of diagrams of the same type. For instance, a diagram of type e can only be folded with diagrams of type d at that point.

Let us denote the sum of all the nodal diagrams as N and the sum of the non-nodal irreducible diagrams as X. Clearly, N and X can be identified by the sub-indexes $dd, de, ee,$ and cc. For the nodal diagrams, we can express the following set of equations

$$N_{dd}(r_{12}) = \int d^3r_3[X_{dd}(r_{13}) + X_{de}(r_{13})]\rho(\mathbf{r}_3)[N_{dd}(r_{32}) + X_{dd}(r_{32})] + \int d^3r_3[X_{dd}(r_{13})]\rho(\mathbf{r}_3)[N_{ed}(r_{32}) + X_{ed}(r_{32})] \, ,$$

$$N_{de}(r_{12}) = \int d^3r_3[X_{dd}(r_{13}) + X_{de}(r_{13})]\rho(\mathbf{r}_3)[N_{de}(r_{32}) + X_{de}(r_{32})] + \int d^3r_3 X_{dd}(r_{13})\rho(\mathbf{r}_3)[N_{ee}(r_{32}) + X_{ee}(r_{32})] \, ,$$

$$N_{ee}(r_{12}) = \int d^3r_3[X_{ed}(r_{13}) + X_{ee}(r_{13})]\rho(\mathbf{r}_3)[N_{de}(r_{32}) + X_{de}(r_{32})] + \int d^3r_3 X_{ed}(r_{13})\rho(\mathbf{r}_3)[N_{ee}(r_{32}) + X_{ee}(r_{32})] \, ,$$

$$N_{cc}(r_{12}) = \int d^3r_3 X_{cc}(r_{13})\rho(\mathbf{r}_3) [N_{cc}(r_{32}) + X_{cc}(r_{32}) - \rho_0(\mathbf{r}_3, \mathbf{r}_2)] \, . \tag{11.61}$$

The equations for the non-nodal diagrams are:

$$\begin{aligned}
X_{dd}(r_{12}) &= g_{dd}(r_{12}) - N_{dd}(r_{12}) - 1 \, , \\
X_{de}(r_{12}) &= g_{dd}(r_{12})[N_{de}(r_{12}) + E_{de}(r_{12})] - N_{de}(r_{12}) \, , \\
X_{ee}(r_{12}) &= g_{dd}(r_{12})\{N_{ee}(r_{12}) + E_{ee}(r_{12}) + [N_{de}(r_{12} + E_{de}(r_{12})]^2 \\
&\quad -\nu[N_{cc}(r_{12}) + E_{cc}(r_{12}) - \rho_0(\mathbf{r}_1, \mathbf{r}_2)]^2\} - N_{ee}(r_{12}) \, , \\
X_{cc}(r_{12}) &= g_{dd}(r_{12})[N_{cc}(r_{12}) + E_{cc}(r_{12}) - \rho_0(\mathbf{r}_1, \mathbf{r}_2)] \\
&\quad - N_{cc}(r_{12}) + \rho_0(\mathbf{r}_1, \mathbf{r}_2) \, .
\end{aligned} \tag{11.62}$$

where we have indicated with ν the particle degeneracy, i.e. $\nu = 2$ for fermions with spin $1/2$ and $\nu = 4$ for nucleons.

The partial definitions of the TBDF are:

$$\begin{aligned}
g_{dd}(r_{12}) &= f^2(r_{12}) \exp[N_{dd}(r_{12}) + E_{dd}(r_{12})] \, , \\
g_{de}(r_{12}) &= N_{de}(r_{12}) + X_{de}(r_{12}) \, , \\
g_{ed}(r_{12}) &= g_{de}(r_{12}) \, , \\
g_{ee}(r_{12}) &= N_{ee}(r_{12}) + X_{ee}(r_{12}) \, , \\
g_{cc}(r_{12}) &= N_{cc}(r_{12}) + X_{cc}(r_{12}) - \rho(\mathbf{r}_1, \mathbf{r}_2) \, .
\end{aligned} \tag{11.63}$$

The total TBDF is obtained by summing the partial terms

$$g(r_{12}) = g_{dd}(r_{12}) + g_{ed}(r_{12}) + g_{de}(r_{12}) + g_{ee}(r_{12}). \tag{11.64}$$

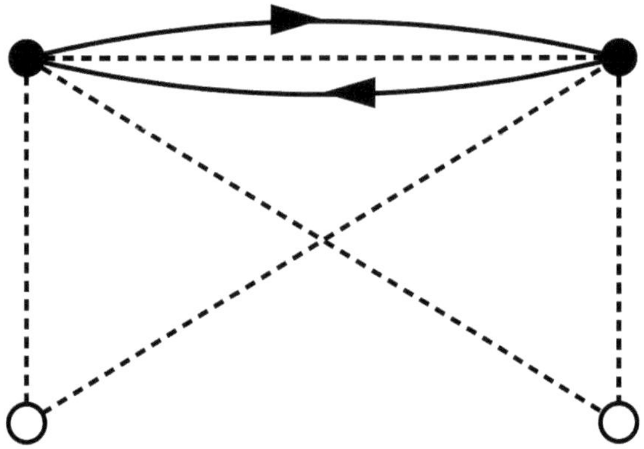

Fig. 11.11 Four-points elementary diagram

The system of Eqs. (11.61), (11.62), (11.63), and (11.64) constitutes the **Fermi HyperNetted Chain** (FHNC) equations. Similar to the bosonic HNC equations, the FHNC framework enables the calculation of all composite and nodal diagrams in a closed form. However, elementary diagrams, such as the one depicted in Fig. 11.11, are excluded and must be calculated individually. In line with the HNC approach, it is common practice to use nomenclature such as FHNC/0, FHNC/4, and so on, to indicate the specific elementary diagrams included in the calculations.

In this case, the FHNC equations are solved using an iterative procedure similar to the one employed for solving the HNC equations. Once the TBDF has been obtained, the energy of the system is calculated following the same methods outlined in Sect. 11.2.1.

The FHNC theory has been applied to fermionic liquid helium [7–9]. In this case, the helium atoms consist of ^{3}He nuclei rather than ^{4}He. The properties of fermionic liquid helium differ significantly from those of the bosonic case, as the two liquids adhere to different quantum statistics. In both scenarios, the interaction is governed by the purely scalar Coulomb interaction, as discussed in Sect. 3.4.

11.3.1 Operator Dependent Correlations

The Jastrow ansatz (11.3) considers purely scalar correlations. However, this is insufficient to capture the complexity of fermionic systems. The polarisation effects in fermionic liquid helium necessitate the use of spin-dependent correlations [8, 9]. The situation becomes even more complex when applying the CBF approach to nuclear systems. As we pointed out in Sect. 3.3, the nuclear interaction can only be accurately described by considering its dependence on various aspects of the two interacting particles, not just their relative distance.

For this reason the Jastrow's ansatz is modified by adding correlation terms depending on the operators used to describe the nuclear hamiltonian (3.2)

$$\mathcal{F}(1, ..., A) = \mathcal{S}\Big(\prod_{j>i=1}^{A} F_{ij}\Big) = \mathcal{S}\Big(\prod_{j>i=1}^{A} \sum_{p=1}^{6} f_p(r_{ij})\hat{O}_{ij}^{p}\Big) . \tag{11.65}$$

with the operators defined as

$$\hat{O}_{ij}^{p=1,6} = \hat{\mathbb{I}}, \hat{\boldsymbol{\tau}}_i \cdot \hat{\boldsymbol{\tau}}_j, \hat{\boldsymbol{\sigma}}_i \cdot \hat{\boldsymbol{\sigma}}_j, (\hat{\boldsymbol{\sigma}}_i \cdot \hat{\boldsymbol{\sigma}}_j)(\hat{\boldsymbol{\tau}}_i \cdot \hat{\boldsymbol{\tau}}_j), \hat{S}_{ij}, \hat{S}_{ij}(\hat{\boldsymbol{\tau}}_i \cdot \hat{\boldsymbol{\tau}}_j) . \tag{11.66}$$

The step (11.12) defining the h-function, and the following expressions (11.13) and (11.14), have a universal value because the correlation was a scalar function. Now it is necessary to consider that the various terms of the correlation do not necessarily commute with the terms of the hamiltonian and neither between them. This fact does not allow the formulation of equations which sum in closed form all the diagrams of a certain type. It is necessary to introduce an approximation.

It is possible to sum in a closed form diagrams containing chains of operator-dependent correlations which have only one operator-dependent correlation line between two points, as in the diagrams of Fig. 11.12. This approximation is called Single Operator Chain (SOC) [10]. The FHNC/SOC theory with the inclusion of the elementary diagram of Fig. 11.11 is the most elaborated theory of CBF type used in nuclear physics.

This theoretical framework is adequate to treat hamiltonians with two-body interactions only. The insertion of three-body forces is not included in the diagrammatic expansion scheme of the FHNC, but, analogously to the elementary diagrams, is inserted by considering the single diagrams.

11.3.2 Infinite Nuclear Matter

As example of application of the CBF theory we discuss here some results obtained for nuclear matter (see [11]). The system under study has translational invariance and constant nucleonic density defined by the sum $\rho = \rho_{\rm p} + \rho_{\rm n}$ of the, constant, proton and neutron densities (see Sect. 8.5). The energy per nucleon $e = E/A$ is traditionally written in power expansion with respect to the asymmetry parameter $\delta = (\rho_{\rm n} - \rho_{\rm p})/\rho$, i.e.

$$e(\rho, \delta) = e(\rho, 0) + e_{\rm sym}(\rho)\, \delta^2 + \mathcal{O}(\delta^4) . \tag{11.67}$$

Around the minimum of symmetric nuclear matter, $\delta = 0$, at the saturation density ρ_0, the two coefficients are expanded in powers of the parameter $\epsilon = (\rho - \rho_0)/(3\,\rho_0)$.

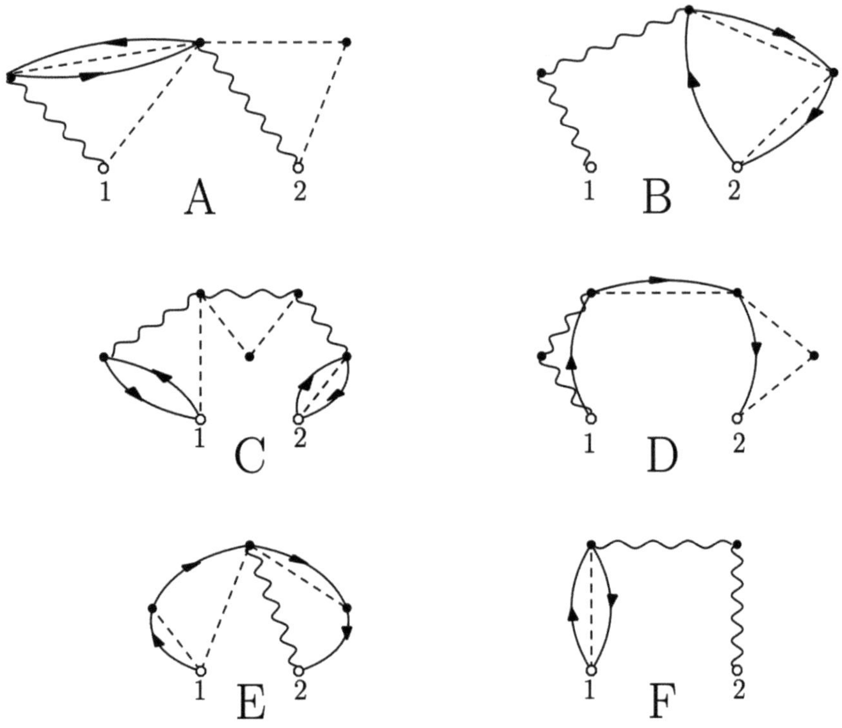

Fig. 11.12 Examples of the FHNC/SOC diagrams. The wavy lines represent operator dependent terms of the correlations

For the symmetric nuclear matter we obtain

$$e(\rho, 0) = a_V + \frac{1}{2} K_V \epsilon^2 + \ldots, \tag{11.68}$$

where the term of first order in ϵ, related to te first derivative with respect to ρ, is zero since $e(\rho, 0)$ has a minimum for $\rho = \rho_0$. In the quadratic term, related to the second derivative, the coeffcient, defined as

$$B = 9\rho_0^2 \left. \frac{\partial^2 e(\rho, 0)}{\partial \rho^2} \right|_{\rho=\rho_0}, \tag{11.69}$$

is called compression modulus, see Eq. (2.59).

The second coefficient of Eq. (11.67), called symmetry energy, is expanded as

$$e_{\text{sym}}(\rho) = a_{\text{sym}} + \mathcal{L}\,\epsilon + \ldots. \tag{11.70}$$

with the coefficient

$$\mathcal{L} = 3\rho_0 \left. \frac{\partial e_{\text{sym}}(\rho)}{\partial \rho} \right|_{\rho=\rho_0}. \tag{11.71}$$

Table 11.1 Properties of infinite nuclear matter obtained with different types of calculations. The saturation density ρ_0 is expressed in fm^{-3}. All the other quantities in MeV. The Monte Carlo results are those of Ref. [13], those of the CBF theory from Ref. [12]. The other results from [14]

	Exp	AFDMC	CBF	D1M	SLy5	DDME2
ρ_0	0.16 ± 0.01	0.16	0.16	0.16	0.16	0.15
$e(\rho_0, 0)$	-16.0 ± 0.1	–16.00	–16.00	–16.01	–15.98	–16.13
B	220 ± 30	276	269	217	228	278
$e_{\text{sym}}(\rho_0)$	30–35	31.3	33.94	29.45	32.66	33.20
$\mathcal{L}$	88 ± 25	60.10	58.08	25.41	48.38	54.74

Table 11.1 shows the values of these quantities calculated with different theories at the value of the saturation density ρ_0. The comparison is done with empirical values (exp). The values obtained with the CBF theory [12] are compared with those obtained by a calculation of Auxiliary Field Diffusion Monte Carlo (AFDMC) [13], both of them are using the same microscopic nucleon-nucleon interaction. The other results are obtained in MF calculation of Hartee-Fock type (see Sect. 15.2) by using different effective nucleon-nucleon interactions.

The values of the saturation densities and of the minimal energy agree within the 2% and 0.4%, respectively. Other variables have similar values in other calculations. The major differences are present in the values of $\mathcal{L}$.

We show in Fig. 11.13 the energies per particle as a function of the density, the so-called equations of state, for pure neutronic matter (a), symmetric nuclear matter (b), and symmetry energy (c). The various calculations agree at the saturation point of symmetric but they show remarkable differences away from this point, especially at higher densities. The agreement between the results of the CBF calculations and those of the AFDM reinforces the validity of the FHNC/SOC calculation scheme, which allows for the resolution of the equations of CBF theory.

11.4 Renormalised FHNC Equations

11.4.1 Finite Systems

The HNC and FHNC equations which we have presented so far have been developed by exploiting the characteristics of infinite and homogenous systems to which they have been applied, mainly bosonic and fermionic liquid helium and nuclear matter. Two of the basic hypotheses done in the previous section, infinite number of particles and translational invariance, are no longer valid in the description finite systems. One of the consequences of the loss of translational requires a change in the evaluation of the OBDM (11.47), and of its diagonal part, the number density, which is not any more constant. Another consequence is related to the cancelation of the reducible

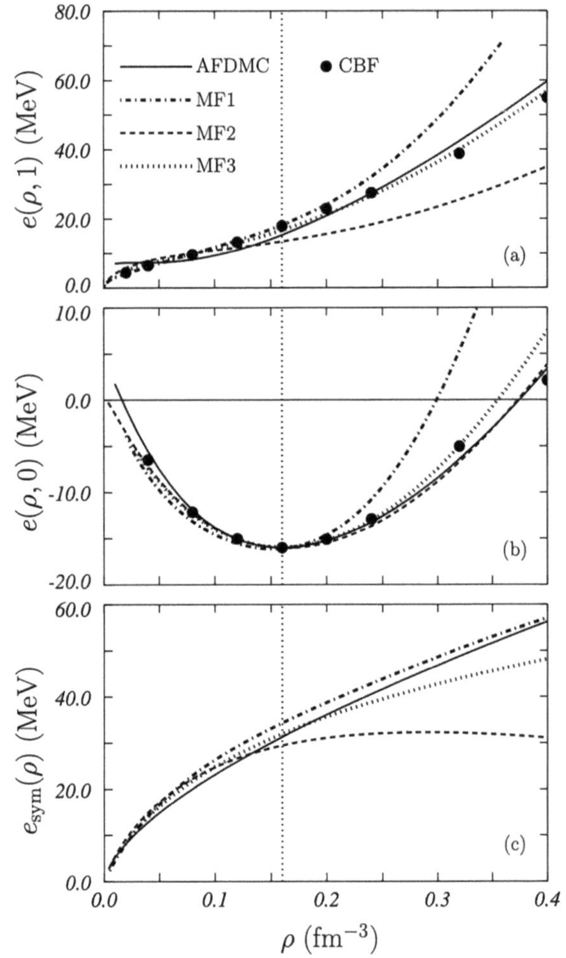

Fig. 11.13 Equation of state for pure neutron matter (a), symmetric neutron matter (b) and symmetry energy (c). The points show the CBF results [12], the full lines those of the AFDMC [13] and the other lines the results of MF calculations carried out with different effective nucleon-nucleon interactions [14]

diagrams which is not any more effective. These differences leads to a modified set of FHNC equations which are namend Renormalized FHNC (RFHNC) equations [15].

In infinite systems the single particle wave functions are plane waves (11.5) eventually multiplied by the Pauli spinors χ_s and χ_t to consider spin and isospin as it has been discussed in Sect. 2.3, see Eq. (2.38). In finite systems the single particle wave functions are obtained by solving one-body Schrödinger equations with a potential having a general symmetry around a fixed point. These single particle wave functions are characterised by angular momentum, and in the case of spin-orbit coupling, by total angular momentum, as we have discussed in Sect. 2.2.4. Since the single particle wave functions depend on the third component of the spin and of the isospin the uncorrelated OBDM (11.47) has to be expressed as

$$\rho_0(x_i, x_j) = \sum_{s,s',t} \rho_0^{ss't}(\mathbf{r}_i, \mathbf{r}_j)\chi_s^+(i)\chi_{s'}(j)\chi_t^+(i)\chi_t(j) \,, \tag{11.72}$$

where by using the symbols of Eq. (2.31) we defined the spatial part as:

$$\rho_0^{ss't}(\mathbf{r}_i, \mathbf{r}_j) = \sum_{nlj} R_{nlj}^t(r_i) R_{nlj}^t(r_j)$$
$$\sum_{\mu\mu' m} \langle l\mu \frac{1}{2} s | jm\rangle \langle l\mu' \frac{1}{2} s' | jm\rangle Y_{l\mu}^*(\Omega_i) Y_{l\mu'}(\Omega_j) \,. \tag{11.73}$$

The uncorrelated density of the system is obtained by considering the diagonal part of the OBDM, and it is not any more constant as it was in the case of the infinite system:

$$\rho_0^t(x_i) \equiv \rho_0^{sst}(\mathbf{r}_i, \mathbf{r}_i) = \frac{1}{4\pi} \sum_{nlj} (2j+1) \left| R_{nlj}^t(r_i) \right|^2 , \tag{11.74}$$

where t is the isopin third component which, in finite nuclear systems, differentiates the proton and neutron densities. In finite systems, the uncorrelated OBDM's describing finite systems do not depend only on the distance between the two particles i and j, $r_{ij} = |\mathbf{r}_i - \mathbf{r}_j|$, as in the infinite systems case. In any case, the properties (11.48), (11.50) and (11.51), relevant for the construction of the FHNC equations, remain valid.

11.4.2 The Vertex Corrections

The construction of the FHNC equations for the finite systems follows the steps used for the infinite systems. The minimization of the energy functional, Eq. (11.1), with the ansatz (11.2) on the wave function and (11.3) on the correlation, leads to the requirement of evaluating the TBDF (11.53).

The arguments used in Sects. 11.2 and 11.3 to show that the contributions of the unlinked diagrams of the numerator are simplified by the denominator, can be repeated also in the finite systems case [15]. The demonstration is done by formally extending up to infinity all the sums of the various cluster terms, since the property (11.51) of the sub-determinant Δ_p ensures that diagrams containing a number of particles greater than the number of particles forming the system, do not contribute.

In the infinite systems case, the next step was the elimination of the reducible diagrams. We have already said that this elimination is only approximated for boson systems, up the $1/A$ order, but it is exact for infinite fermion systems. The basic point of the demonstration for this latter case, was the possibility to associate to each reducible diagram, another diagram containing only one additional exchange loop. The contributions of these two diagrams to the TBDF differ only by a sign, therefore they cancel each other out. This cancellation mechanism is produced by two specific characteristics of the infinite system. The fact that for a given reducible diagram it is always possible to find another diagram having one additional particle, and one additional exchange loop, is ensured by the presence of an infinite number of particles. Thc translational invariance is instead responsible for the fact that the additional exchange loop contributes only with an overall minus sign. In the finite nuclei the number of particles is limited, and the translational invariance is lost, therefore there is no cancellation of the reducible diagrams.

However, even in finite systems it is possible to consider only irreducible diagrams in the cluster expansion by introducing the so-called vertex corrections [15]. A graphical representation of this idea is given in Fig. 11.14. Every reducible diagram can be thought as composed by two parts, as indicated by the diagrams **A** and **B** of the figure. A first part contains the external points and is the irreducible part of the diagram. A second, reducible, part contains only internal points, and it is linked to the irreducible part through the articulation point a. The total contribution of these connected, and reducible, diagrams to the TBDF can be written as the folding integral of the irreducible part with a function taking into account the contribution of all the diagrams connected to the articulation point and it is directly related to the density

$$\rho^{t_1}(\mathbf{r}_1) = \frac{n_{t_1}}{<\Psi|\Psi>} \int dx_2 \ldots dx_A \; \Psi^*(x_1, x_2, \ldots, x_A)\, P_1^{t_1} \times \; \Psi(x_1, x_2, \ldots, x_A) \;, \tag{11.75}$$

where n_t indicates the number of protons ($t = 1/2$) or of neutrons ($t = -1/2$), and the projector operator P^t selects the particle with isospin third component t.

It is necessary to distinguish the case where the irreducible part is linked to the articulation point only by dynamical correlations, as in the **A** diagram of the figure, from the case when there are statistical correlations joining the articulation point, as in the diagram **B**. To simplify the drawing, we show in the **A** diagram only a single dynamical correlation line connecting the irreducible part to the articulation point. In reality, there are no limitations on the number of dynamical correlations. In the case of the **B** diagram we show only the statistical lines connecting the articulation point, but also dynamical correlations may be present.

The fact to be considered is that the Pauli principle allows each point to be reached by no more than two exchange lines. When the articulation point is of type d, i.e. linked to the irreducible part of the diagram only by dynamical correlations, the Pauli principle is not active. In this case, the reducible part of the diagram can reach the articulation point with both dynamical and statistical correlations. We call $C^t(a)$ the sum of all the possible linked diagrams containing the articulation point a which has isospin third component t. We observe that $C^t(a)$ is the correlated density (11.75) of the nucleons with isospin third component t, therefore $C^t(a) = \rho^t(a)$.

The situation changes when the articulation point is of exchange type e, i.e. linked to the irreducible part of the diagram also by statistical correlations. In this case, because of the Pauli principle, the reducible part of the diagram can reach the artic-

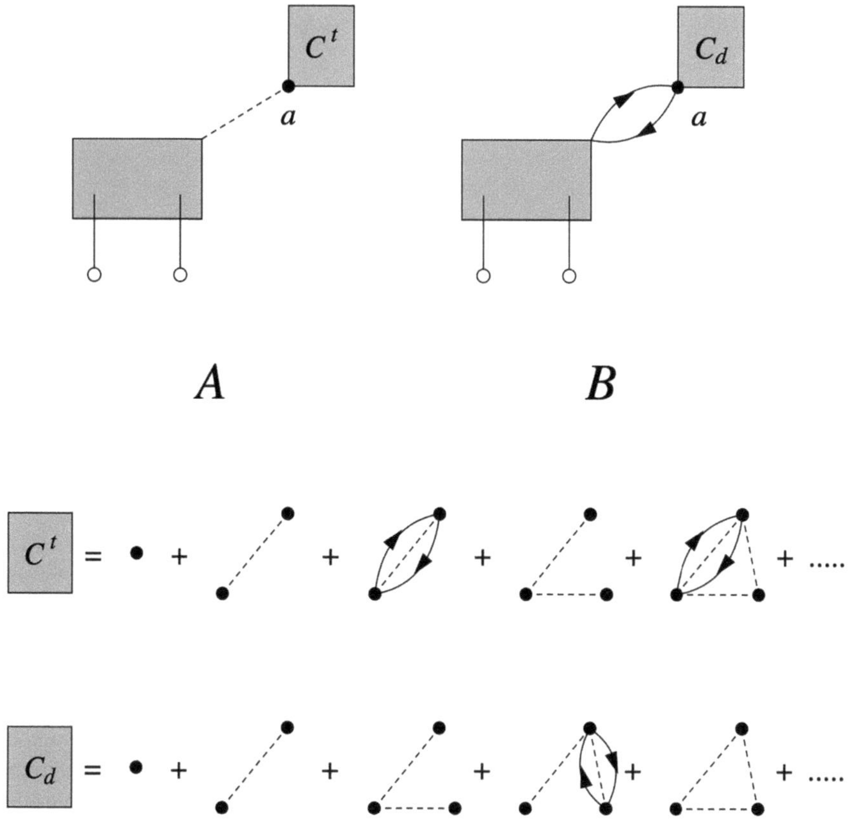

Fig. 11.14 Graphical illustration of the vertex corrections. Since the Pauli exclusion principle allows each point to be reached by no more than two exchange lines, we have to distinguish the reducible parts of the diagrams. In C_d the articulation point a can be reached only by dynamical correlations, while in C^t also by statistical correlations

ulation point exclusively with dynamical correlations. We call $C_d(a)$ the sum of the diagrams reaching the articulation point a with dynamical correlations only.

The evaluation of $C_d(a)$ can be done by extending the diagrams classification done in Sect. 11.2 to the case of a single external point. All the linked diagrams, both simple and composite ones, contribute to $C_d(a)$. As an example, the contribution of the (II) diagram of Fig. 11.14 is obtained by squaring the contribution of the (I) diagrams and dividing by two, in order to avoid double counting. The procedure used in Sect. 11.2 to calculate the contribution of the composite diagrams can be applied also in this case. If we call $U_d(a)$ the sum of all the simple irreducible diagrams connected to the point a by dynamical correlations only, see Fig. 11.15, we can write:

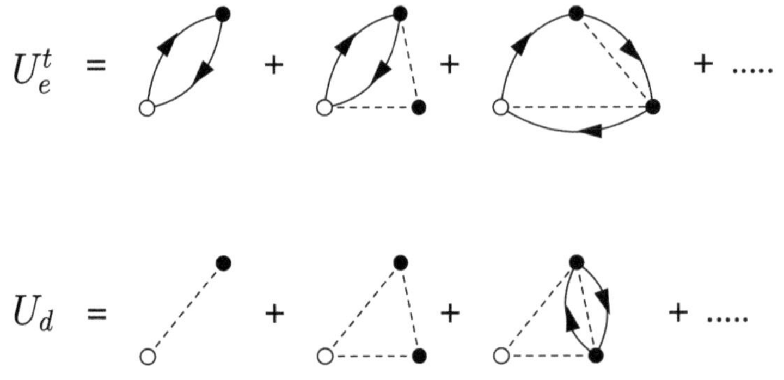

Fig. 11.15 Graphical illustration of the U_e^t and U_d diagrams used to evaluate the vertex corrections $C_d(a)$, Eq. (11.76), and $C^t(a)$, Eq. (11.77)

$$C_d(a) = 1 + U_d(a) + \frac{1}{2!}U_d^2(a) + \frac{1}{3!}U_d^3(a) + \cdots = \exp[U_d(a)]. \tag{11.76}$$

It is understood that all the diagrams contained in $U_d(a)$ are renormalised by the vertex corrections, therefore they must be irreducible in each internal point.

For the calculation of $C^t(a)$ we have to consider also the diagrams linked to the articulation point a with statistical correlations. We call $U_e^t(a)$ the sum of all these simple irreducible diagrams, see Fig. 11.15. Because of the Pauli principle, one can construct composite diagrams with $U_e^t(a)$ combining it only with any number of $U_d(a)$ that produces $C_d(a)$. By definition, $C^t(a)$ is given by all the diagrams contributing to $C_d(a)$, i.e. all those reaching a with dynamical correlations, plus the diagrams constructed by associating those with $U_e^t(a)$:

$$C^t(a) = C_d(a)\left[\rho_0^t(a) + U_e^t(a)\right] = \rho^t(a), \tag{11.77}$$

where $\rho_0^t(a)$ indicates the uncorrelated one-body density for nucleons with isospin t, and $\rho^t(a)$ is the corresponding correlated density (11.75). In the absence of correlations, U_d and U_e^t are zero, therefore C^t is equal to the uncorrelated density, as is expected.

The idea of the vertex corrections allows us to consider, again, the contribution of only the irreducible diagrams in the cluster expansion. Obviously the evaluation of the nodal diagrams is more involved than in the case of infinite systems. However, the basic ideas used to calculate the nodal diagrams in the infinite system are still valid in the present case, and the expressions are a rather straightforward extension of those presented for the infinite systems [16].

The treatment of the operator dependent correlations, in analogy to the infinite system approach, is done by using the SOC approximation. However, the treatment of these correlations presented in Sect. 11.3.1 cannot be straightforwardly extended to finite nuclear systems. First, the systems we want to describe are not saturated in isospin, and this changes the treatment of the isospin dependent terms. Second, the

Table 11.2 Ground state energies, in MeV, for simple models of ^{4}He and ^{16}O nuclei, see text for details. In the column $\langle V \rangle$ the contributions of the two-nucleon interaction are given, and in in the column $\langle T \rangle$ those of the kinetic energy term. The VMC results are from Ref. [18] (the uncertainties are $\sim \pm 0.2$ MeV)

		$\langle V \rangle$	$\langle T \rangle$	E
^{4}He	RFHNC	–101.1	63.4	–37.7
	VMC	-99.0	62.5	–36.5
^{16}O	RFHNC	–467.1	316.7	–150.4
	VMC	–464.9	314.0	–150.9

Table 11.3 Energy per particle, in MeV, for the doubly magic nuclei indicated, obtained with a RFHNC/1 calculation by using the Argonne $V8'$ two-nucleon interaction together with the Urbana IX three-body force [19]. The various raws indicate, respectively, T the contribution of the kinetic energy, $V^6_{2\text{-body}}$ the contribution of the two-body part of the interaction, V_{LS} the contribution of the spin-orbit terms of the interaction, V_{Coul} the contribution of the Coulomb term of the interaction, $V_{3\text{-body}}$ the contribution of the three-body term of the interaction, and E the total contribution. E_{exp} show the experimental values

	^{16}O	^{40}Ca	^{48}Ca	^{208}Pb
T	32.33	41.06	39.64	39.56
$V^6_{2\text{-body}}$	–38.15	–48.97	–46.60	–48.43
V_{LS}	–0.70	–0.85	–0.79	–0.80
V_{Coul}	0.86	1.96	1.57	3.97
$V_{3\text{-body}}$	0.86	1.76	1.61	1.91
E	–4.80	-5.05	–4.62	-3.78
E_{exp}	-7.97	-8.55	-8.66	-7.86

jj coupling of the single-particle wave functions modifies the calculation of the spin traces. A detailed discussion of how to deal with operator dependent correlations is given in Ref. [16].

11.4.3 Application to Doubly Closed Shell Nuclei

The RFHNC scheme has been applied to describe doubly magic nuclei which are shell saturated and spherical. The validity of the RFHNC equations has been tested in a simple model of ^{4}He and ^{16}O nuclei [17]. A nuclear interaction containing only scalar and spin-dependent terms has been used. The trial wave function (11.2) had a purely scalar, Jastrow, correlation function. The single particle wave functions are those generated by a harmonic oscillator well. This simplified model allowed a comparison with Variational Monte Carlo calculations (see Sect. 4.3) carried out with the same input [18].

This comparison is shown in Table 11.2 and the good agreement of the results obtained with the two approaches indicates the validity of the RFHNC calculation.

We present the results of a more realistic calculation in Table11.3 [16]. In this case, the nucleon-nucleon interaction employed is the Argonne V8' [19], which includes the first eight channels of the expansion described in Sect. 3.3 and is specifically designed to reproduce the nucleon-nucleon phase shifts. This two-body interaction is complemented by a three-body interaction known as Urbana-IX, which has been carefully constructed to accurately reproduce the triton binding energy [19].

The RFHNC equations have been solved to describe the set of spherical nuclei indicated in Tab 11.3. The single particle wave functions have been obtained by using a MF potential of Woods-Saxon type, see Eq. (2.10), with spin-orbit potential in a j-j coupling scheme, see Sect. 2.2.4. In the trial wave function a state dependent correlation of type (11.65) has been used.

In Table 11.3 the contribution of the various terms of the hamiltonian are presented. The values indicated are in MeV and they are relative to the contribution per nucleon.

It is interesting to remark that there is a clear indication of the saturation of the various term but for the ^{16}O nucleus which appears to be relatively small. The values of T, the kinetic energy, and $V^6_{2-\text{body}}$, the contribution of the two-body interaction, are very similar for the three heavier nuclei. On the contrary, the repulsive contribution of the Coulomb term V_{Coul} increases with the number of protons. This is the differences between the long and short-range characteristics of the Coulomb and nuclear interaction. The contribution of the tree-body term of the interaction is repulsive.

The search for the minimum has been done by making variations on the various terms of the correlation function (11.65), while the set of single particle wave function has not been optimized. This is probably the reason of the discrepancy between the calculated energies and the experimental ones.

It is interesting to remark that in the process of minimisation it turns out that the various terms of the correlation function are almost independent of the nucleus considered, indicating that the f_p functions of Eq. (11.65) take care of short-range physics where surface effects are neglible. The various f_p functions are shown in Fig. 11.16 as a function of the relative distance between the two correlated nucleons.

The scalar term of the correlation function, f_1, is certainly the largest one, almost one order of magnitude bigger than the other ones. The behaviour of this term is that naively expected. At small distances the correlation function is small restricting the possibility of the the two-nucleons to be too close, this is due to the strongly repulsive core of the interaction. At distances slightly larger than 1 fm it f_1 becomes 1 indicating that the two-nucleons are free, this is a consequence of the short-range nature of the interaction.

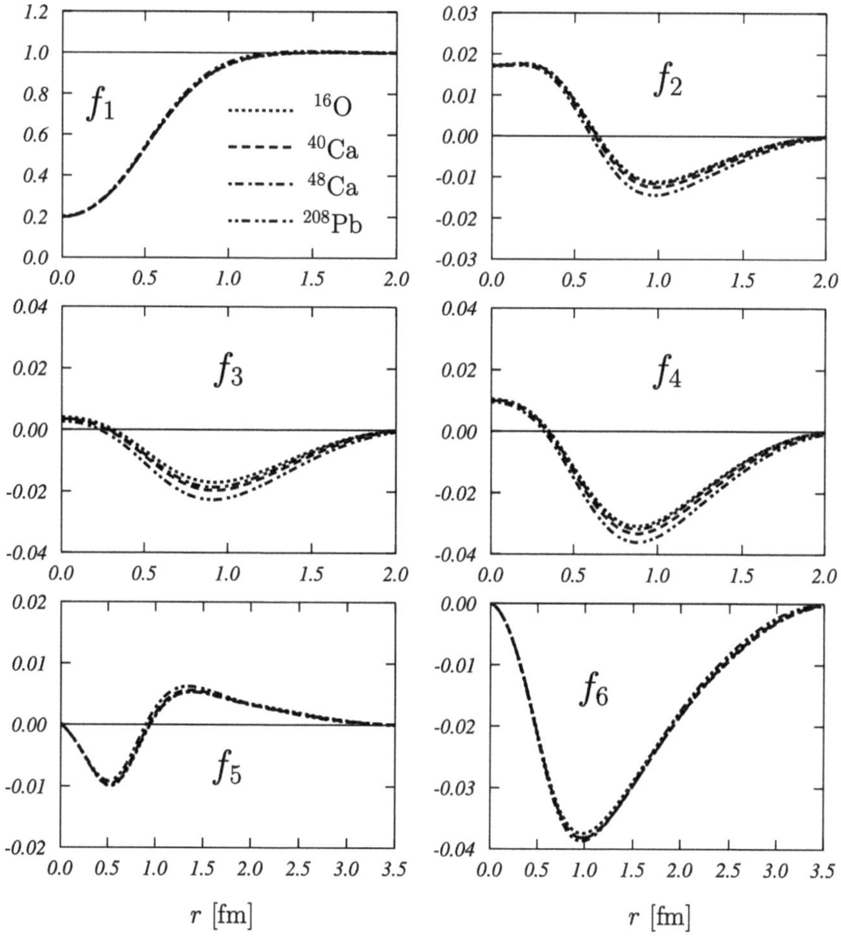

Fig. 11.16 The f_p correlation functions of Eq. (11.65) for the four nuclei considered in Table 11.3 as a function of the relative distance of the two nucleons involved. These functions have been obtained by searching for a minimum of the energy functional (11.1) calculated within the RFHNC/1 computational scheme

References

1. R. Jastrow, Many-body problem with strong forces. Phys. Rev. **98**, 1479 (1955)
2. J.E. Mayer, M.G. Mayer, *Statistical Mechanics* (Wiley, New York, 1940)
3. S.A. Rice, P. Gray, *The statistical mechanics of simple liquids* (Wiley, New York, 1965)
4. D.L. Goodstein, *States of matter* (Dover, New York, 1985)
5. S. Fantoni, S. Rosati, Jastrow correlations and an irreducible cluster expansion for infinite boson or fermion systems. Nuov. Cim. A **20**, 179 (1974)
6. W. Greiner, L. Neise, H. Stöcker, *Thermodynamics and Statistical Mechanics* (Springer, Berlin, 1995)

7. E. Manousakis, S. Fantoni, V.R. Pandharipande, Q.N. Usmani, Microscopic calculations for normal and polarized liquid ^{3}He. Phys. Rev. B **28**, 3770 (1983)
8. M. Viviani, E. Buendía, S. Fantoni, S. Rosati, Spin-dependent correlations in the ground state of liquid ^{3}He. Phys. Rev. B **38**, 4523 (1988)
9. F. Arias de Saavedra, E. Buendía, Variational Fermi-hypernetted-chain approximation [FHNC/$\alpha(r)$] calculations with σ_z-dependent correlations in liquid ^{3}He. Phys. Rev. B **42**, 6018 (1990)
10. V.R. Pandharipande, R.B. Wiringa, Variations on a theme of nuclear matter. Rev. Mod. Phys. **51**, 821 (1979)
11. O. Benhar, S. Fantoni, *Nuclear Matter Theory* (CRC Press, Boca Raton, 1979)
12. A. Akmal, V.R. Pandharipande, D.G. Ravenhall, Equation of state of nucleon matter and neutron star structure. Phys. Rev. C **58**, 1804 (1998)
13. S. Gandolfi, A.Y. Illarionov, S. Fantoni, J.C. Miller, F. Pederiva, K.E. Schmidt, Microscopic calculation of the equation of state of nuclear matter and neutron star structure. Mont. Not. R. Astron. Soc. **404**, L35 (2010)
14. G. Co', V. De Donno, P. Finelli, M. Grasso, M. Anguiano, A.M. Lallena, C. Giusti, A. Meucci, F.D. Pacati, Mean-field calculations of the ground states of exotic nuclei. Phys. Rev. C **85**, 024322 (2012)
15. S. Fantoni, S. Rosati, Extension of the FHNC method to finite systems. Nucl. Phys. A **328**, 478 (1979)
16. F. Arias de Saavedra, C. Bisconti, G. Co', A. Fabrocini, Renormalized Fermi hypernetted chain approach in medium-heavy nuclei. Phys. Rep. **450**, 1 (2007)
17. G. Co', A. Fabrocini, S. Fantoni, I. Lagaris, Model calculations of doubly closed shell nuclei in CBF theory (I). Nucl. Phys. A **549**, 439 (1992)
18. M.C. Boscá, E. Buendía, R. Guardiola, Monte Carlo test of the convergence of cluster expansions in Jastrow correlated nuclei. Phys. Lett. B **198**, 312 (1987)
19. B. S. Pudliner, V. R. Pandharipande, J. Carlson, S. C. Pieper, R. B. Wiringa: *Quantum Monte Carlo calculations of nuclei with* $A <\sim 7$, Phys. Rev. C 56 (1997) 1720

Chapter 12
Unitary Correlation Operator Method

12.1 Introduction

In the 1990s, particularly within the field of theoretical nuclear physics, significant attention was focused on short-range correlations. Among the various approaches which focus their attention to these types of correlations, the Unitary Correlation Operator Method (UCOM) [1] stands out as particularly interesting, as it highlights the duality in the interpretation of these correlations.

12.2 The Unitary Correlation Operator

We consider the Schrödinger equation with the full hamiltonian containing a microscopic interaction,

$$\hat{H}\,|\Psi_n\rangle = E_n\,|\Psi_n\rangle \quad . \tag{12.1}$$

and define an operator $\hat{C}$ such as its action on a MF state generates the eigenstate of the hamiltonian $\hat{H}$,

$$|\Psi_n\rangle = \hat{C}(\hat{H}, \{\Phi\})\,|\Phi_n\rangle \quad , \tag{12.2}$$

where we made explicit the dependence of the $\hat{C}$ operator on both the hamiltonian $\hat{H}$ and also on the single-particle wave functions generated by a MF hamiltonian and forming the Slater determinant $|\Phi_n\rangle$.

Since both the set of $|\Phi_n\rangle$ states and that of the eigenstates of $\hat{H}$, $|\Psi_n\rangle$, form complete bases, the $\hat{C}(\hat{H}, \{\Phi\})$ operator can be formally expressed as

$$\hat{C}(\hat{H}, \{\Phi\}) = \sum_k |\Psi_k\rangle\,\langle\Phi_k|\,, \tag{12.3}$$

© The Author(s), under exclusive license to Springer Nature Switzerland AG 2026

G. Co', *Concepts in Quantum Many-Body Physics*, UNITEXT for Physics,
https://doi.org/10.1007/978-3-032-08920-5_12

showing that $\hat{C}$ is a unitary operator which diagonalizes the hamiltonian $\hat{H}$. In fact we can write

$$\left\langle \Psi_k | \hat{H} | \Psi_n \right\rangle = \left\langle \Phi_k | \hat{C}^+ \hat{H} \hat{C} | \Phi_n \right\rangle = \langle \Phi_k | \sum_i |\Phi_i\rangle \langle \Psi_i | \hat{H} \sum_j |\Psi_j\rangle \langle \Phi_j | \Phi_n\rangle$$

$$= \sum_{ij} \langle \Phi_k | \Phi_i \rangle E_i \delta_{ij} \langle \Phi_j | \Phi_n \rangle = \sum_{ij} E_i \delta_{ij} \delta_{ik} \delta_{jn} = E_n \delta_{kn} \ . \tag{12.4}$$

Since $\hat{C}$ is unitary, the state $|\Psi_n\rangle$, as defined in Eq. (12.2) has the same normalisation as the MF state $|\Phi_n\rangle$. The structure of the equations in (12.2) and (11.2), which define the correlations in UCOM and CBF theories respectively, is similar. However, Jastrow's ansatz does not yield a unitary operator. Therefore, a specific treatment is required to achieve the proper normalisation of the wave function. In contrast, the correlation in UCOM is defined in a way that ensures the property of unitarity is maintained.

The second term of Eq. (12.4) can be interpreted in two ways: either by considering that the correlation affects the state $|\Phi_n\rangle$ or by viewing it as acting on the operator $\hat{H}$

$$\left\langle \Psi_k | \hat{H} | \Psi_n \right\rangle = \left\langle \Phi_k | \hat{C}^+ \hat{H} \hat{C} | \Phi_n \right\rangle = \left\langle \Phi_k | \hat{H}^{\text{eff}} | \Phi_n \right\rangle \ , \tag{12.5}$$

with the obvious definition of the effective hamiltonian $\hat{H}^{\text{eff}}$. This is what we referred to in the introduction regarding the duality in interpreting the correlation effects.

We pointed out how the unitarity of the correlation operator is an essential requirement of the model. A unitary operator can be described as

$$\hat{C} = e^{-i\hat{G}}, \tag{12.6}$$

where $\hat{G}$ is a hermitian operator. Since we are considering a fermionic system, the number of particles is conserved under the action of the operator. The operator $\hat{G}$ is invariant under the exchange of two particles due to the antisymmetry inherent in the Slater determinant $|\Phi\rangle$. Furthermore, $\hat{G}$ must be at least a two-body operator, or more complex, because a unitary transformation induced by a one-body operator would yield another Slater determinant that is completely equivalent in terms of calculations of observables.

The operator $\hat{G}$ can be expressed as

$$\hat{G} = \sum_{i<j} \hat{g}(i,j) + \cdots \ , \tag{12.7}$$

where $\hat{g}(i,j)$ is an operator that acts on the fermions i and j, and the dots indicates the possible presence of operators acting on more than two particles. Each correlated operator

$$\underline{\hat{B}} = \hat{C}^{+}\hat{B}\hat{C} = \hat{C}^{-1}\hat{B}\hat{C} = e^{i\hat{G}}\hat{B}e^{-i\hat{G}} \quad , \tag{12.8}$$

is an overlap of zero-, one-, two-, three-, $\cdots$, $A-$body operators.

It is useful to employ a notation that enables the identification of an n-body operator that cannot be expressed as a sum of simpler operators; for this reason, it is referred to as irreducible:

$$\hat{B}_A^{[n]} = \sum_{i_1<\cdots<i_n}^{A} \hat{b}^{[n]}(i_1, i_2, \cdots, i_n) \quad , \tag{12.9}$$

where $\hat{b}^{[n]}(i_1, i_2, \cdots, i_n)$ is an operator acting on n of the A particles forming the system. If $n > A$ then $\hat{B}_A^{[n]} = 0$. In this notation, the hamiltonian composed by the kinetic energy and by a two-body potential is expressed as

$$\hat{H} \equiv \hat{H}_A = \hat{T}_A^{[1]} + \hat{V}_A^{[2]} = \sum_{i}^{A} \hat{t}^{[1]}(i) + \sum_{i<j}^{A} \hat{v}^{[2]}(i, j) \quad . \tag{12.10}$$

The absence of the upper index $[n]$ indicates that the operator is sum of operators acting on different numbers of particles.

The expression of a correlated operator in terms of irreducible operators can be written as

$$\underline{\hat{B}} \equiv \underline{\hat{B}}_A = \hat{C}_A^{-1}\hat{B}_A\hat{C}_A = \sum_{n=1}^{A} \underline{\hat{B}}_A^{[n]} \quad , \tag{12.11}$$

where

$$\underline{\hat{B}}_A^{[n]} = \sum_{i_1<\cdots<i_n}^{A} \underline{\hat{b}}^{[n]}(i_1, i_2, \cdots, i_n) \quad . \tag{12.12}$$

From the above definitions we obtain the expression

$$\underline{\hat{b}}^{[n]} \equiv \underline{\hat{B}}_n - \sum_{k=1}^{n-1} \underline{\hat{B}}_n^{[k]} = \hat{C}_n^{-1}\hat{B}_n\hat{C}_n - \sum_{k=1}^{n-1} \sum_{i_1<\cdots<i_k}^{n} \underline{\hat{b}}^{[k]}(i_1, \cdots, i_k) \quad . \tag{12.13}$$

The first $\underline{\hat{b}}^{[n]}$ term different from zero, is related to the number of particles connected by the $\hat{B}$ operator. For one-body operators $n = 1$, for two-body operators $n = 2$, etc..

We shall consider, henceforth, a generator $\hat{G}$ composed by two-body operators only. This means that we neglect the terms indicated by the dots in the expression (12.7).

The correlated kinetic energy term is described as

$$\hat{\underline{T}}_A = \hat{C}_A^+ \hat{T}_A \hat{C}_A = \sum_{n=1}^{A} \hat{\underline{T}}_A^{[n]} \ . \tag{12.14}$$

The first term, $n = 1$, is the MF kinetic energy

$$\hat{\underline{T}}_A^{[1]} = \sum_i^A \hat{\underline{t}}^{[1]}(i) = \sum_i^A \hat{t}^{[1]}(i) = \hat{T}_A^{[1]} \ . \tag{12.15}$$

The second term, $n = 2$, is obtained by considering Eqs. (12.12) and (12.13),

$$\hat{\underline{T}}_A^{[2]} = \sum_{i<j} \hat{\underline{t}}^{[2]}(i, j) \ , \tag{12.16}$$

where

$$\hat{\underline{t}}^{[2]}(i, j) = \hat{c}(i, j)^{-1}\Big(\hat{t}^{[1]}(i) + \hat{t}^{[1]}(j)\Big)\hat{c}(i, j) - \Big(\hat{t}^{[1]}(i) + \hat{t}^{[1]}(j)\Big) \ , \tag{12.17}$$

with

$$\hat{c}(i, j) = \exp\{-i\hat{g}^{[2]}(i, j)\} \ . \tag{12.18}$$

This means that, for each pair i and j, the two-body part is the difference between the correlated term and uncorrelated one.

For the third term of Eq. (12.14), $n = 3$, we have that

$$\hat{\underline{T}}_A^{[3]} = \sum_{i<j<k} \hat{\underline{t}}^{[3]}(i, j, k) \ , \tag{12.19}$$

where

$$\begin{aligned} \hat{\underline{t}}^{[3]}(i, j) = {} & \hat{c}(i, j, k)^{-1}\Big(\hat{t}^{[1]}(i) + \hat{t}^{[1]}(j) + \hat{t}^{[1]}(k)\Big)\hat{c}(i, j, k) \\ & - \Big(\hat{\underline{t}}^{[2]}(i, j) + \hat{\underline{t}}^{[2]}(i, k) + \hat{\underline{t}}^{[2]}(j, k)\Big) \\ & - \Big(\hat{t}^{[1]}(i) + \hat{t}^{[1]}(j) + \hat{t}^{[1]}(k)\Big) \ , \end{aligned} \tag{12.20}$$

with the definition

$$\hat{c}(i, j, k) = \exp\Big\{ - i\Big(\hat{g}^{[2]}(i, j) + \hat{g}^{[2]}(i, k) + \hat{g}^{[2]}(j, k)\Big)\Big\} \ . \tag{12.21}$$

Also in this case, the correlated contribution of the three-body term is obtained by subtracting the two-body and one-body terms.

We use the same approach to describe the two-body interaction term. In this case, the first term different from zero is $n = 2$, therefore

$$\underline{\hat{V}}_A \equiv \hat{C}_A^+ \hat{V}_A \hat{C}_A = \sum_{n=2}^{A} \underline{\hat{V}}_A^{[n]} , \tag{12.22}$$

where

$$\underline{\hat{V}}_A^{[2]} = \sum_{i<j} \underline{\hat{v}}^{[2]}(i, j) , \tag{12.23}$$

with

$$\underline{\hat{v}}^{[2]}(i, j) = \hat{c}(i, j)^{-1} \hat{v}^{[2]}(i, j) \hat{c}(i, j) , \tag{12.24}$$

and

$$\underline{\hat{V}}_A^{[3]} = \sum_{i<j<k} \underline{\hat{v}}^{[3]}(i, j, k) , \tag{12.25}$$

where

$$\begin{aligned} \underline{\hat{v}}^{[3]}(i, j) = {} & \hat{c}(i, j, k)^{-1} \Big(\hat{v}^{[2]}(i, j) + \hat{v}^{[2]}(i, k) + \hat{v}^{[2]}(j, k) \Big) \hat{c}(i, j, k) \\ & - \Big(\underline{\hat{v}}^{[2]}(i, j) + \underline{\hat{v}}^{[2]}(i, k) + \underline{\hat{v}}^{[2]}(j, k) \Big). \end{aligned} \tag{12.26}$$

12.3 Representation as Coordinates Transformation

In this section, we express the action of the operator $\hat{c}(i, j)$, Eq. (12.18), in coordinate space. The relative coordinate and momentum of the identical fermions i and j are defined as

$$\mathbf{r} = \mathbf{r}(i) - \mathbf{r}(j) \text{ and } \hat{\mathbf{q}} = \frac{1}{2}(\hat{\mathbf{p}}(i) - \hat{\mathbf{p}}(j)) . \tag{12.27}$$

where the momenta are operators in coordinate space. We define the generator $\hat{g}(i, j) \equiv \hat{g}(\mathbf{r}, \hat{\mathbf{q}})$, Eq. (12.7), in a symmetric and hermitian form as

$$\hat{g}(\mathbf{r}, \hat{\mathbf{q}}) = \frac{1}{2} \left\{ \left(\hat{\mathbf{q}} \cdot \frac{\mathbf{r}}{r} \right) s(r) + s(r) \left(\frac{\mathbf{r}}{r} \cdot \hat{\mathbf{q}} \right) \right\} \text{ with } r \equiv |\mathbf{r}| , \tag{12.28}$$

where $s(r)$ is a suitable scalar function. The unitary operator $\hat{c}(\mathbf{r}, \mathbf{q})$ moves the relative position of the interacting pair of particles from $\mathbf{r}$ at about $\mathbf{r} + s(r)\mathbf{r}/r$. The function $s(r)$ is defined in such a way that the particles remain always far from the region of the strongly repulsive core of the potential.

In coordinates representation, the action of the generator $g(\mathbf{r}, \hat{\mathbf{q}})$ on the wave function $\langle \mathbf{r} | \phi \rangle$ describing the relative motion of the two-independent particles is

$$\frac{1}{\hbar}\langle \mathbf{r}|\hat{g}|\phi\rangle = -i\left(\frac{1}{2}\frac{\partial s(r)}{\partial r} + \frac{s(r)}{r} + s(r)\frac{\partial}{\partial r}\right)\langle \mathbf{r}|\phi\rangle$$
$$= -i\frac{1}{r\sqrt{s(r)}}\,s(r)\frac{\partial}{\partial r}r\sqrt{s(r)}\,\langle \mathbf{r}|\phi\rangle \quad . \tag{12.29}$$

where we used the explicit expression of the operator $\hat{\mathbf{q}}$. For the correlation operator we obtain

$$\frac{1}{\hbar}\langle \mathbf{r}|\hat{c}|\phi\rangle = \exp\left(\frac{-1}{r\sqrt{s(r)}}\,s(r)\frac{\partial}{\partial r}r\sqrt{s(r)}\right)\langle \mathbf{r}|\phi\rangle$$
$$= \frac{1}{r\sqrt{s(r)}}\exp\left(-s(r)\frac{\partial}{\partial r}\right)r\sqrt{s(r)}\,\langle \mathbf{r}|\phi\rangle \quad . \tag{12.30}$$

The last step can be verified by making a power expansion of the exponential. It is convenient to make a change of variables

$$s(r)\frac{\partial}{\partial r} \to \frac{\partial}{\partial y} \quad . \tag{12.31}$$

We call $Y(r)$ and $R(y)$, respectively, the transformations between y e r, and viceversa,

$$r \xrightarrow{Y} y = Y(r) \;\; ; \;\; y \xrightarrow{R} r = R(y) \quad , \tag{12.32}$$

and, by construction

$$R\big(Y(r)\big) = r \;\; ; \;\; Y\big(R(y)\big) = y \quad . \tag{12.33}$$

From the definition (12.31) we obtain

$$\frac{\partial}{\partial y}R(y) = s\big(R(y)\big). \tag{12.34}$$

As already stated, the function $s(r)$ is defined such as the particles always remain far from the strongly repulsive core region of the interaction.

The correlated relative distance can be expressed as

$$\underline{\mathbf{r}} = \hat{c}^{-1}\,\mathbf{r}\,\hat{c} = R(r)\frac{\mathbf{r}}{r} \quad . \tag{12.35}$$

In coordinate space, the action of every correlated operator $\underline{\hat{b}}(\mathbf{r})$ can be described as that of uncorrelated operator evaluated at a different distance

$$\underline{\hat{b}}(\mathbf{r}) \equiv \hat{c}^{+}\,\hat{b}(\mathbf{r})\,\hat{c} = \hat{c}^{-1}\,\hat{b}(\mathbf{r})\,\hat{c} = \hat{b}\left(R(r)\frac{\mathbf{r}}{r}\right). \tag{12.36}$$

This general description must be adapted to the physical situation under study. This means to find a convenient coordinate transformation to describe the physical problem under investigation. There are properties which $R(r)$ must have: it has to increase monotonically and must be differentiable in the full dominion. Since the repulsive core of the potential is limited to short relative distances, the effect of the short-range correlation is that of modifying the structure of the uncorrelated wave functions only in the surroundings of the repulsive core.

By considering terms up to the second order in the correlation, we can express the hamiltonian as

$$\underline{\hat{H}}^{[1]} + \underline{\hat{H}}^{[2]} = \hat{T}_A^{[1]} + \underline{\hat{T}}_A^{[2]} + \underline{\hat{V}}_A^{[2]}, \tag{12.37}$$

where the three terms are defined, respectively, in Eqs. (12.15), (12.16) and (12.22). By using the expression of the correlation operators in the coordinate space we write the various terms of the correlated hamiltonian. The easiest term is the two-body term (12.24). In case of purely scalar two-body interaction we obtain

$$\underline{\hat{v}}^{[2]}(i, j) = v\big(R(r_{ij})\big) \ , \tag{12.38}$$

where r_{ij} is the distance between the i and j particles. The expression of the two-body correlated term of the kinetic energy $\underline{\hat{T}}_A^{[2]}$ is more involved and depends on $R(r)$ and its first, second and even third derivative [1].

Actual calculations are performed using the variational principle. One selects expressions for $R(r)$ that contain parameters, the values of which are determined by searching for the minimum of the energy functional.

12.4 Spin, and Isospin, Dependent Correlations

The presentation done so far has used purely scalar expressions of the correlation. The complexity of the nuclear interaction requires that the correlations depend also on the spin and the isospin of the interacting nucleons. For this purpose the generator of the unitary correlation operator is decomposed in four different channels.

$$\hat{g}^{[2]}\big(\mathbf{r}, \hat{\mathbf{q}}, \hat{\boldsymbol{\sigma}}(i), \hat{\boldsymbol{\sigma}}(j), \hat{\boldsymbol{\tau}}(i), \hat{\boldsymbol{\tau}}(j)\big) = \sum_{S=0,1} \sum_{T=0,1} \hat{g}_{ST}^{[2]}(\mathbf{r}, \hat{\mathbf{q}}) \, \hat{\Pi}_S \otimes \hat{\Pi}_T \ , \tag{12.39}$$

where we have indicated with the symbol $\hat{\Pi}$ the spin and isospin projection operators

$$\hat{\Pi}_{S=0} = \frac{1}{4}\big(1 - \hat{\boldsymbol{\sigma}}(i) \cdot \hat{\boldsymbol{\sigma}}(j)\big) \ ; \ \hat{\Pi}_{S=1} = \frac{1}{4}\big(3 + \hat{\boldsymbol{\sigma}}(i) \cdot \hat{\boldsymbol{\sigma}}(j)\big) \ ; \tag{12.40}$$

$$\hat{\Pi}_{T=0} = \frac{1}{4}\big(1 - \hat{\boldsymbol{\tau}}(i) \cdot \hat{\boldsymbol{\tau}}(j)\big) \ ; \ \hat{\Pi}_{T=1} = \frac{1}{4}\big(3 + \hat{\boldsymbol{\tau}}(i) \cdot \hat{\boldsymbol{\tau}}(j)\big) \ . \tag{12.41}$$

The unitary correlation operator is

$$\hat{C}_A = \exp\left(-i\sum_{i<j}\hat{g}^{[2]}(\mathbf{r}_{ij}, \hat{\mathbf{q}}_{ij}, \hat{\boldsymbol{\sigma}}(i), \hat{\boldsymbol{\sigma}}(j), \hat{\boldsymbol{\tau}}(i), \hat{\boldsymbol{\tau}}(j)\right) \quad . \tag{12.42}$$

Also in this case, a truncation to the two-body terms is adopted $\hat{c} \equiv \hat{C}_2$, therefore

$$\hat{c} = \exp\left(-i\sum_{ST}^{\{0,1\}}\hat{g}_{ST}^{[2]}(\mathbf{r}, \hat{\mathbf{q}})\hat{\Pi}_S \otimes \hat{\Pi}_T\right) = \sum_{ST}^{\{0,1\}}\exp\left(-i\hat{g}_{ST}^{[2]}(\mathbf{r}, \hat{\mathbf{q}})\right)\hat{\Pi}_S \otimes \hat{\Pi}_T \quad . \tag{12.43}$$

By using the projectors $\hat{\Pi}_{S,T}$ it is possible to express the correlation operator as a sum of four, commuting, correlation operators, one for each channel. With respect to the case of purely scalar correlation, there are four correlation functions $R(r)$ to be determined.

In Table 12.1 we show the binding energies of some doubly closed shell nuclei [1]. The calculations have been carried out by using an interaction, called Afnan-Tang S3 [2, 3], with only the four central channels, see Sect. 3.3.1. In the table, the contribution of the kinetic energy term T and that of the interaction potential V are separately presented. The results of the column labelled MF have been obtained by switching off the correlations, i.e. by setting $\hat{C} = 1$.

We observe that the effect of the correlations is analogous in all nuclei considered. The contribution of the kinetic energy is increased. At the same time, the potential energy is made more attractive. This potential does not bind nuclei in MF calculations, but when the correlations are activated it becomes enough attractive to bind all the nuclei.

Table 12.1 Energy per nucleon in MeV for some doubly closed shell nuclei [1]. The column labelled as T, V and E indicate, respectively, the contribution of the kinetic energy, the interaction energy and the total energy. The columns labelled as MF show the uncorrelated energies, those labelled as UCOM show the result of the full calculation. The last column, labelled E(th) report the results of other microscopic calculations carried out with the same input. Specifically Yakuobvsky for ^{4}He and CBF for ^{16}O, ^{40}Ca and ^{48}Ca. The interaction used is a relatively simple interaction of central type [2, 3]

Nucleus	T		V		E		E(th)
	MF	UCOM	MF	UCOM	MF	UCOM	
^{4}He	14.89	19.63	–13.51	–26.61	1.38	–6.96	–7.2
^{12}C	17.68	22.37	–15.15	–27.48	2.53	–5.11	
^{16}O	18.48	23.81	–16.67	–30.56	1.82	–6.74	–6.58
^{40}Ca	21.21	27.72	–19.47	–36.09	1.74	–8.37	–8.65
^{48}Ca	20.99	27.05	–18.67	–34.09	2.31	–7.04	–7.57

The global binding energies are compared with those obtained by other microscopic theories, CBF for example. There is a reasonable agreement between these results, especially if one consider the remarkable differences of the various theoretical approaches.

These are results of toy model calculations useful to understand the role of the various ingredients of the theory. More involved calculations which consider more realistic interactions, three-body and tensor correlation terms have been carried out [4]. The comparison with experimental data is very satisfying.

References

1. H. Feldmeier, T. Neff, R. Roth, J. Schnack, A unitary correlation operator method. Nucl. Phys. A **632**, 61 (1998)
2. I.R. Afnan, Y.C. Tang, Investigation of nuclear three- and four-body systems with soft-core nucleon-nucleon potentials. Phys. Rev. **175**, 1337 (1968)
3. R. Guardiola, A. Faessler, H. Müther, A. Polls, Brueckner theory and Jastrow approach for finite nuclei. Nucl. Phys. A **371**, 79 (1981)
4. R. Roth, T. Neff, H. Feldmeier, Nuclear structure in the framework of the Unitary Correlation Operator Method. Prog. Part. Nucl. Phys. **65**, 50 (2010)

Chapter 13
The Coupled Cluster Method

13.1 Introduction

In Chaps. 11 and 12, the CBF and UCOM theories address the challenge posed by the strongly repulsive core of microscopic interactions, focusing on the definition of short-range correlations. The terminology used in these two theories differs significantly from that presented in Part II. From a linguistic perspective, the Coupled Cluster Method (CCM) is a bridge between the two distinct languages employed thus far. Physically, it is an approach closely related to Statistical Mechanics, even though it incorporates concepts of particle and hole states derived from Field Theory. The key point is that, in this context, the fundamental quantity of the theory remains the correlation function, which is described in terms of particle-hole excitations.

13.2 The Many-Body State in the CCM

In the description of a many-body system, each eigenstate $|\Psi\rangle$ of the total Hamiltonian $\hat{H}$ can be expressed as a linear combination of Slater determinants formed from single-particle states obtained within the IPM framework (see Chap. 2). Using the terminology introduced in Chap. 5, each Slater determinant is characterised by a specific number of particle-hole $(p-h)$ excitations. Notably, the Slater determinant representing the ground state of the IPM contains no excitations of this type. There are Slater determinants that include one-particle one-hole $(1p-1h)$ excitations, states with two-particles two-holes $(2p-2h)$, and so forth, extending up to excitations involving A particles and A holes $(Ap-Ah)$.

One way to describe this situation is by defining operators that indicate the number of $p-h$ excitations. For instance, for a one-particle one-hole $1p-1h$ excitation, we have the operator

$$\hat{S}_1 = \sum_{ph} Z_{ph} \hat{a}_p^+ \hat{a}_h \quad , \tag{13.1}$$

© The Author(s), under exclusive license to Springer Nature Switzerland AG 2026

G. Co', *Concepts in Quantum Many-Body Physics*, UNITEXT for Physics,
https://doi.org/10.1007/978-3-032-08920-5_13

for $2p - 2h$ the operator

$$\hat{S}_2 = \frac{1}{(2!)^2} \sum_{p_1 h_1 p_2 h_2} Z_{p_1 h_1 p_2 h_2} \hat{a}^+_{p_1} \hat{a}^+_{p_2} \hat{a}_{h_2} \hat{a}_{h_1} , \tag{13.2}$$

and for $3p - 3h$ the operator

$$\hat{S}_3 = \frac{1}{(3!)^2} \sum_{p_1 h_1 p_2 h_2} Z_{p_1 h_1 p_2 h_2 p_3 h_3} \hat{a}^+_{p_1} \hat{a}^+_{p_2} \hat{a}^+_{p3} \hat{a}_{h_3} \hat{a}_{h_2} \hat{a}_{h_1} , \tag{13.3}$$

and so on. The goal of the CCM is to determine the values of the Z coefficients using the variational principle.

Since the Hamiltonian includes a two-body interaction term, the simplest type of excitation induced by the Hamiltonian arises from the action of $\hat{S}_2$ on the IPM ground state $|\Phi_0\rangle$. This action generates a $\hat{S}_2 |\Phi\rangle$ component of the eigenstate $|\Psi\rangle$ of the total Hamiltonian $\hat{H}$. This component accounts for states in which two fermions occupying the h_1 and h_2 states below the Fermi surface are promoted to occupy the p_1 and p_2 states above the Fermi surface.

It is also possible to conceive of states where this promotion occurs without the influence of the interaction term in the Hamiltonian. This implies that the excitations of $2p - 2h$ pairs occur independently of one another. The simplest example of this is a Slater determinant with $4p - 4h$ excitations, where the two pairs of $2p - 2h$ excitations are not connected to each other. In general, the components of the total wave function that involve these types of excitations are described by applying the $\hat{S}_2$ operator m times. In this case, the component of the total wave function is

$$\frac{1}{m!} \hat{S}_2^m |\Phi_0\rangle \quad ,$$

where the $1/m!$ term takes care of the multiple counting. The total contribution of $2m$-particles and $2m-$holes is:

$$\sum_{m=0}^{\infty} \frac{1}{m!} \hat{S}_2^m |\Phi_0\rangle = e^{\hat{S}_2} |\Phi_0\rangle \quad . \tag{13.4}$$

Obviously, the sum is truncated at $m = A/2$ which gives the maximum number of independent $2p - 2h$ excitations. However, it is convenient to formally express the contribution of the various $p - h$ excitations by using exponential expressions.

The contribution to the total wave function of $3p - 3h$ independent excitations is

$$\frac{1}{p!} \hat{S}_3^p |\Phi_0\rangle \quad .$$

The contribution of a set of m $2p-2h$ and n $3p-3h$ excitations all independent of each other is

$$\frac{1}{m!n!}\hat{S}_3^n\hat{S}_2^m\,|\Phi_0\rangle \quad .$$

Since the excitations are independent of each other, the operators $\hat{S}_2$ and $\hat{S}_3$ commutes. This means that all the indexes p and h identifying the creation and destruction operators are different from each other. By expressing everything in terms of infinite sums we obtain for the total contribution the expression

$$e^{\hat{S}_2+\hat{S}_3}\,|\Phi_0\rangle \quad ,$$

and systematically continuing with $np-nh$ excitations we arrive to the expression

$$e^{\hat{S}_2+\hat{S}_3+\cdots+\hat{S}_A}\,|\Phi_0\rangle \quad .$$

Since it is possible to excite also $1p-1h$ states we can write

$$|\Psi\rangle = e^{\hat{S}}\,|\Phi_0\rangle \quad \text{with} \quad \hat{S}=\sum_{n=1}^{A}\hat{S}_n \quad . \tag{13.5}$$

A different method to describe the same kind of physics is that of expressing the eigenstate of $\hat{H}$ as

$$|\Psi\rangle = \left(1+\sum_{n=1}^{A}\hat{F}_n\right)|\Phi_0\rangle \quad , \tag{13.6}$$

with the definition

$$\hat{F}_n \equiv \frac{1}{(n!)^2}\sum_{p_1\cdots p_n,h_1\cdots h_n}\Lambda_{p_1\cdots p_n,h_1\cdots h_n}\hat{a}^+_{p_1}\cdots\hat{a}^+_{p_n}\hat{a}_{h_n}\cdots\hat{a}_{h_1} \quad , \tag{13.7}$$

where the Λ amplitudes are numbers.

The two operators $\hat{S}_n$ and $\hat{F}_n$ describe the same state in terms excitations of $np-nh$ type. The relation between these two pictures is

$$\begin{aligned}
\hat{F}_1 &= \hat{S}_1, \\
\hat{F}_2 &= \hat{S}_2+\frac{1}{2}\hat{S}_1^2, \\
\hat{F}_3 &= \hat{S}_3+\hat{S}_2\hat{S}_1+\frac{1}{6}\hat{S}_1^3, \\
\hat{F}_4 &= \hat{S}_4+\hat{S}_3\hat{S}_1+\frac{1}{2}\hat{S}_2^2+\frac{1}{2}\hat{S}_2\hat{S}_1^2+\frac{1}{24}\hat{S}_1^4, \\
\hat{F}_5 &= \cdots \, .
\end{aligned}$$

These expressions clarify the difference between these two types of operators. The operators $\hat{S}_n$ describe connected clusters of $np - nh$ excitations, while the operators $\hat{F}_n$ describe all the possible $np - nh$ excitations, also those induced by non-connected excitations clusters.

13.3 The CCM Equations

Here, we derive a set of equations that enables us to determine the amplitudes Z from Eqs. (13.1, 13.2, 13.3). We will write the Schrödinger equation and express the wave function using Eq. (13.5).

$$E_0 |\Psi_0\rangle = \hat{H} |\Psi_0\rangle = \hat{H} e^{\hat{S}} |\Phi_0\rangle \quad . \tag{13.8}$$

In the following, we shall consider the overlap of this equation on states of the type

$$\begin{aligned} &\langle \Phi_0 | \hat{a}^+_{h_1} \hat{a}_{p_1} \quad , \\ &\langle \Phi_0 | \hat{a}^+_{h_2} \hat{a}^+_{h_1} \hat{a}_{p_1} \hat{a}_{p_2} \quad , \\ &\langle \Phi_0 | \hat{a}^+_{h_3} \hat{a}^+_{h_2} \hat{a}^+_{h_1} \hat{a}_{p_1} \hat{a}_{p_2} \hat{a}_{p_3} \quad . \end{aligned} \tag{13.9}$$

We find convenient to rewrite the $\hat{S}$ operators by using the expression

$$\hat{S} = \sum_{n \neq 0} \mathcal{S}_n \hat{C}^+_n , \tag{13.10}$$

where the $\mathcal{S}_n$ are numbers, and the $\hat{C}^+_n$ operator represents the set of the n pairs of creation and destruction operators

$$\hat{C}^+_n = \hat{a}^+_{p_1} \cdots \hat{a}^+_{p_n} \hat{a}_{h_n} \cdots \hat{a}_{h_1} \quad . \tag{13.11}$$

Let us multiply the Schrödinger equation, on the left hand side, by $\exp(-\hat{S})$

$$e^{-\hat{S}} \hat{H} e^{\hat{S}} |\Phi_0\rangle = e^{-\hat{S}} \hat{H} |\Psi_0\rangle = e^{-\hat{S}} E_0 |\Psi_0\rangle = e^{-\hat{S}} E_0 e^{\hat{S}} |\Phi_0\rangle = E_0 |\Phi_0\rangle \quad , \tag{13.12}$$

Projecting on the $\langle \Phi_0 |$ state we have

$$E_0 = \left\langle \Phi_0 | e^{-\hat{S}} \hat{H} e^{\hat{S}} | \Phi_0 \right\rangle = \left\langle \Phi_0 | \hat{H} e^{\hat{S}} | \Phi_0 \right\rangle \quad . \tag{13.13}$$

The last equality has been obtained by considering that

$$\langle\Phi_0|\, e^{-\hat{S}} = \langle\Phi_0| \sum_{n=0}^{\infty} \frac{(-\hat{S}_n)^n}{n!} = \langle\Phi_0|\, \hat{S}_0 = \langle\Phi_0| \quad , \tag{13.14}$$

because $\hat{C}_{n>0}$ contains creation operators above the Fermi surface such as

$$\langle\Phi_0|\, \hat{a}_p^+ = 0 \ .$$

By projecting the eigenvalue equation (13.12) onto states that include particles and holes, as indicated in Eq. (13.9), we obtain equations of the following form:

$$\left\langle\Phi_0|\hat{C}_n e^{-\hat{S}}\hat{H}e^{\hat{S}}|\Phi_0\right\rangle = 0 \text{ with } n > 0 \ . \tag{13.15}$$

The expressions in (13.15) indicate a set of interconnected equations related to different numbers of particle-hole excitations, as represented by the operator $\hat{C}_n$.

We express the operator term of Eq. (13.12) as a set of connected commutators

$$\begin{aligned} e^{-\hat{S}}\hat{H}e^{\hat{S}} &= \hat{H} + [\hat{H},\hat{S}] + \frac{1}{2!}[[\hat{H},\hat{S}],\hat{S}] + \frac{1}{3!}\Big[[[\hat{H},\hat{S}],\hat{S}],\hat{S}\Big] + \cdots \\ &= \hat{H} + \sum_n \hat{\mathcal{S}}_n[\hat{H},\hat{C}_n^+] + \frac{1}{2}\sum_{n,m}\hat{\mathcal{S}}_n\hat{\mathcal{S}}_m\big[[\hat{H},\hat{C}_n^+],\hat{C}_m^+\big] + \cdots \end{aligned} \tag{13.16}$$

This expression inserted in Eq. (13.15) generates a system of equations whose unknown are the amplitudes $\mathcal{S}_n$.

The expression (13.16) contains an infinite number of terms. When calculating the expectation value with respect to $|\Phi_0\rangle$, a natural truncation of the sum occurs. This can be understood by noting that $[\hat{C}_n^+, \hat{C}_m^+] = 0$. This implies that, in the evaluation of the contractions, only those indexes that relate the free indexes of the Hamiltonian to the fixed indexes of the C_n^+ operators contribute non-zero values. Consequently, the sum is naturally truncated. For instance, in a Hamiltonian that includes only two-body interaction terms, the sum considers commutators only up to the fourth order:

$$\begin{aligned} &\left\langle\Phi_0|\hat{C}_n\hat{H}|\Phi_0\right\rangle + \sum_m \mathcal{S}_m\left\langle\Phi_0|\hat{C}_n[\hat{H},\hat{C}_m^+]|\Phi_0\right\rangle \\ &+\frac{1}{2!}\sum_{m,p}\mathcal{S}_m\mathcal{S}_p\left\langle\Phi_0|\hat{C}_n\big[[\hat{H},\hat{C}_m^+],\hat{C}_p^+\big]|\Phi_0\right\rangle \\ &+\frac{1}{3!}\sum_{m,p,q}\mathcal{S}_m\mathcal{S}_p\mathcal{S}_q\left\langle\Phi_0|\hat{C}_n\Big[\big[[\hat{H},\hat{C}_m^+],\hat{C}_p^+\big],\hat{C}_q^+\Big]|\Phi_0\right\rangle \\ &+\frac{1}{4!}\sum_{m,p,q,r}\mathcal{S}_m\mathcal{S}_p\mathcal{S}_q\mathcal{S}_r\left\langle\Phi_0|\hat{C}_n\Big[\Big[\big[[\hat{H},\hat{C}_m^+],\hat{C}_p^+\big],\hat{C}_q^+\Big],\hat{C}_r^+\Big]|\Phi_0\right\rangle = 0 \ . \end{aligned} \tag{13.17}$$

This is because the two-body term of the Hamiltonian contains four creation and annihilation operators, and it is necessary to include a $\hat{C}^+$ operator to fully saturate the creation and annihilation operators of the $\hat{C}$ applied to $\langle\Phi_0|$.

Once the number of particle-hole $p-h$ excitations defining the bra state, represented by $\hat{C}_n$, has been established, the expression (13.17) indicates a closed and finite set of interconnected equations. The unknowns in these equations are the $\mathcal{S}_n$, which are directly related to the Z amplitudes in Eqs. (13.1, 13.2, 13.3).

13.4 Approximations

The computational scheme outlined in the previous sections necessitates the calculation of Eq. (13.13) to determine the ground state energy. This calculation can be performed by solving Eqs. (13.17), which yield the amplitudes $\mathcal{S}_n$. If only the $\hat{C}_1$ and $\hat{C}_2$ operators are selected in Eq. (13.10), then the following equations must be solved:

$$\sum_{ph} \left\langle \Phi_0 | \hat{a}_h^+ \hat{a}_p e^{-\hat{S}} \hat{H} e^{\hat{S}} | \Phi_0 \right\rangle = 0 \ , \tag{13.18}$$

and

$$\sum_{p_1 h_1 p_2 h_2} \left\langle \Phi_0 | \hat{a}_{h_2}^+ \hat{a}_{h_1}^+ \hat{a}_{p_2} \hat{a}_{p_1} e^{-\hat{S}} \hat{H} e^{\hat{S}} | \Phi_0 \right\rangle = 0 \ . \tag{13.19}$$

The computational cost of this calculation scales with the square of the number of hole states and the fourth power of the number of particle states [1]. The solutions obtained in this manner are not satisfactory, necessitating an improvement in the approximation used in the truncation of the sum in Eq. (13.10). On the other hand, incorporating C_3 incurs a significantly higher computational cost, scaling with the cube of the number of hole states and the fifth power of the number of particle states. For this reason, in actual calculation it is adopted a different strategy.

Let us define an eigenvalue problem related to the bra state as

$$\langle \Phi_0 | \, \hat{\Lambda} e^{-\hat{S}} \hat{H} e^{\hat{S}} = E_0 \, \langle \Phi_0 | \, \hat{\Lambda} \ , \tag{13.20}$$

where $\hat{\Lambda}$ is an operator, called of de-excitation, defined as:

$$\hat{\Lambda} = \mathbb{I} + \sum_{I>0} \hat{\Lambda}_n \hat{C}_n = \mathbb{I} + \sum_{ph} \lambda_{ph} \hat{a}_h^+ \hat{a}_p + \sum_{p_1 h_1 p_2 h_2} \lambda_{p_1 p_2 h_1 h_2} \hat{a}_{h_2}^+ \hat{a}_{h_1}^+ \hat{a}_{p_2} \hat{a}_{p_1} \cdots \ , \tag{13.21}$$

where the λ are numbers.

The idea is to consider the eigenvalue equation (13.20) and to project it on $\hat{C}_n^+ \, |\Phi_0\rangle$ states

$$\left\langle \Phi_0 | \Lambda (e^{-\hat{S}} \hat{H} e^{\hat{S}} - E_0) \hat{C}_n^+ | \Phi_0 \right\rangle = 0 \;\; \text{with} \;\; n > 0 \; . \tag{13.22}$$

At this point, the problem is addressed in two steps. First, the CCM equations (13.17) are solved, specifically Eqs. (13.18) and (13.19) in the case of truncation up to C_2. Once the amplitudes $\mathcal{S}_1$ and $\mathcal{S}_2$ have been determined, the equations in (13.22) are solved to obtain the Λ_n. In this case, we only need to consider Λ_1 and Λ_2.

13.5 Applications

The CCM calculations are quite complex, particularly from a numerical standpoint. Currently, the application of this method is limited to relatively simple systems that exhibit well-defined symmetries.

The parameters to consider when defining the problems include the choice of C_n, specifically the number of particle-hole $p - h$ excitations to be included. This choice determines the configuration space, particularly the number of particle states above the Fermi surface. Based on these selections, CCM calculations are categorised using various acronyms. Unfortunately, each research field has its own method of classification, which creates challenges when comparing calculations conducted within different frameworks.

Below, we summarise some important points from the results obtained, which are continually evolving.

Systems interacting with Coulomb interaction
In this case, the Hamiltonian is well-defined and includes only the two-body scalar term of Coulomb type. The electronic properties of simple atoms and molecules are evaluated. The results obtained by restricting the analysis to the $\hat{C}_2$ choice align closely with those derived from the GFMC method. The differences in binding energies, as well as ionisation energies, for some of the studied molecules are within 1–2%. Furthermore, the comparison with experimental values reveals differences of a similar order of magnitude [1].

Nuclear systems
The comparison between CCM results and those obtained from other microscopic calculations has been conducted for spherical systems and nuclear interactions that exclude tensor and spin-orbit terms. For the few-body system ^{4}He, the differences in binding energies calculated using GFMC and CCM with approximations restricted to $\hat{C}_2$ are approximately 6%. The other system studied is the ^{16}O nucleus, where the theoretical results serve as a reference obtained with FHNC-1. The differences in this case are closely related to the approximations used for the $\hat{C}_2$ choice. With one approximation, the differences are about 20%, while with another approximation, the differences are less than 1%. This indicates that both FHNC-1 and the approximation chosen for the CCM calculations describe the

Table 13.1 Binding energies in K, for bosonic systems, He_4, and fermionic systems, He_3, of helium atoms [1]. N indicates the number of atoms composing the system

	N	VMC	CCM
He_4	20	–30.44	–32.73
	40	–93.44	–98.17
He_3	20	4.12	3.44
	40	–1.44	–2.55

same underlying physics. The CCM has been applied to spherical nuclei up to ^{208}Pb [2], and more recently, an extension of the theory has been proposed to address open-shell nuclei [3].

He atoms drops

A very useful system for verifying the validity of many-body theories is that of helium droplets. This involves constructing many-body systems where the fundamental particles are helium atoms.

The system can be bosonic if the nuclei of the helium atoms are ^{4}He, or fermionic if they are ^{3}He. The interaction is purely scalar and includes a repulsive core, as discussed in Sect. 3.4. It is intriguing to study how these systems become bound as a function of the number of particles.

In Table 13.1, we present a comparison between the CCM results and those obtained using VMC. The relative differences between the results of the two calculations are smaller for the bosonic systems than for the fermionic ones. The bosonic systems are more tightly bound than the fermionic systems. Notably, the fermionic system with 20 atoms is unbound, while the system with 40 particles is bound.

References

1. J. Navarro, R. Guardiola, I. Moliner, *The Coupled Cluster Method and its applications*, in Introduction to modern methods of quantum many-body theory and their applications, A. Fabrocini, S. Fantoni, E. Krotsheck Eds., World Scientific, Singapore (2002), p. 121
2. B. Hu et al., Ab initio predictions link the neutron skin of ^{208} Pb to nuclear forces. Nat. Phys. **18**, 1196 (2022)
3. A. Tichai, P. Demol, T. Duguet, Towards heavy-mass ab initio nuclear structure: Open-shell Ca, Ni and Sn isotopes from Bogoliubov coupled-cluster theory. Phys. Lett. B **851**, 138571 (2024)

Part IV
Phenomenological Theories

Chapter 14
Effective Theories

The aim of the theories presented so far has been to solve the Schrödinger equation using microscopic Hamiltonians, as discussed in Chap. 3. These Hamiltonians incorporate interactions specifically designed to describe the properties of two-body and, eventually, three-body systems. Due to the presence of a strongly repulsive core at short interaction distances, these theories have developed various non-perturbative methods to address this characteristic of the interaction.

In theories inspired by Quantum Field Theory, the approach involves modifying the microscopic interaction to eliminate the repulsive core. For instance, in Brueckner's theory, this objective is achieved by summing all the ladder diagrams, as detailed in Chap. 8. In contrast, theories influenced by Statistical Mechanics, such as the CBF approach, tackle the issue of the repulsive core by employing correlation functions that prevent the two interacting particles from coming too close together. In all these approaches, the Hamiltonian is treated as an external quantity, developed independently of theory where it will be applied.

The effective theories approach the many-body problem from a different perspective. The goal is to identify an effective Hamiltonian that operates within a subspace of the full Hilbert space and can reproduce the same eigenvalues as the true Hamiltonian. This implies that the equation

$$\hat{H}|\Psi_n\rangle = E_n|\Psi_n\rangle \ , \tag{14.1}$$

is substituted by the equation

$$\hat{H}^{\text{eff}}|\Psi_n^{\text{eff}}\rangle = E_n|\Psi_n^{\text{eff}}\rangle \ . \tag{14.2}$$

Formally, we can find the relation between $\hat{H}$ and $\hat{H}^{\text{eff}}$ by separating the hamiltonian in two parts $\hat{H} = \hat{H}_0 + \hat{H}_1$, and by considering the projection operator $\hat{P}$ on the $\hat{H}_0$ eigenstates, and the operator $\hat{Q}$ which is its complement

© The Author(s), under exclusive license to Springer Nature Switzerland AG 2026

G. Co', *Concepts in Quantum Many-Body Physics*, UNITEXT for Physics,
https://doi.org/10.1007/978-3-032-08920-5_14

$$\hat{P}^2 = \hat{P} \ ; \ \ \hat{Q}^2 = \hat{Q} \ ; \ \ \hat{Q} = \hat{\mathbb{I}} - \hat{P} \ ; \ \ \hat{P}\hat{Q} = 0 \ , \tag{14.3}$$

where

$$\hat{H}_0 \hat{P}|\Psi_n\rangle = E_n^0 \hat{P}|\Psi_n\rangle \Rightarrow \hat{P}\hat{H}_0|\Psi_n\rangle = E_n^0 \hat{P}|\Psi_n\rangle \ , \tag{14.4}$$

since $\hat{P}$ and $\hat{H}_0$ commute. Equation (14.1) can be rewritten as follows

$$\hat{H}_1|\Psi_n\rangle = (E_n - \hat{H}_0)|\Psi_n\rangle \ , \tag{14.5}$$

and by multiplying to the left hand side by $\hat{P}$, we obtain

$$\begin{aligned}
\hat{P}\hat{H}_1|\Psi_n\rangle &= \hat{P}(E_n - \hat{H}_0)|\Psi_n\rangle = (E_n - \hat{H}_0)\hat{P}|\Psi_n\rangle \\
\hat{P}\hat{H}_1(\hat{P} + \hat{Q})|\Psi_n\rangle &= (E_n - \hat{H}_0)\hat{P}|\Psi_n\rangle \\
\hat{P}\hat{H}_1\hat{Q}|\Psi_n\rangle &= (E_n - \hat{H}_0 - \hat{P}\hat{H}_1)\hat{P}|\Psi_n\rangle \ .
\end{aligned} \tag{14.6}$$

By working in a similar manner and multiplying both sides by $\hat{Q}$ on the left, we obtain

$$\hat{Q}\hat{H}_1\hat{P}|\Psi_n\rangle = (E_n - \hat{H}_0 - \hat{Q}\hat{H}_1)Q|\Psi_n\rangle \ . \tag{14.7}$$

By using the properties of these two operators, we can rewrite the two equations as

$$(\hat{P}\hat{H}_1\hat{Q})\hat{Q}|\Psi_n\rangle = (E_n - \hat{H}_0 - \hat{P}\hat{H}_1\hat{P})\hat{P}|\Psi_n\rangle \tag{14.8}$$

$$(\hat{Q}\hat{H}_1\hat{P})\hat{P}|\Psi_n\rangle = (E_n - \hat{H}_0 - \hat{Q}\hat{H}_1\hat{Q})\hat{Q}|\Psi_n\rangle \ . \tag{14.9}$$

We obtain $Q|\Psi_n\rangle$ from Eq. (14.9) and we substitute it in Eq. (14.8), the result is

$$\begin{aligned}
(\hat{P}\hat{H}_1\hat{Q})\frac{1}{E_n - \hat{H}_0 - \hat{Q}\hat{H}_1\hat{Q}}(\hat{Q}\hat{H}_1\hat{P})\hat{P}|\Psi_n\rangle &= (E_n - \hat{H}_0 - \hat{P}\hat{H}_1\hat{P})\hat{P}|\Psi_n\rangle \\
\left[\hat{H}_0 + \hat{P}\left(\hat{H}_1 + \hat{H}_1\hat{Q}\frac{1}{E_n - \hat{H}_0 - \hat{Q}\hat{H}_1\hat{Q}}\hat{Q}\hat{H}_1\right)\hat{P}\right]\hat{P}|\Psi_n\rangle &= E_n\hat{P}|\Psi_n\rangle \ .
\end{aligned} \tag{14.10}$$

This equation has the structure of Eq. (14.2) and it is possible to identify the effective interaction with the term

$$\hat{V}^{\text{eff}} = \hat{P}\left(\hat{H}_1 + \hat{H}_1\hat{Q}\frac{1}{E_n - \hat{H}_0 - \hat{Q}\hat{H}_1\hat{Q}}\hat{Q}\hat{H}_1\right)\hat{P} \ . \tag{14.11}$$

First, it is important to note that the effective interaction depends on the energy E_n. Secondly, and more crucially in this context, the effective interaction is influenced by $\hat{P}$, which refers to how the Hamiltonian is separated. The effective interaction is closely tied to the specific theory in which it is applied, as each theory has its own unique effective interaction.

The derivation presented above holds purely formal value, demonstrating how it is theoretically possible to rigorously define effective theories and relate them to microscopic theories. However, from a practical standpoint, solving these equations is as complex as directly solving the original Schrödinger equation.

To address this challenge, researchers employ phenomenological approaches that utilise interactions characterised by parameters chosen to ensure the theory can reproduce specific empirical data from the many-body system under investigation. The primary goal of these theories is to make predictions about various quantities that differ from those used to define the interaction.

These phenomenological effective interactions exhibit different characteristics compared to their microscopic counterparts, particularly in that they lack the strongly repulsive core at short distances. Generally, effective theories address and solve the many-body problem in a significantly simpler manner than microscopic theories. This simplicity makes them widely used for comparing with experimental data, especially in cases where the systems being described are highly complex and lack evident symmetries. Effective interactions are commonly applied in MF approaches, such as Hartree-Fock and Density Functional Theories. The relationship between effective and microscopic theories remains an active area of research.

Chapter 15
Mean-Field Applications of the Variational Principle

15.1 Introduction

In the previous chapters, we utilized the variational principle to obtain approximate solutions to the Schrödinger equation. This method is based on the idea that the wave function that minimises the energy corresponds to the correct eigenfunction of the Hamiltonian. This holds true if the search for the minimum is conducted over the entire Hilbert space. However, in practice, the problem is simplified by assuming a specific form for the wave function, and the search for the minimum is performed within the subspace spanned by all wave functions that conform to this chosen form. Consequently, the energy value obtained serves as an upper bound for the true energy eigenvalue of the Hamiltonian. The formal properties of the variational principle are discussed in various Quantum Mechanics textbooks, such as [1]. A summary of the relevant properties is provided in Appendix A.

We have previously presented an application of the variational principle in Sect. 4.3. In that instance, the multi-dimensional integrals necessary for calculating the energy were performed using Monte Carlo techniques. The trial wave functions in that calculation were expressed by Eq. (4.12), which represents the product of a Slater determinant and a correlation function. A similar trial wave function has been employed in the CBF theory discussed in Chap. 11. Additionally, the variational principle has been applied in the UCOM and CCM theories outlined in Chaps. 12 and 13.

In this chapter, we will focus on applying the variational principle to much simpler trial wave functions, specifically those represented by a single Slater determinant. This choice of trial wave function gives rise to theories that are classified as mean-field (MF) theories in the context of many-body physics, namely Hartree-Fock (HF) and Density Functional Theory (DFT).

The microscopic theories that describe many-body systems begin with the assumption that the MF problem has been solved, utilizing the single-particle basis derived from this solution. In Chap. 2, we discussed how the MF model is solved for various

© The Author(s), under exclusive license to Springer Nature Switzerland AG 2026

G. Co', *Concepts in Quantum Many-Body Physics*, UNITEXT for Physics,
https://doi.org/10.1007/978-3-032-08920-5_15

types of many-body systems. These approaches were primarily phenomenological, with the average potentials containing parameters chosen based on the need to reproduce specific experimental data. In this chapter, we introduce an approach that, by taking the particle interactions into account, provides a solid theoretical foundation for the phenomenological models presented in Chap. 2.

15.2 Hartree-Fock

15.2.1 The Mean-Field Hamiltonian

In this section, we derive a form of the Hamiltonian that is particularly useful for HF calculations. We are not introducing any new physics, we are simply reformulating the Hamiltonian expression. This new representation will clearly highlight the approximations made by MF theories.

Let us consider the expression of the hamiltonian in ONR Eq. (5.53):

$$
\begin{aligned}
\hat{H} &= \sum_{\nu\nu'} T_{\nu\nu'} \hat{a}^+_\nu \hat{a}_{\nu'} + \frac{1}{2} \sum_{\nu\mu\nu'\mu'} V_{\nu\mu\nu'\mu'} \hat{a}^+_\nu \hat{a}^+_\mu \hat{a}_{\mu'} \hat{a}_{\nu'} \\
&= \sum_{\nu\nu'} T_{\nu\nu'} \hat{a}^+_\nu \hat{a}_{\nu'} + \frac{1}{4} \sum_{\nu\nu'\mu\mu'} \overline{V}_{\nu\mu\nu'\mu'} \hat{a}^+_\nu \hat{a}^+_\mu \hat{a}_{\mu'} \hat{a}_{\nu'} \ .
\end{aligned}
\tag{15.1}
$$

where only two-body interactions have been considered. We have defined the antisymmetrised matrix element as:

$$
\overline{V}_{\nu\mu\nu'\mu'} \equiv \langle \nu\mu | V | \nu'\mu' \rangle - \langle \nu\mu | V | \mu'\nu' \rangle \ . \tag{15.2}
$$

From the definition of contraction (see Sect. 6.3), for single particle states below the Fermi energy we have that

$$
\overbracket{\hat{a}^+_\nu \, \hat{a}_{\nu'}} = \delta_{\nu\nu'} \delta_{\nu' i} \ ; \quad \overbracket{\hat{a}_\nu \, \hat{a}^+_{\nu'}} = 0 \ ; \quad \overbracket{\hat{a}_\nu \, \hat{a}_{\nu'}} = 0 \ ; \quad \overbracket{\hat{a}^+_\nu \, \hat{a}^+_{\nu'}} = 0 \ . \tag{15.3}
$$

By considering the definition of normal ordered product $\hat{\mathbb{N}}$ we obtain

$$
\hat{a}^+_\nu \, \hat{a}_{\nu'} = \hat{\mathbb{N}}[\hat{a}^+_\nu \, \hat{a}_{\nu'}] + \overbracket{\hat{a}^+_\nu \, \hat{a}_{\nu'}} \ , \tag{15.4}
$$

and, for the Wick's theorem,

$$
\begin{aligned}
\hat{a}^+_\nu \hat{a}^+_\mu \, \hat{a}_{\mu'} \, \hat{a}_{\nu'} = {} & \hat{\mathbb{N}}[\hat{a}^+_\nu \hat{a}^+_\mu \, \hat{a}_{\mu'} \, \hat{a}_{\nu'}] \\
& + \hat{\mathbb{N}}[\hat{a}^+_\mu \hat{a}_{\mu'}] \overbracket{\hat{a}^+_\nu \, \hat{a}_{\nu'}} + \hat{\mathbb{N}}[\hat{a}^+_\nu \hat{a}_{\nu'}] \overbracket{\hat{a}^+_\mu \, \hat{a}_{\mu'}} \\
& - \hat{\mathbb{N}}[\hat{a}^+_\mu \hat{a}_{\nu'}] \overbracket{\hat{a}^+_\nu \, \hat{a}_{\mu'}} - \hat{\mathbb{N}}[\hat{a}^+_\nu \hat{a}_{\mu'}] \overbracket{\hat{a}^+_\mu \, \hat{a}_{\nu'}} \\
& + \overbracket{\hat{a}^+_\mu \hat{a}_{\mu'}} \overbracket{\hat{a}^+_\nu \, \hat{a}_{\nu'}} - \overbracket{\hat{a}^+_\nu \hat{a}_{\mu'}} \overbracket{\hat{a}^+_\mu \, \hat{a}_{\nu'}} \; .
\end{aligned}
\tag{15.5}
$$

Let us insert the above expression in Eq. (15.1)

$$
\begin{aligned}
\hat{H} = {} & \sum_{\nu\nu'} T_{\nu\nu'} \hat{a}^+_\nu \hat{a}_{\nu'} + \frac{1}{4} \sum_{\mu\mu'\nu\nu'} \overline{V}_{\nu\mu\nu'\mu'} \Big\{ \hat{\mathbb{N}}[\hat{a}^+_\nu \hat{a}^+_\mu \, \hat{a}_{\mu'} \, \hat{a}_{\nu'}] \\
& + \hat{\mathbb{N}}[\hat{a}^+_\mu \hat{a}_{\mu'}] \delta_{\nu\nu'} \delta_{\nu i} + \hat{\mathbb{N}}[\hat{a}^+_\nu \hat{a}_{\nu'}] \delta_{\mu\mu'} \delta_{\mu i} \\
& - \hat{\mathbb{N}}[\hat{a}^+_\mu \hat{a}_{\nu'}] \delta_{\nu\mu'} \delta_{\nu i} - \hat{\mathbb{N}}[\hat{a}^+_\nu \hat{a}_{\mu'}] \delta_{\mu\nu'} \delta_{\mu i} \\
& + \delta_{\nu\nu'} \delta_{\nu i} \delta_{\mu\mu'} \delta_{\mu j} - \delta_{\nu\mu'} \delta_{\nu i} \delta_{\mu\nu'} \delta_{\mu j} \Big\},
\end{aligned}
\tag{15.6}
$$

where we have already considered the fact that a contraction is different from zero only if the single-particle state is of hole type, i.e. if its energy is below the Fermi surface. We used the common convention of indicating with i, j, k, l the hole states and with m, n, p, q, r the particle states.

By considering the restrictions imposed by the indexes of the Kronecker δ, we obtain

$$
\begin{aligned}
\hat{H} = {} & \sum_{\nu\nu'} T_{\nu\nu'} \hat{a}^+_\nu \hat{a}_{\nu'} + \frac{1}{4} \sum_{\mu\mu'\nu\nu'} \overline{V}_{\nu\mu\nu'\mu'} \hat{\mathbb{N}}[\hat{a}^+_\nu \hat{a}^+_\mu \, \hat{a}_{\mu'} \, \hat{a}_{\nu'}] \\
& + \frac{1}{4} \sum_{\mu\mu' i} \overline{V}_{\mu i \mu' i} \hat{\mathbb{N}}[\hat{a}^+_\mu \hat{a}_{\mu'}] + \frac{1}{4} \sum_{\nu\nu' i} \overline{V}_{i \nu i \nu'} \hat{\mathbb{N}}[\hat{a}^+_\nu \hat{a}_{\nu'}] \\
& - \frac{1}{4} \sum_{\mu\nu' i} \overline{V}_{i \mu \nu' i} \hat{\mathbb{N}}[\hat{a}^+_\mu \hat{a}_{\nu'}] - \frac{1}{4} \sum_{\nu\mu' i} \overline{V}_{\nu i i \mu'} \hat{\mathbb{N}}[\hat{a}^+_\nu \hat{a}_{\mu'}] \\
& + \frac{1}{4} \sum_{ij} \overline{V}_{ijij} - \frac{1}{4} \sum_{ij} \overline{V}_{ijji} \; .
\end{aligned}
\tag{15.7}
$$

The definition (15.2) of the antisymmetric matrix element implies the following relations:

$$
\overline{V}_{\nu\mu\nu'\mu'} = -\overline{V}_{\mu\nu\nu'\mu'} = \overline{V}_{\mu\nu\mu'\nu'} = -\overline{V}_{\nu\mu\mu'\nu'} \; , \tag{15.8}
$$

therefore

$$\hat{H} = \sum_{\nu\nu'} T_{\nu\nu'} \hat{a}^+_\nu \hat{a}_{\nu'} + \frac{1}{4} \sum_{\mu\mu'\nu\nu'} \overline{V}_{\nu\mu\nu'\mu'} \hat{\mathbb{N}}[\hat{a}^+_\nu \hat{a}^+_\mu \hat{a}_{\mu'} \hat{a}_{\nu'}]$$
$$+ \sum_{\nu\nu' i} \overline{V}_{\nu i \nu' i} \hat{\mathbb{N}}[\hat{a}^+_\nu \hat{a}_{\nu'}] + \frac{1}{2} \sum_{ij} \overline{V}_{ijij} \ . \quad (15.9)$$

Let us consider the normal ordered product of two operators and write it following the definition of contraction

$$\hat{\mathbb{N}}[\hat{a}^+_\nu \hat{a}_{\nu'}] = \hat{a}^+_\nu \hat{a}_{\nu'} - \overbrace{\hat{a}^+_\nu \hat{a}_{\nu'}} \ . \quad (15.10)$$

The last but one term of Eq. (15.9) becomes

$$\sum_{\nu\nu' i} \overline{V}_{\nu i \nu' i} \hat{\mathbb{N}}[\hat{a}^+_\nu \hat{a}_{\nu'}] = \sum_{\nu\nu' i} \overline{V}_{\nu i \nu' i} \hat{a}^+_\nu \hat{a}_{\nu'} - \sum_{ij} \overline{V}_{ijij} \ , \quad (15.11)$$

therefore, the hamiltonian can be expressed as

$$\hat{H} = \sum_{\nu\nu'} \left(T_{\nu\nu'} + \sum_i \overline{V}_{\nu i \nu' i} \right) \hat{a}^+_\nu \hat{a}_{\nu'}$$
$$+ \frac{1}{4} \sum_{\mu\mu'\nu\nu'} \overline{V}_{\nu\mu\nu'\mu'} \hat{\mathbb{N}}[\hat{a}^+_\nu \hat{a}^+_\mu \hat{a}_{\mu'} \hat{a}_{\nu'}] - \frac{1}{2} \sum_{ij} \overline{V}_{ijij} \ . \quad (15.12)$$

This expression clearly indicates the presence of a one-body operator in the Hamiltonian, specifically the term that multiplies $\hat{a}^+_\nu \hat{a}_{\nu'}$. It is noteworthy that part of the interaction $\overline{V}$ contributes to this one-body term. Until now, we have not made any assumptions about the structure of the basis of single-particle wave functions that compose the Slater determinant upon which the creation and annihilation operators act. Therefore, we can select the single-particle basis that diagonalizes the one-body term in Eq. (15.12)

$$h_{\nu\nu'} = T_{\nu\nu'} + \sum_i \overline{V}_{\nu i \nu' i} \ , \quad (15.13)$$

therefore

$$\langle \nu | h | \nu \rangle = \epsilon_\nu \ . \quad (15.14)$$

The expression of the hamiltonian in this basis is

$$
\hat{H} = \underbrace{\sum_\nu \epsilon_\nu \hat{a}^+_\nu \hat{a}_\nu - \frac{1}{2}\sum_{ij} \overline{V}_{ijij}}_{\hat{H}_0}
+ \underbrace{\frac{1}{4}\sum_{\mu\mu'\nu\nu'} \overline{V}_{\nu\mu\nu'\mu'} \hat{\mathbb{N}}[\hat{a}^+_\nu \hat{a}^+_\mu \hat{a}_{\mu'} \hat{a}_{\nu'}]}_{\hat{V}_{\rm res}} \equiv \hat{H}_0 + \hat{V}_{\rm res} \ . \tag{15.15}
$$

In the expression above, $\hat{H}_0$ is defined as the sum of the one-body term and the constant term. The operator $\hat{V}_{\rm res}$, referred to as the residual interaction, is the final term that includes the normal-ordered product of four creation and annihilation operators.

The expectation value of the Hamiltonian (15.15) calculated using the Slater determinant constructed from the eigenstates of $\hat{h}$, which describes the ground state of the system, is

$$
\begin{aligned}
\langle \Phi_0 | \hat{H} | \Phi_0 \rangle &= \langle \Phi_0 | \hat{H}_0 | \Phi_0 \rangle + \langle \Phi_0 | \hat{V}_{\rm res} | \Phi_0 \rangle \\
&= \sum_\nu \epsilon_\nu \langle \Phi_0 | \hat{a}^+_\nu \hat{a}_\nu | \Phi_0 \rangle - \frac{1}{2}\sum_{ij} \overline{V}_{ijij} \langle \Phi_0 | \Phi_0 \rangle \\
&+ \frac{1}{4}\sum_{\mu\mu'\nu\nu'} \overline{V}_{\nu\mu\nu'\mu'} \langle \Phi_0 | \hat{\mathbb{N}}[\hat{a}^+_\nu \hat{a}^+_\mu \hat{a}_{\mu'} \hat{a}_{\nu'}] | \Phi_0 \rangle \\
&= \sum_i \epsilon_i - \frac{1}{2}\sum_{ij} \overline{V}_{ijij} = \mathcal{E}_0 \ .
\end{aligned} \tag{15.16}
$$

Let us summaries here below some important points.

- The expression (15.15) of the hamiltonian is only a different way of writing Eq. (15.1), but in the former expression the normal ordered product is highlighted. There is not difference in the physics contents of the two expressions.
- The expectation value of the Hamiltonian calculated for the ground state Slater determinant neglects one term of the Hamiltonian. In the description of a many-body system using a single Slater determinant, or in other words, within the MF model, the contribution from the term related to the normal ordered product is zero by definition. This represents a significant approximation inherent in any mean-field description of a many-body system. The term of the Hamiltonian that is not included in this description is referred to as the residual interaction. The one-body part, $\hat{H}_0$, of the Hamiltonian is responsible for the IPM or MF model. Phenomena that go beyond this description arise from the presence of the residual interaction $\hat{V}_{\rm res}$, and are generally referred to as *long-range correlations*. This terminology is used to distinguish them from other effects related to the strongly repulsive core of the interaction, which are instead called *short-range correlations*.
- The total energy of the many-body system in the framework of a MF model can be calculated by knowing only the two-body interaction $\hat{V}$. The expression (15.16)

indicates that the total energy is not simply the sum of the single-particle energies but it explicitly contains an interaction term.

It is clear from the discussion in Chap. 8 that the interaction employed in the HF calculations is not one of the microscopic interactions outlined in Chap. 3. The strongly repulsive core of these interactions is not adequately treated in these calculations, necessitating the use of effective interactions, such as those presented in Chap. 8.

15.2.2 Hartree-Fock Equations

The variational calculation of the system energy is performed using Eq. (15.16). In HF theory, the goal is to identify the Slater determinant that minimises this energy value. This involves searching for the minimum of the ground state energy, which is considered a functional of the many-body wave function, within the subspace of the Hilbert space spanned by all possible Slater determinants that can be constructed.

The variational principle is applied by selecting a set of single-particle states to form the Slater determinant. It is essential that these single-particle wave functions are orthonormalised to create a valid basis. This requirement imposes an additional external condition that must be satisfied. Consequently, the problem becomes one of finding a constrained minimum, which is addressed using the technique of Lagrange multipliers.

The variational principle with the orthonormalisation conditions can be expressed as

$$\delta\left[E(\Phi) - \sum_{ij} \lambda_{ij} \langle i|j\rangle\right] = 0 \ , \tag{15.17}$$

and, by using the hamiltonian operator $\hat{H}$

$$\delta\langle\Phi|\hat{H}|\Phi\rangle - \sum_{ij} \lambda_{ij} \delta\langle i|j\rangle = 0 \ , \tag{15.18}$$

where $|\Phi\rangle$ is the Slater determinant formed by the functions $|i\rangle$, and λ_{ij} is the Lagrange multiplier.

By using the expression (15.12) of the hamiltonian operator, and recalling that $\left\langle\Phi|\hat{\mathbb{N}}|\Phi\right\rangle = 0$, we obtain

$$\sum_i \delta\langle i|\hat{T}|i\rangle + \frac{1}{2}\sum_{ij}\left[\delta\langle ij|\hat{V}|ij\rangle - \delta\langle ij|\hat{V}|ji\rangle\right] - \sum_{ij}\lambda_{ij}\delta\langle i|j\rangle = 0 \ . \tag{15.19}$$

Since i and j are dummy indexes indicating hole states, we have that

$$\sum_{ij} \delta \langle ij| = \sum_{ij} [\langle (\delta i) j| + \langle i(\delta j)|] = 2 \sum_{ij} \langle (\delta i) j| \ , \tag{15.20}$$

and therefore,

$$\sum_{i} \langle \delta i|\hat{T}|i\rangle + \sum_{ij} \left[\langle (\delta i) j|\hat{V}|ij\rangle - \langle (\delta i) j|\hat{V}|ji\rangle \right] - \sum_{ij} \lambda_{ij} \langle (\delta i)|j\rangle = 0 \ . \tag{15.21}$$

Since the variations $\langle \delta i|$ are independent of each other, each term of the above sum on i must be zero. This implies that for a generic term with $i = k$ we have

$$\langle \delta k|\hat{T}|k\rangle + \sum_{j} \left[\langle \delta k|\langle j|\hat{V}|j\rangle|k\rangle - \langle \delta k|\langle j|\hat{V}|k\rangle|j\rangle \right] = \sum_{j} \lambda_{kj} \langle \delta k|j\rangle \ . \tag{15.22}$$

Since $|\delta k\rangle$ is different from zero, we can simplify the above expression by writing

$$\hat{T}|k\rangle + \sum_{j} \left[\langle j|\hat{V}|j\rangle|k\rangle - \langle j|\hat{V}|k\rangle|j\rangle \right] = \sum_{j} \lambda_{kj}|j\rangle \ . \tag{15.23}$$

The Lagrange multiplier λ_{ij} is a real number, therefore it can be considered as the expectation value of a single-particle hamiltonian $\hat{h}$

$$\lambda_{kj} = \langle k|\hat{h}|j\rangle \ . \tag{15.24}$$

By using a unitary transformation, it is possible to find a single-particle basis where $\hat{h}$ is diagonal:

$$\langle \tilde{k}|\hat{h}|\tilde{j}\rangle = \epsilon_k \delta_{kj} \ . \tag{15.25}$$

The operator $\hat{\mathbb{U}}$ which makes the unitary transformation from a single-particle basis to another one is given by

$$|\tilde{k}\rangle = \sum_{k'} \hat{\mathbb{U}}_{kk'}|k'\rangle \quad \text{with} \quad \sum_{kk'} \hat{\mathbb{U}}^{\dagger}_{kk'} \hat{\mathbb{U}}_{k'k} = \hat{\mathbb{I}} \ . \tag{15.26}$$

The Slater determinant in the new basis is

$$|\tilde{\Phi}\rangle = \det(\hat{\mathbb{U}})|\Phi\rangle \ , \tag{15.27}$$

The $\hat{\mathbb{U}}$ is unitary operator which means that it satisfies the relation $|\det(\hat{\mathbb{U}})| = 1$. This implies that the value of the determinant in the two bases is the same, therefore the value of the functional $E(\Phi)$ is invariant under a unitary transformation of the basis, and, clearly, also its variation is invariant.

In the new basis (and here we substituted k to $\tilde{k}$ to simplify the writing), we obtain the expression

$$\hat{h}|k\rangle = \hat{T}|k\rangle + \sum_j \Big[\langle j|\hat{V}|j\rangle|k\rangle - \langle j|\hat{V}|k\rangle|j\rangle \Big] = \epsilon_k|k\rangle \ . \tag{15.28}$$

We define the average potential as

$$\hat{U}(\mathbf{r}) = \sum_j \langle j|\hat{V}|j\rangle = \sum_j \int d^3r' \phi_j^*(\mathbf{r}')\hat{V}(\mathbf{r},\mathbf{r}')\phi_j(\mathbf{r}') \ , \tag{15.29}$$

which is also called Hartree term [2]. The sum is carried on all the states below the Fermi surface. This term describes the interaction of the k particle with all the other ones.

The second term containing the interaction is non-local, and is called Fock - Dirac term [3]

$$\hat{W}(\mathbf{r},\mathbf{r}') = \sum_j \phi_j^*(\mathbf{r}')\hat{V}(\mathbf{r},\mathbf{r}')\phi_j(\mathbf{r}) \ . \tag{15.30}$$

In configuration space, the expression of Eq. (15.28) is

$$\hat{h}\phi_k(\mathbf{r}) = -\frac{\hbar^2\nabla^2}{2m}\phi_k(\mathbf{r}) + \underbrace{\hat{U}(\mathbf{r})\phi_k(\mathbf{r})}_{\text{Hartree}} - \underbrace{\int d^3r' \hat{W}(\mathbf{r},\mathbf{r}')\phi_k(\mathbf{r}')}_{\text{Fock-Dirac}} = \epsilon_k\phi_k(\mathbf{r}) \ . \tag{15.31}$$

By neglecting the Fock-Dirac term, we arrive at a differential equation of MF type. The Fock-Dirac term, also known as the exchange term, modifies the bare MF equation by incorporating the effects of the Pauli exclusion principle. This set of equations is the same as the one we derived in Sect. 10.4 using the first-order perturbation terms of the Dyson equation (see Eq. 10.57 for comparison).

It is important to note that the sums in Eqs. (15.29) and (15.30) are limited to states below the Fermi surface and do not have any additional restrictions. When considered separately in Eq. (15.28), these sums include self-interaction terms when $j = k$. In the HF case, the two terms are identical but have opposite signs; therefore, in the expression (15.28), there are no self-interaction contributions.

The non-linear-integro-differential equations (15.31) are solved numerically using an iterative procedure. The process begins with a set of trial wave functions $|k\rangle_{(1)}$, which are constructed using MF potentials, such as harmonic oscillator or Woods-Saxon. These trial wave functions are then used to calculate the Hartree (15.29) and Fock-Dirac (15.30) terms, which are incorporated into Eq. (15.31).

By applying standard numerical methods to solve these equations, a new set of wave functions $|k\rangle_{(2)}$ is generated. This new set can be utilized to compute updated $\hat{U}$ and $\hat{W}$ potentials. The iterative process continues until convergence is achieved. Typically, the convergence criterion is established by examining the differences between the total energies of the system (15.16) calculated in the i and $i+1$ iterations.

Koopmans' theorem
The physical significance of the Lagrange multiplier ϵ_k is elucidated by what is known as Koopmans' theorem [4], which states that this quantity represents the energy required to add a particle to a system of A particles.

We call $|\Phi_A\rangle$ the HF state describing the A nucleon system, and with $\mathcal{E}_A$ the energy of this system calculated with the HF theory. In the framework of the HF theory, we define the state of the system with an additional particle as

$$|\Phi_{A+1}\rangle = \hat{a}_k^+ |\Phi_A\rangle \quad . \tag{15.32}$$

The difference between the energies of the two systems is given by:

$$\begin{aligned}
\Delta\mathcal{E}_k^{\rm HF} &\equiv \left\langle \Phi_{A+1}|\hat{H}_0 + \hat{V}_{\rm res}|\Phi_{A+1}\right\rangle - \left\langle \Phi_A|\hat{H}_0|\Phi_A\right\rangle \\
&= \left\langle \Phi_A|\hat{a}_k \hat{H}_0 \hat{a}_k^+|\Phi_A\right\rangle + \left\langle \Phi_A|\hat{a}_k \hat{V}_{\rm res}\hat{a}_k^+|\Phi_A\right\rangle - \left\langle \Phi_A|\hat{H}_0|\Phi_A\right\rangle \\
&= \left\langle \Phi_A|\hat{a}_k \hat{H}_0 \hat{a}_k^+|\Phi_A\right\rangle + \left\langle \Phi_A|\hat{a}_k \hat{V}_{\rm res}\hat{a}_k^+|\Phi_A\right\rangle - \mathcal{E}_A^{\rm HF} \equiv \mathcal{E}_{A+1}^{\rm HF} - \mathcal{E}_A^{\rm HF} ,
\end{aligned} \tag{15.33}$$

where we used the fact that the property

$$\left\langle \Phi_A|\hat{\mathbb{N}}[\cdots]|\Phi_A\right\rangle = 0 \ , \tag{15.34}$$

is defined for the ground state of the A fermion system, but, in general, one has that

$$\left\langle \Phi_{A+1}|\hat{\mathbb{N}}[\cdots]|\Phi_{A+1}\right\rangle \neq 0 \ . \tag{15.35}$$

Let us calculate the first term of Eq. (15.33)

$$\left\langle \Phi_A|\hat{a}_k \hat{H}_0 \hat{a}_k^+|\Phi_A\right\rangle = \sum_{\nu=1}^{A+1} \epsilon_\nu - \frac{1}{2}\sum_{i,j=1}^{A+1} \overline{V}_{ijij}$$

$$= \sum_{\nu=1}^{A} \epsilon_\nu + \epsilon_k - \frac{1}{2}\sum_{i,j=1}^{A} \overline{V}_{ijij} - \sum_{i=1}^{A}\left(\frac{1}{2}\overline{V}_{ikik} + \frac{1}{2}\overline{V}_{kiki}\right) - \frac{1}{2}\overline{V}_{kkkk} \tag{15.36}$$

$$= \mathcal{E}_A^{\rm HF} + \epsilon_k - \sum_{i=1}^{A}\overline{V}_{ikik} \ , \tag{15.37}$$

since for the definition of $\overline{V}$ we have that

$$\overline{V}_{ikik} = \overline{V}_{kiki} \quad \text{and} \quad \overline{V}_{kkkk} = 0 \ . \tag{15.38}$$

Here ϵ_k is the Lagrange multiplier associated to the k single-particle state. The term of Eq. (15.33) related to the residual interaction is

$$\left\langle \Phi_A|\hat{a}_k \hat{V}_{\rm res}\hat{a}_k^+|\Phi_A\right\rangle = \frac{1}{4}\sum_{\mu\mu'\nu\nu'} \overline{V}_{\nu\mu\nu'\mu'}\langle \Phi_A|\hat{a}_k \hat{\mathbb{N}}[\hat{a}_\nu^+ \hat{a}_\mu^+ \hat{a}_{\mu'} \hat{a}_{\nu'}]\hat{a}_k^+|\Phi_A\rangle \ . \tag{15.39}$$

The expectation value of the term containing the creation and destruction operators is

$$
\begin{aligned}
&\langle \Phi_A | \hat{a}_k \hat{\mathbb{N}}[\hat{a}^+_\nu \hat{a}^+_\mu \, \hat{a}_{\mu'} \, \hat{a}_{\nu'}] \hat{a}^+_k | \Phi_A \rangle = \\
&\langle \Phi_A | \hat{a}_k \hat{a}^+_\nu \hat{a}^+_\mu \hat{a}_{\mu'} \hat{a}_{\nu'} \hat{a}^+_k | \Phi_A \rangle + \langle \Phi_A | \hat{a}_k \hat{a}^+_\nu \hat{a}^+_\mu \hat{a}_{\mu'} \hat{a}_{\nu'} \hat{a}^+_k | \Phi_A \rangle \\
+ \; &\langle \Phi_A | \hat{a}_k \hat{a}^+_\nu \hat{a}^+_\mu \hat{a}_{\mu'} \hat{a}_{\nu'} \hat{a}^+_k | \Phi_A \rangle + \langle \Phi_A | \hat{a}_k \hat{a}^+_\nu \hat{a}^+_\mu \hat{a}_{\mu'} \hat{a}_{\nu'} \hat{a}^+_k | \Phi_A \rangle \\
= \; &\delta_{k\nu} \delta_{k\nu'} \delta_{\mu\mu'} \delta_{\mu,i} - \delta_{k\nu} \delta_{k\mu'} \delta_{\mu\nu'} \delta_{\mu,i} \\
- \; &\delta_{k\mu} \delta_{k\nu'} \delta_{\nu\mu'} \delta_{\mu',i} + \delta_{k\mu} \delta_{k\mu'} \delta_{\nu\nu'} \delta_{\nu,i} \;\; . \qquad (15.40)
\end{aligned}
$$

Using the properties of $\overline{V}$ we can write

$$
\left\langle \Phi_A | a_k \hat{V}_{\rm res} a^+_k | \Phi_A \right\rangle = \sum_i \overline{V}_{kiki} \;\; . \qquad (15.41)
$$

Therefore

$$
\Delta \mathcal{E}^{\rm HF}_k = \mathcal{E}^{\rm HF}_A + \epsilon_k - \sum_{i=1}^{A} \overline{V}_{ikik} + \sum_{i=1}^{A} \overline{V}_{kiki} - \mathcal{E}^{\rm HF}_A = \epsilon_k \;\; , \qquad (15.42)
$$

which is Koopmans' theorem. The Lagrange multipliers ϵ_k of the constrained minimisation problem correspond to the energy differences between the MF energies of the $A + 1$ and A fermion systems.

This calculation can be repeated for the $A - 1$ case

$$
|\Phi_{A-1}\rangle = \hat{a}_k \, |\Phi_A\rangle \, ,
$$

and the result corresponds to the MF energy difference between the MF energies of the A and $A - 1$ fermion systems.

15.2.3 Hartree-Fock in Fermi Gas

A simple application of the HF theory is the description of an infinite and homogeneous system of fermions. This is the system that in Sect. 2.3 we called *Fermi gas.*

We rewrite the HF equations in the coordinate space

$$
\begin{aligned}
&-\frac{\hbar^2}{2m} \nabla^2 \phi_k(\mathbf{r}) + \sum_{k' \le k_{\rm F}} \int d^3 r' \, |\phi_{k'}(\mathbf{r}')|^2 \hat{V}(\mathbf{r}, \mathbf{r}') \phi_k(\mathbf{r}) \\
- &\sum_{k' \le k_{\rm F}} \int d^3 r' \phi^*_{k'}(\mathbf{r}') \hat{V}(\mathbf{r}, \mathbf{r}') \phi_k(\mathbf{r}') \phi_{k'}(\mathbf{r}) = \epsilon_k \phi_k(\mathbf{r}) \;\; , \qquad (15.43)
\end{aligned}
$$

where the sums on the occupied states are indicated as sums on the wave number k whose values are smaller than the Fermi value, k_F. The translational invariance implies that $\hat{V}(\mathbf{r}, \mathbf{r}') = \hat{V}(\mathbf{r} - \mathbf{r}')$. The single-particle wave functions are eigenstates of the momentum $\mathbf{p} = \hbar\mathbf{k}$, that is the plane waves defined in Eq. (2.38), and satisfy the Schrödinger equation

$$-\frac{\hbar^2}{2m}\nabla^2\phi_k(\mathbf{r}) = \epsilon_k^{(0)}\phi_k(\mathbf{r}) \ . \tag{15.44}$$

The term $\hat{V}(\mathbf{r} - \mathbf{r}')$ represents the interaction between two of the fermions of the system. In Fermi gas case, there is not interaction among the particles composing the system. Each particle moves independently of the presence of the other particles in a uniform, and constant, potential which can also be eliminated by defining the zero of the energy in appropriated way.

The HF theory proposes an alternative approximate solution to the many-body problem. The total wave function is still represented as a Slater determinant, but the interactions between fermions are now taken into account.

We define the Fourier transform of the interaction between two fermions as

$$\tilde{V}(\mathbf{k}) = \int d^3x \, \hat{V}(\mathbf{x}) \, e^{i\mathbf{k}\cdot\mathbf{x}} \ . \tag{15.45}$$

Let us consider the third term of Eq. (15.43), the Fock–Dirac term, by including the normalization factor (2.46)

$$\begin{aligned}
&\frac{\mathcal{V}}{(2\pi)^3}\int d^3k'\Theta(k_\mathrm{F}-k)\int d^3\mathbf{r}'\frac{e^{-i\mathbf{k}'\cdot\mathbf{r}'}}{\mathcal{V}^{1/2}}\left[e^{i\mathbf{k}\cdot\mathbf{r}}e^{-i\mathbf{k}\cdot\mathbf{r}}\right]\hat{V}(\mathbf{r}-\mathbf{r}')\frac{e^{i\mathbf{k}\cdot\mathbf{r}'}}{\mathcal{V}^{1/2}}\frac{e^{-i\mathbf{k}'\cdot\mathbf{r}}}{\mathcal{V}^{1/2}} \\
&= \frac{1}{(2\pi)^3}\int d^3k'\Theta(k_\mathrm{F}-k)\int d^3(r'-r)e^{i(\mathbf{k}'-\mathbf{k})\cdot(\mathbf{r}'-\mathbf{r})}\hat{V}(\mathbf{r}-\mathbf{r}')\frac{e^{i\mathbf{k}\cdot\mathbf{r}}}{\mathcal{V}^{1/2}} \\
&= \frac{1}{(2\pi)^3}\int d^3k'\Theta(k_\mathrm{F}-k)\tilde{V}(\mathbf{k}-\mathbf{k}')\frac{e^{i\mathbf{k}\cdot\mathbf{r}}}{\mathcal{V}^{1/2}} \ .
\end{aligned} \tag{15.46}$$

Since the density of the system is defined as (2.49),

$$\rho(r) = \sum_{k\leq k_\mathrm{F}} |\phi_a(\mathbf{r})|^2 \ , \tag{15.47}$$

by using (15.44) we can write the HF equation (15.43) as

$$\left[\epsilon_k^{(0)} + \rho\,\tilde{V}(0) - \frac{1}{(2\pi)^3}\int d^3k'\Theta(k_\mathrm{F}-k)\tilde{V}(\mathbf{k}-\mathbf{k}')\right]\phi_k(\mathbf{r}) = \epsilon_k\phi_k(\mathbf{r}) \ . \tag{15.48}$$

where

$$\tilde{V}(0) = \int d^3x \, \hat{V}(\mathbf{x}), \tag{15.49}$$

is called volume integral of the interaction.

The presence of the interaction modifies the energy of the particle

$$\epsilon_k = \epsilon_k^{(0)} + \mathcal{U}_{\mathrm{HF}}(\mathbf{k}) = \frac{\hbar^2 k^2}{2m} + \mathcal{U}_{\mathrm{HF}}(\mathbf{k}) \ , \tag{15.50}$$

where

$$\mathcal{U}_{\mathrm{HF}}(\mathbf{k}) = \rho\, \tilde{V}(0) - \frac{1}{(2\pi)^3} \int d^3k' \Theta(k_{\mathrm{F}} - k') \tilde{V}(\mathbf{k} - \mathbf{k}') \ . \tag{15.51}$$

15.3 Density Functional Theory

The Hartree-Fock (HF) model is widely utilized in nuclear and atomic physics, but its application presents two significant challenges. The first issue pertains to the theoretical framework, particularly within nuclear physics. Since HF is a MF theory, it necessitates the use of effective interactions, specifically, interactions that exclude the strongly repulsive core present at short distances between interacting particles.

In nuclear physics, the commonly employed interactions are phenomenological in nature and include terms that explicitly depend on the system density. Without these density-dependent terms, HF calculations struggle to accurately reproduce the binding energies and densities of nuclei. By incorporating these terms, researchers can develop interactions that yield high-quality results across the entire nuclide table.

However, the specific components of the bare nucleon-nucleon interaction, as well as the many-body effects that renormalise it, which are simulated by the density dependence of the effective interaction, remain subjects of ongoing research. A key formal consideration is that the variational principle, as previously defined, does not hold if the interaction $\hat{V}$ explicitly depends on density.

The second issue lies in the difficulty of evaluating the Fock-Dirac term in Eq. (15.31). Nuclei and atoms, even those exhibiting deformation, are systems that develop around a central point, which can conveniently serve as the origin of the coordinate system. While evaluating the Fock-Dirac term in these systems is not straightforward, it is still feasible. However, the situation becomes significantly more complex for intricate molecules that lack a well-defined central reference point or exhibit no translational invariance.

Density Functional Theory (DFT) addresses both of these issues. This theory is grounded in a theorem formulated in the latter half of the 1960s. Based on this theoretical foundation, a set of equations has been derived. These equations closely resemble the HF equations but incorporate an effective local term that accounts for the Pauli exclusion principle and includes correlations beyond the MF approach, effectively replacing the Fock-Dirac term.

15.3.1 Theorem of Hoenberg-Kohn

The starting point of the DFT is the theorem of Hohenberg-Kohn [5] stating that the ground state of a many-particle system can be completely characterised by the density of the particles and by correlated quantities. Let us consider the hamiltonian of a system of A fermions of mass m, and spin 1/2. We write it as

$$\hat{H} = \hat{T} + \hat{U}_{ext} + \hat{V} \ , \tag{15.52}$$

where

$$\hat{T} = \sum_{i=1} -\hbar^2 \frac{\nabla_i^2}{2m} \ , \quad \hat{U}_{ext} = \sum_{i=1} \hat{u}_{ext}(i) \ , \quad \hat{V} = \frac{1}{2} \sum_{ij} \hat{v}(i,j) \ , \tag{15.53}$$

where all the sums run on all the A fermions. The kinetic energy term, $\hat{T}$, and the external potential $\hat{U}_{ext}$, are one-body operators, while the interaction term $\hat{V}$ is a two-body potential. Other many-body interactions are not considered. The kinetic energy term plus $\hat{V}$ are characteristic of the many-fermion system, while $\hat{U}_{ext}$ depends on external situations, and therefore, in principle, it can be modified. For example, in an atomic system $\hat{U}_{ext}$ is given by the interaction of the electrons with the nucleus and it can be modified if the neutron number changes, modifying in this manner the charge distribution of the nucleus, even though the global electric charge of the nucleus remains the same. This is the origin of the so-called isotope shift effects on the atomic spectra. In an electron gas of a crystal the external field can be modified by changing the position of the ions in the crystal, or by considering the system of positive charges as uniform, and homogeneous, distribution with a defined charge density. In the case of the nucleus, the external field can be the MF where the nucleons are immersed and it can be described by potentials of harmonic oscillator or Woods-Saxon type.

Let us consider the set of all the hamiltonians of type (15.52) having non degenerate ground states, i.e. the set of external potentials $\hat{U}_{ext}$ generating a specific ground state $|\Psi_0\rangle$. This set of hamiltonians contains not only the physically reasonable potentials, but also an infinite number of potentials which have only a mathematical relevance. In addition, for each $\hat{U}_{ext}$ there is an infinite number of copies obtained by adding a constant. These copies generate the same ground state, therefore, from the physics point of view they are equivalent. The presence of degenerate states can be removed by inserting a small perturbation which removes the symmetry of the system.

The theorem states that there is a *bijective correspondence between the external potential* $\hat{U}_{ext}$*, the ground state* $|\Psi_0\rangle$ *and the number density*

$$\rho_0(\mathbf{r}) = \langle \Psi_0 | \sum_{i=1}^{N} \delta(\mathbf{r} - \mathbf{r}_i) | \Psi_0 \rangle \ , \tag{15.54}$$

of the system. This implies that a single potential cannot give rise to more than one ground state, nor can a ground state be produced by multiple external potentials. Furthermore, each ground state corresponds to only one density ρ_0, and each density is associated with a unique Ψ_0 state. A key aspect of the theorem is that both mappings are injective, ensuring their uniqueness.

The proof of the theorem requires two steps:
(i) for each $\hat{U}_{ext}$ only one Ψ_0 exists,
(ii) there is not Ψ_0 which is simultaneously the ground state of two potentials $\hat{U}_{ext}$ and $\hat{U}'_{ext}$ which can differ by more than a constant value.

(i) Since we consider a non degenerate system, by definition for each hamiltonian there is only one Ψ_0.
(ii) The proof of the second point is done by absurd. Let us assume that the same state $|\Psi_0\rangle$ is eigenstate of two hamiltonians differing by more than a constant term.

$$\hat{H}|\Psi_0\rangle = [\hat{T} + \hat{V} + U_{ext}]|\Psi_0\rangle = E_0|\Psi_0\rangle$$
$$\hat{H}'|\Psi_0\rangle = [\hat{T} + \hat{V} + U'_{ext}]|\Psi_0\rangle = E'_0|\Psi_0\rangle \ .$$

By subtracting the two terms of these equations we obtain

$$[\hat{U}_{ext} - \hat{U}'_{ext}]|\Psi_0\rangle = (E_0 - E'_0)|\Psi_0\rangle \ ,$$

We divide by $|\Psi_0\rangle$ and calculate the expectation value between $|r\rangle$

$$\sum_i \left\langle r_i|[\hat{U}_{ext} - \hat{U}'_{ext}]|r_i\right\rangle \equiv \sum_i [U_{ext}(\mathbf{r}_i) - U'_{ext}(\mathbf{r}_i)] = E_0 - E'_0 \ .$$

The previous equation implies that $U_{ext}(\mathbf{r}_i) - U'_{ext}(\mathbf{r}_i)$ is constant for each value of $\mathbf{r}_i$ and for each i. This means that the two potentials differ by a constant value. If we exclude this situation, the above equation leads to a contradiction. The left hand side changes if $\mathbf{r}_i$ changes, while the right hand side remains constant. Therefore every $\hat{U}_{ext}$, apart from a constant value, defines a hamiltonian with only one eigenstate describing the ground state of the system.

Also the second part of the statement, each density ρ_0 is generated by a single state $|\Psi_0\rangle$, is demonstrated by absurd. Let us assume that the same density is produced both by the $|\Psi_0\rangle$ and the $|\Psi'_0\rangle$ eigenstates of the hamiltonians H and H' above defined. For the variational principle we obtain the inequality

$$E_0 = \langle\Psi_0|\hat{H}|\Psi_0\rangle < \langle\Psi'_0|\hat{H}|\Psi'_0\rangle \ , \qquad (15.55)$$

where $\hat{H}$ is the hamiltonian of which $|\Psi_0\rangle$ is eigenstate, and the inequality comes from the fact that the system is not degenerate. We can rewrite the right hand side by adding and subtracting $\hat{U}'_{ext}$

$$\begin{aligned} E_0 &< \langle\Psi'_0|[(\hat{T} + \hat{V} + \hat{U}_{ext}) + \hat{U}'_{ext} - \hat{U}'_{ext}]|\Psi'_0\rangle \\ &= \langle\Psi'_0|[(\hat{T} + \hat{V} + \hat{U}'_{ext}) + \hat{U}_{ext} - \hat{U}'_{ext}]|\Psi'_0\rangle \\ &= E'_0 + \langle\Psi'_0|\hat{U}_{ext} - \hat{U}'_{ext}|\Psi'_0\rangle \ , \end{aligned}$$

The contribution of the one-body external potential can be written as:

$$\langle \Psi_0' | \sum_i U_{ext}(\mathbf{r}_i) | \Psi_0' \rangle = \int d^3r\, U_{ext}(\mathbf{r}) \langle \Psi_0' | \sum_i \delta(\mathbf{r} - \mathbf{r}_i) | \Psi_0' \rangle$$
$$= \int d^3r\, U_{ext}(\mathbf{r})\, \rho_0(\mathbf{r}) \ ,$$

where we assumed that

$$\langle \Psi_0' | \sum_i \delta(\mathbf{r} - \mathbf{r}_i) | \Psi_0' \rangle = \rho_0(\mathbf{r}) = \langle \Psi_0 | \sum_i \delta(\mathbf{r} - \mathbf{r}_i) | \Psi_0 \rangle \ .$$

We can write the inequality as

$$E_0 < E_0' + \int d^3r\, [U_{ext}(\mathbf{r}) - U_{ext}'(\mathbf{r})]\, \rho_0(\mathbf{r}).$$

All the discussion has been done by assuming Eq. (15.55) and it could be repeated by inverting the role of U and U' and we obtain

$$E_0' < E_0 + \int d^3r\, [\hat{U}_{ext}'(\mathbf{r}) - \hat{U}_{ext}(\mathbf{r})]\, \rho_0'(\mathbf{r}) \ .$$

Since $\rho_0 = \rho_0'$, by summing the two equations member to member

$$E_0 + E_0' < E_0 + E_0' \ ,$$

which is absurd. Therefore there is unique mapping between $|\Psi_0\rangle$ and ρ_0.

The Hoenberg-Kohn theorem has the following implications.

(a) There is a bijective mapping between external potential $\hat{U}_{ext}$ and the, not degenerate, ground state $|\Psi_0\rangle$ obtained by the solution of the Schrödinger equation, and the ground state density ρ_0

$$\hat{U}_{ext} \Longleftrightarrow |\Psi_0\rangle \Longleftrightarrow \rho_0 \ . \tag{15.56}$$

Since the three quantities are related by bijective mappings, we can consider the states as functionals of the density $|\Psi_0[\rho]\rangle$.

(b) Because of (a), $|\Psi_0[\rho]\rangle$ is a functional of ρ, every observable is also a functional of ρ: $O[\rho]$. Specifically, this is true for the energy of the system

$$E[\rho] = \langle \Psi[\rho] | \hat{H} | \Psi[\rho] \rangle = F[\rho] + \int d^3r\, \hat{U}_{ext}(\mathbf{r})\, \rho(\mathbf{r}) \ , \tag{15.57}$$

where the universal part, the part independent of the external potential, is defined as

$$F[\rho] \equiv \langle \Psi[\rho] | \hat{T} + \hat{V} | \Psi[\rho] \rangle \ . \tag{15.58}$$

(c) There is a principle related to the minimum of E. If ρ_0 is the ground state density corresponding to a specific value of $\hat{U}_{ext}$, then, for each $\rho \neq \rho_0$ the following relation holds:

$$E_0 \equiv E[\rho_0] < E[\rho] . \tag{15.59}$$

This is a consequence of the unicity of the relation between density, eigenstate and external potential, and of the variational principle.

15.3.2 Khon and Sham Equations

The application of the Hoenberg-Khon theorem is based on the idea of building the ground state density ρ_0 of a system of interacting fermions by using a fictitious system of non-interacting fermions by changing the hamiltonian. The idea is graphically described in Fig. 15.1.

The left part of the figure represents the system to be described. The particles are immersed in a potential and interact with each other through an interaction $\hat{V}$. The density of this system is represented by the dashed line.

The idea of Khon and Sham is to describe this system by changing the MF potential in such a way that, while not altering the density of the system, the particles no longer interact with each other.

The idea of describing a system of interacting fermions by using an effective system of non interacting fermions is analogous to that used by Landau to describe the Fermi liquids (see Chap. 17). In this latter case the properties of the fermions are modified, they acquire effective masses and charges. In the Khon e Sham approach the

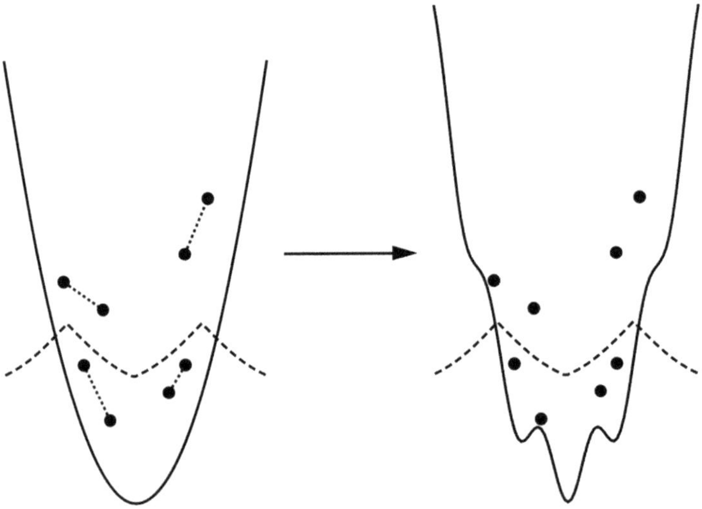

Fig. 15.1 Graphical representation of the Khon-Sham approach

properties of the particles remains the same as in the interacting case. The hamiltonian is modified by changing the external potential and neglecting the interaction term $\hat{V}$.

The starting point consists in writing the density (15.54) as a sum of orthonormalized single-particle wave functions

$$\rho_0(\mathbf{r}) = \rho_0^{\mathrm{KS}}(\mathbf{r}) = \sum_{i<\epsilon_{\mathrm{F}}} |\phi_i^{\mathrm{KS}}(\mathbf{r})|^2 \ , \tag{15.60}$$

where KS indicates Kohn-Sham. The density (15.60) is generated by a one-body hamiltonian whose eigenstate is a Slater determinant Φ^{KS}.

The energy functional built in this system is usually expressed as:

$$E[\rho_0] = T^{\mathrm{KS}}[\rho_0] + E_{\mathrm{H}}^{\mathrm{KS}}[\rho_0] + E_{\mathrm{ext}}^{\mathrm{KS}}[\rho_0] + E_{\mathrm{xc}}^{\mathrm{KS}}[\rho_0] \ , \tag{15.61}$$

where there is a kinetic energy term,

$$T^{\mathrm{KS}}[\rho_0] = \langle \Phi^{\mathrm{KS}} | \hat{T} | \Phi^{\mathrm{KS}} \rangle = \int d^3r \sum_i \phi_i^{*\mathrm{KS}}(\mathbf{r}) \left(-\frac{\hbar^2 \nabla^2}{2m} \right) \phi_i^{\mathrm{KS}}(\mathbf{r}) \ , \tag{15.62}$$

a Hartree term,

$$E_{\mathrm{H}}^{\mathrm{KS}}[\rho_0] = \int d^3r_i \int d^3r_j \rho_0(\mathbf{r}_i) \hat{V}(\mathbf{r}_i, \mathbf{r}_j) \rho_0(\mathbf{r}_j) \ , \tag{15.63}$$

and an external MF term

$$E_{\mathrm{ext}}^{\mathrm{KS}}[\rho_0] = \int d^3r \rho_0(\mathbf{r}_i) \hat{U}_{\mathrm{ext}}^{\mathrm{KS}}(\mathbf{r}_i) \ . \tag{15.64}$$

The additional term is called of exchange and correlation $E_{\mathrm{xc}}^{\mathrm{KS}}$.

The application of the variational principle follows the path outlined in the case of HF. The final result is again a set of equations which allows the evaluation of the KS single-particle wave functions [6]

$$\left\{ -\frac{\hbar^2 \nabla^2}{2m} + \int d^3r_j \hat{V}(\mathbf{r}, \mathbf{r}_j) \rho_0(\mathbf{r}_j) + \hat{U}_{\mathrm{ext}}^{\mathrm{KS}}(\mathbf{r}) + \hat{U}_{\mathrm{xc}}^{\mathrm{KS}}(\mathbf{r}) \right\} \phi_i^{\mathrm{KS}}(\mathbf{r}) = \epsilon_i \phi_i^{\mathrm{KS}}(\mathbf{r}) \ . \tag{15.65}$$

This non-linear-integro-differential equation is numerically solved by using iterative techniques analogous to those used in the HF case.

- In Eq. (15.65) only local terms appear, contrary to the HF equation containing the non-local Fock-Dirac term. This makes the numerical solution of the KS equations much simpler than that of the HF equations.
- The expression of the operators of kinetic energy and in the Hartree term of the KS functional are the same as in the interacting system. This does not mean that

from the quantitative point of view the values of the kinetic energy and of the Hartree term are equal in the two systems. In effect, here the expectation values are evaluated between the Slater determinant for the KS functional, and not between $|\Psi_0\rangle$ as in the case of the interacting systems.

- In the KS energy functional (15.61), the components that account for the contributions from kinetic energy, the Hartree term and the external potential, are significantly larger than the contribution from the exchange and correlation term.
- The Slater determinant $|\Phi^{\rm KS}\rangle$ is not the exact eigenvalue $|\Psi_0\rangle$ of the many-body hamiltonian. The DFT is based on the equality between densities (15.60), called one-body densities. The one-body densities contain poorer information than the states $|\Psi_0\rangle$. We shall discuss this point in more detail in the next section.
- The single-particle wave functions $\phi_i^{\rm KS}(\mathbf{r})$ do not possess a precise physical interpretation; rather, they should be viewed as a mathematical tool for deriving the particle density. In DFT, the density is the only quantity that holds physical significance for calculating observable values. Additionally, the eigenvalues ϵ_i should be regarded solely as Lagrange multipliers, rather than as single-particle energies, as indicated by Koopmans' theorem in the HF framework.
- There are no theoretical prescriptions or limits to guide the selection of the exchange and correlation term. Instead, this choice is made on a pragmatic basis. A significant portion of the theoretical work within the DFT framework focuses on the development of this term.
- The DFT is an independent particle model.

15.4 Density and Single-Particle Wave Functions

In this section, we discuss the definitions of particle densities and their significance within the framework of HF and DFT. We define the density matrix of a many-body system as

$$\rho(\mathbf{r}_1, \mathbf{r}_1') = \frac{A}{\langle\Psi|\Psi\rangle} \int d^3r_2\, d^3r_3 \cdots d^3r_A \Psi^*(\mathbf{r}_1, \mathbf{r}_2, \ldots, \mathbf{r}_A)\Psi(\mathbf{r}_1', \mathbf{r}_2, \ldots, \mathbf{r}_A) \ , \tag{15.66}$$

where Ψ is the wave function describing the system.

For the sake of clarity, in Eq. (15.66) we define the one-body density matrix, where all coordinates, except one, are integrated. The density matrix (15.66) is normalized to the particle number A, which can be obtained by integrating over the coordinates $\mathbf{r}_1$ and $\mathbf{r}_1'$, while also incorporating a $\delta(\mathbf{r}_1 - \mathbf{r}_1')$ factor. The density considered in DFT is the diagonal component of the one-body density matrix.

Evidently, the information contained in the one-body density matrix is less comprehensive than that included in the wave function Ψ, as integrating over $A-1$ coordinates results in a significant loss of information. Moreover, this loss is exacerbated when only the diagonal elements of the density matrix are considered. This limitation becomes particularly apparent when studying observables that are of a two-body nature or are sensitive to the off-diagonal elements of the density matrix.

In the IPM, the state $|\Psi\rangle$ is a Slater determinant $|\Phi\rangle$ composed by a set of orthonormal single-particle states $|\phi_i\rangle$. By inserting in Eq. (15.66) a Slater determinant we obtain a density matrix given by

$$\rho^{\mathrm{IPM}}(\mathbf{r}_1, \mathbf{r}'_1) = \sum_{i \leq \epsilon_{\mathrm{F}}} \phi_i^*(\mathbf{r}_1)\phi_i(\mathbf{r}'_1). \tag{15.67}$$

The **mean-field single particle wave functions** are those generated in a IPM and produce densities of the type (15.67).

The **natural orbits** are defined as those single particle wave functions which diagonalise the density matrix (15.66) which can be described as

$$\rho(\mathbf{r}_1, \mathbf{r}'_1) = \sum_{\alpha} c_\alpha \phi_\alpha^{*\mathrm{NO}}(\mathbf{r}_1)\phi_\alpha^{\mathrm{NO}}(\mathbf{r}'_1). \tag{15.68}$$

The difference between the expressions (15.67) and (15.68) consists in the fact that, in the second expression, the sum is unlimited, i.e. extended up to infinity, and the coefficients c_α are the diagonal terms of the density matrix, and indicate the occupation of the natural orbit. The two densities are normalised such as

$$\begin{aligned} A &= \int d^3r_1 \rho^{\mathrm{IPM}}(\mathbf{r}_1, \mathbf{r}'_1)\delta(\mathbf{r}_1 - \mathbf{r}'_1) \\ &= \int d^3r_1 \rho^{\mathrm{NO}}(\mathbf{r}_1, \mathbf{r}'_1)\delta(\mathbf{r}_1 - \mathbf{r}'_1) = \sum_{\alpha} c_\alpha \,. \end{aligned} \tag{15.69}$$

This allows the interpretation of the expression (15.67) in analogy to (15.68) where the occupation numbers are 1 for the states below the Fermi surface, and 0 for those above it.

In Fig. 15.2 we compare the occupation numbers of an IPM with those obtained by a microscopic calculations (FHNC-SOC/1 framework see Sect. 11.3.1) carried out with two different correlation functions [7]. The many-body system under investigation is the ^{48}Ca.

The IPM predicts full occupation for the states below the Fermi surface, while above it the occupation is zero. The calculations which include correlations show that the natural orbits are only partially occupied below the Fermi surface, even though the occupation numbers are very close to 1. Above the Fermi surface, the occupation numbers are very small but not zero.

In [7] it is shown that there is a great similarity between single-particle wave functions and natural orbits. The effects of correlations are mainly related to the changes of the occupation numbers rather than on the shape of the wave functions.

Another type of wave functions referred to the individual fermion is that of the **quasi-particle wave functions** defined as the superposition of the wave functions

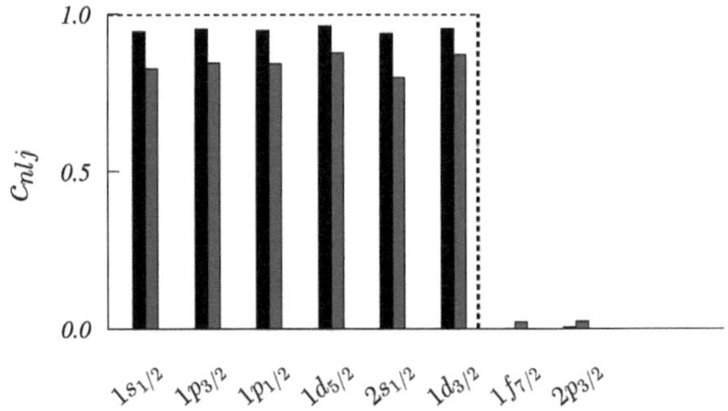

Fig. 15.2 Comparison between the occupation numbers of MF single-particle wave functions, represented by dashed lines, with those of natural orbits indicated by the histograms [7]. Each wave function is identified by its principal quantum number, orbital angular momentum, and total angular momentum. The natural orbit occupation numbers c_{nlj} have been calculated within a FHNC-SOC/1 framework (see Sect. 11.3.1) using two different correlation functions. The black bars represent the values obtained with the scalar correlation function, while the grey bars indicate the values derived from a correlation that includes operator-dependent terms. The system under investigation is the ^{48}Ca nucleus, and the results presented here pertain to the protons

Fig. 15.3 Differences between the charge distributions of ^{206}Pb and ^{205}Tl. IPM indicates the result obtained with MF wave functions. The other lines have been obtained by considering different types of correlations, long-range correlations LRC, and short-range correlations, SRC. The results have been taken from Ref. [8]

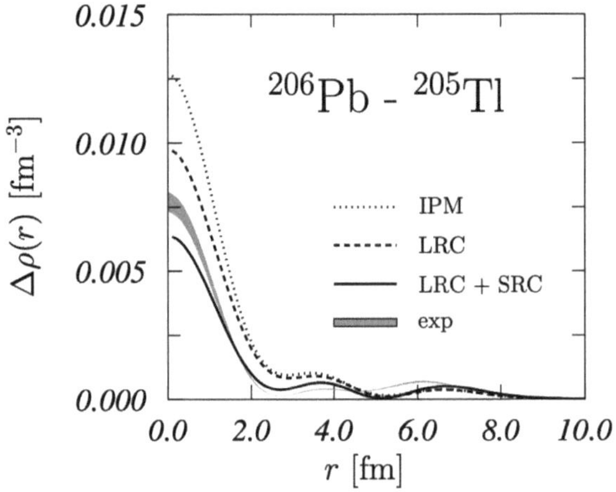

describing systems with A and $A-1$ particles

$$\psi_\alpha(\mathbf{r}) = \frac{\sqrt{A}\langle\Psi(A-1)|\delta(\mathbf{r}-\mathbf{r}_A)|\Psi(A)\rangle}{\langle\Psi(A-1)|\Psi(A-1)\rangle^{1/2}\langle\Psi(A)|\Psi(A)\rangle^{1/2}} \; . \tag{15.70}$$

As example of quasi-particle wave function, we show in Fig. 15.3 the difference between the charge distributions of the ^{206}Pb and ^{205}Tl nuclei. The shell structure of these nuclei indicates that the difference between these two charge distributions is due to the absence of a proton in the $3s_{1/2}$ state in ^{205}Tl.

The line labeled as IPM represents the square modulus of the single-particle wave function within the IPM framework. The form of this wave function closely resembles that observed in experiments. However, the results from the IPM are significantly higher than the experimental curve. To accurately describe this charge distribution, it is essential to incorporate various effects, such as the collective surface vibrations of the nuclei, referred to as Long-Range Correlations (LRC). Additionally, it is important to consider the strong repulsive core present in the interaction between two nucleons, which is denoted as Short-Range Correlations (SRC).

In the IPM, the single-particle wave functions, natural orbits, and quasi-particle wave functions are equivalent. However, in calculations that extend beyond the IPM, the subtle differences among these three types of wave functions become apparent only in very specific cases. This is what makes MF calculations, such HF and DFT, particularly valuable. Within the framework of DFT, which is formulated in terms of a functional of the density, correlation effects are only observable in quantities that are sensitive to the off-diagonal elements of the one-body density matrix (15.66).

One observable of this type is the momentum distribution, traditionally indicated as $n(\mathbf{k})$, and defined as:

$$n(\mathbf{k}) = \frac{1}{(2\pi)^3} \frac{1}{\langle \Psi | \Psi \rangle} \int d^3(r - r')\, e^{i\mathbf{k}\cdot(\mathbf{r}-\mathbf{r}')} \rho(\mathbf{r}, \mathbf{r}') \ . \tag{15.71}$$

The momentum distribution addresses the question of the probability of finding a particle within a many-body system that has a momentum (or wave number) value between $\mathbf{k}$ and $\mathbf{k} + d\mathbf{k}$.

In Fig. 15.4, we present the momentum distributions for five doubly magic nuclei, which are spherical in shape. As a result, the momentum distributions depend on k, the absolute value of $\mathbf{k}$

The full lines have been obtained by considering an IPM, or, in other words, a diagonal one-body density matrix. The dashed lines show the results of CBF calculations carried on within a FHNC-SOC/1 computational scheme (see Chap. 11) [7].

The difference between the IPM results and the correlated ones is remarkable at large values of k, which indicate that the off-diagonal parts of the density matrix generates high-momentum components.

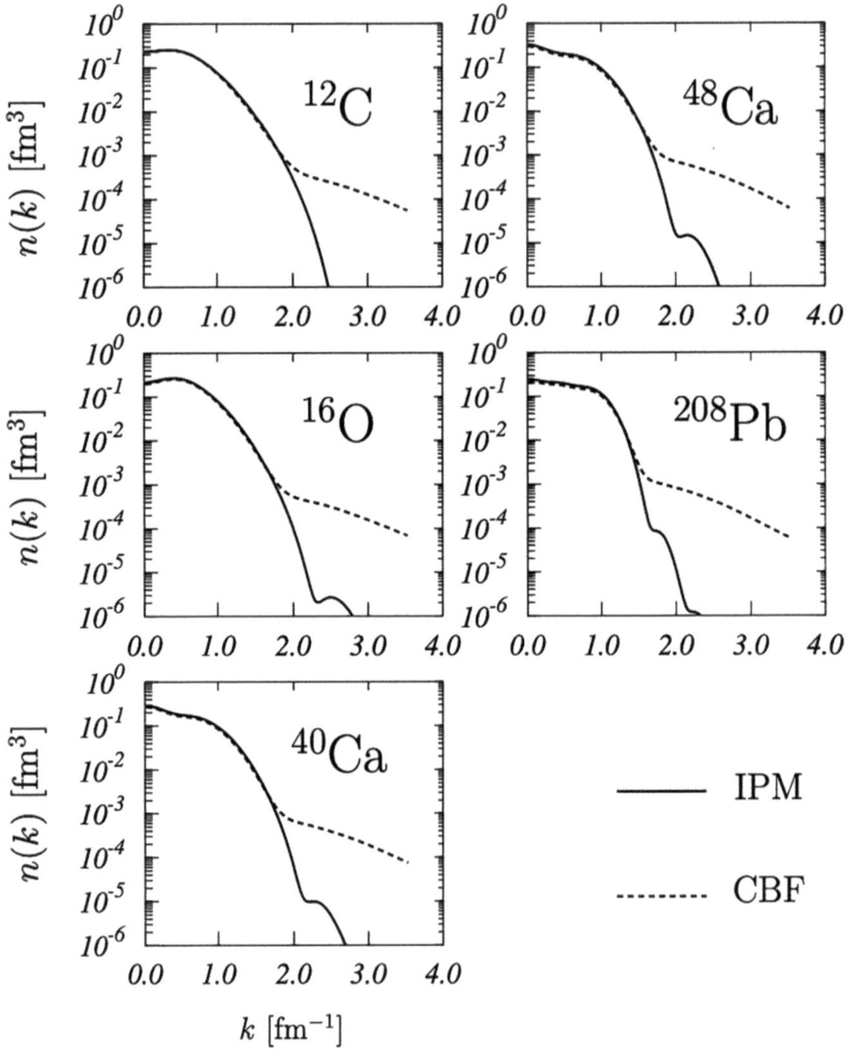

Fig. 15.4 Momentum distributions for five doubly-magic nuclei obtained in IPM and with CBF calculations (see Chap. 11). The effect of the correlations appears at high values of the momentum [7]

References

1. A. Messiah, *Quantum Mechanics* (North Holland, Amsterdam, 1961)
2. D.R. Hartree, The wave mechanics of an atom with a non-coulomb central field. Math. Proc. Camb. Philos. Soc. **24**, 89 (1928)
3. V. Fock, Näherungsmethode zur Lösung der Quantennahen Probleme. Zeit. für Phys. **61**, 126 (1930)

4. B.O. Koopman, Hamiltonian systems and the theory of integral equations. Proc. Natl. Acad. Sci. U. S. A. **17**, 315 (1931)
5. P. Hohenberg, W. Kohn, Inhomogeneous electron gas. Phys. Rev. **136**, B864 (1964)
6. E. Engel, R.M. Dreizler, *Density Functional Theory* (Springer, Berlin, 2011)
7. C. Bisconti, G. Co', F. Arias de Saavedra, Momentum distributions and spectroscopic factors of doubly-closed shell nuclei in correlated basis function theory. Phys. Rev. C **75**, 054302 (2007)
8. M. Anguiano, G. Co', Correlations and charge distributions of medium heavy nuclei. J. Phys. G **27**, 2109 (2001)

Chapter 16
Excited States

In the previous chapters, we focused on describing the ground state of many-fermion systems. In this chapter, we introduce the Random Phase Approximation (RPA), a theory designed to describe the excitation spectrum of these systems. The RPA extends beyond the Independent Particle Model (IPM) since it considees effects generated by the residual interaction, as defined in Eq. (15.16).

The RPA theory was proposed in the mid-1950s to describe plasma excitations [1]. The name "Random Phase Approximation" has historical significance, as the authors of Ref. [1] assumed that the coupling of plasma oscillations with different momenta could be neglected. This assumption pertains to the system under investigation rather than the theory itself, which does not incorporate any random phase.

We have already presented the RPA theory in Sect. 10.6 using two-body Green's functions. The formulation presented in this chapter, known as the Equation of Motion Method, is less general but it offers a simpler interpretation. There is also a third method for deriving the RPA equations, based on time-dependent Hartree-Fock (HF) theory. The basic steps of this latter approach are outlined in Appendix D. All these formulations ultimately lead to the same set of equations that must be solved numerically, each highlighting different aspects that may be obscured in other approaches. Typically, for infinite systems with translational invariance, the Green's function formulation is preferred, while the Equation of Motion method is often chosen for systems with rotational invariance.

16.1 The Equations of Motion Method

This method is inspired to the Heisenberg picture of the Quantum Mechanics, see Sect. 6.1. Also in this case, the commutation properties of a hamiltonian with an operator are exploited.

© The Author(s), under exclusive license to Springer Nature Switzerland AG 2026

G. Co', *Concepts in Quantum Many-Body Physics*, UNITEXT for Physics,
https://doi.org/10.1007/978-3-032-08920-5_16

Let us consider the Schrödinger equation

$$\hat{H} |\Psi_\nu\rangle = E_\nu |\Psi_\nu\rangle \quad , \tag{16.1}$$

where $|\Psi_\nu\rangle$ describe an excited state of the system. We define the $\hat{Q}^+_\nu$ operator whose action on the system ground state defines the excited states:

$$\hat{Q}^+_\nu |\Psi_0\rangle = |\Psi_\nu\rangle \quad ; \quad \hat{Q}_\nu |\Psi_0\rangle = 0 \ . \tag{16.2}$$

The choice of $\hat{Q}^+_\nu$ defines completely the problem to be solved, and also the ground state of the system. Let us calculate the commutator of the $\hat{Q}^+_\nu$ operator with the hamiltonian

$$\begin{aligned}\left[\hat{H}, \hat{Q}^+_\nu\right] |\Psi_0\rangle &= \left(\hat{H}\hat{Q}^+_\nu - \hat{Q}^+_\nu \hat{H}\right) |\Psi_0\rangle = \hat{H} |\Psi_\nu\rangle - \hat{Q}^+_\nu E_0 |\Psi_0\rangle \\ &= E_\nu |\Psi_\nu\rangle - \hat{Q}^+_\nu E_0 |\Psi_0\rangle = (E_\nu - E_0)\, \hat{Q}^+_\nu |\Psi_0\rangle \quad , \end{aligned} \tag{16.3}$$

and for the operator $\hat{Q}_\nu$ we obtain

$$\left[\hat{H}, \hat{Q}_\nu\right] |\Psi_0\rangle = \left(\hat{H}\hat{Q}_\nu - \hat{Q}_\nu \hat{H}\right) |\Psi_0\rangle = \hat{H}\hat{Q}_\nu |\Psi_0\rangle - E_0 \hat{Q}_\nu |\Psi_0\rangle = 0 \ . \tag{16.4}$$

We multiply Eq. (16.3) by a generic operator $\hat{\mathbb{O}}$ and by $\langle\Psi_0|$, and we subtract the complex conjugate

$$\begin{aligned}&\left\langle\Psi_0\right|\left[\hat{\mathbb{O}}, [\hat{H}, \hat{Q}^+_\nu]\right]\left|\Psi_0\right\rangle = \left\langle\Psi_0|\hat{\mathbb{O}}[\hat{H}, \hat{Q}^+_\nu]|\Psi_0\right\rangle - \left\langle\Psi_0|[\hat{H}, \hat{Q}^+_\nu]\hat{\mathbb{O}}|\Psi_0\right\rangle \\ =\ &\left\langle\Psi_0\right|\left(\hat{\mathbb{O}}[\hat{H}, \hat{Q}^+_\nu] - [\hat{H}, \hat{Q}^+_\nu]\hat{\mathbb{O}}\right)\left|\Psi_0\right\rangle \ \text{for Eqs. (16.3) and (16.4)} \\ =\ &(E_\nu - E_0)\left\langle\Psi_0|\hat{\mathbb{O}}\hat{Q}^+_\nu|\Psi_0\right\rangle - \left\langle\Psi_0|\hat{H}\hat{Q}^+_\nu\hat{\mathbb{O}}|\Psi_0\right\rangle + \left\langle\Psi_0|\hat{Q}^+_\nu\hat{H}\hat{\mathbb{O}}|\Psi_0\right\rangle \\ =\ &(E_\nu - E_0)\left\langle\Psi_0|\hat{\mathbb{O}}\hat{Q}^+_\nu|\Psi_0\right\rangle - E_0\left\langle\Psi_0|\hat{Q}^+_\nu\hat{\mathbb{O}}|\Psi_0\right\rangle + \left\langle\Psi_0|\hat{Q}^+_\nu\hat{H}\hat{\mathbb{O}}|\Psi_0\right\rangle \ .\end{aligned}$$

Since $\langle\Psi_0|\,\hat{Q}^+_\nu = 0$, we can write

$$\begin{aligned}\left\langle\Psi_0\right|\left[\hat{\mathbb{O}}, [\hat{H}, \hat{Q}^+_\nu]\right]\left|\Psi_0\right\rangle &= (E_\nu - E_0)\left\langle\Psi_0|\hat{\mathbb{O}}\hat{Q}^+_\nu|\Psi_0\right\rangle \\ &= (E_\nu - E_0)\left\langle\Psi_0\right|\left[\hat{\mathbb{O}}, \hat{Q}^+_\nu\right]\left|\Psi_0\right\rangle \ .\end{aligned} \tag{16.5}$$

This result is independent of the expression of the operator $\hat{\mathbb{O}}$. In the construction of the various theories describing the system excited states, the $\hat{\mathbb{O}}$ operator is substituted by the $\delta\hat{Q}_\nu$ operator representing an infinitesimal variation of the excitation operator defined by the Eq. (16.2).

16.2 Tamm-Dankoff Approximation (TDA)

A first choice the $\hat{Q}^+_\nu$ consists in considering the excited state as linear combination of particle-hole excitations. Clearly, this implies that the MF problem has already been solved, and a set of orthonormal single-particle wave functions, with the relative single-particle energies, is available. The ground state is characterised by fully occupied hole states, those below the Fermi energy, and totally empty particle states, those above the Fermi energy. The set of single-particle wave functions can be obtained by using the IPM techniques presented in Chap. 2 or by solving the HF or the KS equations as indicated in Chap. 15.

The definition of $\hat{Q}^+_\nu$ in terms of creation and destruction operators is

$$\hat{Q}^+_\nu = \sum_{ph} X^\nu_{ph} \hat{a}^+_p \hat{a}_h, \tag{16.6}$$

and then

$$\delta \hat{Q}_\nu = \sum_{ph} \hat{a}^+_h \hat{a}_p \delta X^{*\nu}_{ph} \ . \tag{16.7}$$

In the equation above, the X^ν_{ph} are a numbers. We have adopted the conventional notation of using the letters h, i, j, k, l to denote hole states, and m, n, p, q, r to represent particle states. The ground state $|\Psi_0\rangle$ satisfying the Eqs. (16.3) and (16.4) is the MF ground state $|\Phi_0\rangle$, in effect

$$\hat{Q}_\nu \, |\Phi_0\rangle = \sum_{ph} X^\nu_{ph} \hat{a}^+_h \hat{a}_p \, |\Phi_0\rangle = 0 \ , \tag{16.8}$$

since it is not possible to remove particles above the Fermi surface or to add particles below it (see Chap. 5).

By substituting Eq. (16.7) in Eq. (16.5) we obtain

$$\begin{aligned} &\left\langle \Phi_0 \Big| \Big[\sum_{ph} \hat{a}^+_h \hat{a}_p \delta X^{*\nu}_{ph}, \big[\hat{H}, \sum_{p'h'} X^\nu_{p'h'} \hat{a}^+_{p'} \hat{a}_{h'} \big] \Big] \Big| \Phi_0 \right\rangle \\ &= (E_\nu - E_0) \left\langle \Phi_0 \Big| \Big[\sum_{mi} \hat{a}^+_i \hat{a}_m \delta X^{*\nu}_{mi}, \sum_{nj} X^\nu_{nj} \hat{a}^+_n \hat{a}_j \Big] \Big| \Phi_0 \right\rangle \ . \end{aligned} \tag{16.9}$$

Every variation $\delta X^{*\nu}_{ph}$ is independent of the other ones. For this reason, the above equation is a sum of terms independent of each other. The equation is satisfied if all the terms related to the same variation of X^ν_{ph} satisfy the relation. Let us consider a single term of the sum, and, since $\delta X^{*\nu}_{ph} \neq 0$, we can divide it by this factor to obtain a system of connected equations which can be expressed as follow

$$
\begin{aligned}
&\left\langle \Phi_0 \middle| \left[\hat{a}_i^+ \hat{a}_m , [\hat{H}, \sum_{nj} X^\nu_{nj} \hat{a}_n^+ \hat{a}_j] \right] \middle| \Phi_0 \right\rangle \\
&= (E_\nu - E_0) \sum_{qk} X^\nu_{qk} \left\langle \Phi_0 \right| \left[a_i^+ a_m , a_q^+ a_k \right] \left| \Phi_0 \right\rangle \ . \qquad (16.10)
\end{aligned}
$$

Let us calculate the right hand side of Eq. (16.10):

$$
\langle \Phi_0 | \left[\hat{a}_i^+ \hat{a}_m , \hat{a}_q^+ \hat{a}_k \right] | \Phi_0 \rangle = \langle \Phi_0 | \hat{a}_i^+ \hat{a}_m \hat{a}_q^+ \hat{a}_k | \Phi_0 \rangle - \langle \Phi_0 | \hat{a}_q^+ \hat{a}_k , \hat{a}_i^+ \hat{a}_m | \Phi_0 \rangle \ .
$$

We apply the Wick theorem to the first term

$$
\langle \Phi_0 | \hat{a}_i^+ \hat{a}_m \hat{a}_q^+ \hat{a}_k | \Phi_0 \rangle = \delta_{mq} \delta_{ik} \ .
$$

The second term is zero since

$$
\hat{a}_m \, | \Phi_0 \rangle = 0 \ .
$$

By using this result in Eq. (16.10) we obtain

$$
\left\langle \Phi_0 \middle| \left[\hat{a}_i^+ \hat{a}_m , [\hat{H}, \sum_{nj} X^\nu_{nj} \hat{a}_n^+ \hat{a}_j] \right] \middle| \Phi_0 \right\rangle = (E_\nu - E_0) \, X^\nu_{mi} \ .
$$

For the calculation of the left hand side of Eq. (16.2) we use the expression of the hamiltonian given in Eq. (15.15)

$$
\hat{H} = \sum_{\alpha\beta} h_{\alpha\beta} \hat{a}_\alpha^+ \hat{a}_\beta - \frac{1}{2} \sum_{ij} \overline{V}_{ijij} + \frac{1}{4} \sum_{\mu\mu'\nu\nu'} \overline{V}_{\nu\mu\nu'\mu'} \hat{\mathbb{N}}[\hat{a}_\nu^+ \hat{a}_\mu^+ \, \hat{a}_{\mu'} \, \hat{a}_{\nu'}] \ , \qquad (16.11)
$$

where the greek indexes can assume values above or below the Fermi surface, while for the latin indexes we used the common convention. The calculation of the double commutator is carried out here below.

We calculate the double commutators of Eq. (16.2) by considering one term at the time. For the first commutator of the first term of Eq. (16.11) we obtain, by considering the anti-commutation rules of the creation and destruction operators,

$$
[\hat{a}_\alpha^+ \hat{a}_\beta , \hat{a}_n^+ \hat{a}_j] = \delta_{n\beta} \hat{a}_\alpha^+ \hat{a}_j - \delta_{j\alpha} \hat{a}_n^+ \hat{a}_\beta \ ,
$$

therefore, the commutator of the hamiltonian can be written as

$$
\begin{aligned}
[\hat{H}, \hat{a}_n^+ \hat{a}_j] &= \sum_{\alpha\beta} h_{\alpha\beta} \left(\delta_{n\beta} \hat{a}_\alpha^+ \hat{a}_j - \delta_{j\alpha} \hat{a}_n^+ \hat{a}_\beta \right) \\
&\quad + \frac{1}{4} \sum_{\alpha\alpha'\beta\beta'} \overline{V}_{\alpha\beta\alpha'\beta'} \left[\hat{\mathbb{N}}[\hat{a}_\alpha^+ \hat{a}_\beta^+ \, \hat{a}_{\beta'} \, \hat{a}_{\alpha'}], \hat{a}_n^+ \hat{a}_j \right] \ .
\end{aligned}
$$

We did not considered the second term of the hamiltonian (16.11) since it is a number, therefore commuting with every operator. We consider the expectation value of the double commutator

$$\left\langle \Phi_0 | \left[\hat{a}_i^+ \hat{a}_m, [\hat{H}, \hat{a}_n^+ \hat{a}_j] \right] | \Phi_0 \right\rangle .$$

The contribution of the first term of the hamiltonian can be rewritten as

$$\begin{aligned} & h_{\alpha\beta} \left\langle \Phi_0 | \left[\hat{a}_i^+ \hat{a}_m, \left(\delta_{n\beta} \hat{a}_\alpha^+ \hat{a}_j - \delta_{j\alpha} \hat{a}_n^+ \hat{a}_\beta \right) \right] | \Phi_0 \right\rangle \\ = \; & h_{\alpha\beta} \left\langle \Phi_0 \right| \left(\hat{a}_i^+ \hat{a}_m \delta_{n\beta} \hat{a}_\alpha^+ \hat{a}_j - \hat{a}_i^+ \hat{a}_m \delta_{j\alpha} \hat{a}_n^+ \hat{a}_\beta \right) \left| \Phi_0 \right\rangle \\ = \; & h_{\alpha\beta} \left\langle \Phi_0 | \hat{a}_i^+ \hat{a}_m \hat{a}_\alpha^+ \hat{a}_j | \Phi_0 \right\rangle \delta_{n\beta} - h_{\alpha\beta} \left\langle \Phi_0 | \hat{a}_i^+ \hat{a}_m \hat{a}_n^+ \hat{a}_\beta | \Phi_0 \right\rangle \delta_{j\alpha} \\ = \; & h_{\alpha\beta} \delta_{ij} \delta_{m\alpha} \delta_{n\beta} - h_{\alpha\beta} \delta_{i\beta} \delta_{mn} \delta_{j\alpha} \\ = \; & (\epsilon_m - \epsilon_i) \delta_{ij} \delta_{mn} \; , \end{aligned} \tag{16.12}$$

where in the last step we considered the diagonal expression of $h_{\alpha,\beta}$. For the calculation of the second term of the hamiltonian we consider that

$$\left[\hat{\mathbb{N}}[\hat{a}_\alpha^+ \hat{a}_\beta^+ \hat{a}_{\beta'} \hat{a}_{\alpha'}], \hat{a}_n^+ \hat{a}_j \right] = \hat{\mathbb{N}}[\hat{a}_\alpha^+ \hat{a}_\beta^+ \hat{a}_{\beta'} \hat{a}_{\alpha'}] \hat{a}_n^+ \hat{a}_j - \hat{a}_n^+ \hat{a}_j \hat{\mathbb{N}}[\hat{a}_\alpha^+ \hat{a}_\beta^+ \hat{a}_{\beta'} \hat{a}_{\alpha'}] \; ,$$

therefore

$$\begin{aligned} & \left\langle \Phi_0 | \left[\hat{a}_i^+ \hat{a}_m, \hat{\mathbb{N}}[\hat{a}_\alpha^+ \hat{a}_\beta^+ \hat{a}_{\beta'} \hat{a}_{\alpha'}] \hat{a}_n^+ \hat{a}_j - \hat{a}_n^+ \hat{a}_j \hat{\mathbb{N}}[\hat{a}_\alpha^+ \hat{a}_\beta^+ \hat{a}_{\beta'} \hat{a}_{\alpha'}] \right] | \Phi_0 \right\rangle \\ = \; & \left\langle \Phi_0 | \hat{a}_i^+ \hat{a}_m \hat{\mathbb{N}}[\hat{a}_\alpha^+ \hat{a}_\beta^+ \hat{a}_{\beta'} \hat{a}_{\alpha'}] \hat{a}_n^+ \hat{a}_j | \Phi_0 \right\rangle && (16.13) \\ - \; & \left\langle \Phi_0 | \hat{\mathbb{N}}[\hat{a}_\alpha^+ \hat{a}_\beta^+ \hat{a}_{\beta'} \hat{a}_{\alpha'}] \hat{a}_n^+ \hat{a}_j \hat{a}_i^+ \hat{a}_m | \Phi_0 \right\rangle && (16.14) \\ - \; & \left\langle \Phi_0 | \hat{a}_i^+ \hat{a}_m \hat{a}_n^+ \hat{a}_j \hat{\mathbb{N}}[\hat{a}_\alpha^+ \hat{a}_\beta^+ \hat{a}_{\beta'} \hat{a}_{\alpha'}] | \Phi_0 \right\rangle && (16.15) \\ + \; & \left\langle \Phi_0 | \hat{a}_n^+ \hat{a}_j \hat{\mathbb{N}}[\hat{a}_\alpha^+ \hat{a}_\beta^+ \hat{a}_{\beta'} \hat{a}_{\alpha'}] \hat{a}_i^+ \hat{a}_m | \Phi_0 \right\rangle \; . && (16.16) \end{aligned}$$

The terms (16.14) and (16.16) are zero since $a_m \left| \Phi_0 \right\rangle = 0$. The situation for the term (16.15) is more involved. In applying Wick's theorem, it becomes evident that, between all possible sets of contractions, there are always terms where $\hat{a}_n^+$ is contracted with $\hat{a}_{\alpha'}$ or $\hat{a}_{\beta'}$ and these contractions yield a value of zero. Only the term (16.13) is different from zero, and, by applying the Wick's theorem we obtain

$$\begin{aligned} & \left\langle \Phi_0 | a_i^+ a_m N[a_\alpha^+ a_\beta^+ a_{\beta'} a_{\alpha'}] a_n^+ a_j | \Phi_0 \right\rangle \\ = \; & \delta_{i\alpha'} \delta_{m\alpha} \delta_{\beta' n} \delta_{\beta j} - \delta_{i\alpha'} \delta_{m\beta} \delta_{\beta' n} \delta_{\alpha j} \\ - \; & \delta_{i\beta'} \delta_{m\alpha} \delta_{\alpha' n} \delta_{\beta j} + \delta_{i\beta'} \delta_{m\beta} \delta_{\alpha' n} \delta_{\alpha j} . \end{aligned} \tag{16.17}$$

By substituting the above result into Eq. (16.2) and taking into account the symmetry properties of $\overline{V}_{\alpha,\beta,\alpha',\beta'}$ we obtain the TDA equations

$$\sum_{nj} X_{nj}^\nu \left[(\epsilon_n - \epsilon_j) \delta_{mn} \delta_{ij} + \overline{V}_{mjin} \right] = (E_\nu - E_0) X_{mi}^\nu \; , \tag{16.18}$$

where

$$\overline{V}_{mjin} \equiv \langle mj|V|in\rangle - \langle mj|V|ni\rangle \quad . \tag{16.19}$$

The expression (16.18) represents a homogenous system of linear equations whose unknown are the X^{ν}_{mi} amplitudes. The number of unknowns, and therefore that of the solutions, is given by the number of the particle-hole pairs which truncates the sum.

The normalization condition of the excited state induces a relation between the X^{ν}_{mi} amplitudes:

$$1 = \langle \Psi_\nu|\Psi_\nu\rangle = \left\langle \Phi_0|\hat{Q}_\nu \hat{Q}^+_\nu|\Phi_0\right\rangle = \left\langle \Phi_0|\sum_{ph} \hat{a}^+_h \hat{a}_p X^{*\nu}_{ph} \sum_{p'h'} X^{\nu}_{p'h'} \hat{a}^+_{p'} \hat{a}_{h'}|\Phi_0\right\rangle$$
$$= \sum_{ph}\sum_{p'h'} X^{*\nu}_{ph} X^{\nu}_{p'h'} \left\langle \Phi_0|\hat{a}^+_h \hat{a}_p \hat{a}^+_{p'} \hat{a}_{h'}|\Phi_0\right\rangle = \sum_{ph} |X^{\nu}_{ph}|^2 \quad , \tag{16.20}$$

which defines without ambiguity the values of the X^{ν}_{ph} and offers a probabilistic interpretation. Here below we present an example of TDA equations applied to the case of only two particle-hole pairs.

Let us consider the case where there are only two possible particle-hole excitation pairs. The TDA equations are then

$$\begin{pmatrix} \epsilon_{p_1} - \epsilon_{h_1} - \omega + \overline{V}_{p_1h_1h_1p_1} & \overline{V}_{p_2h_2h_1p_1} \\ \overline{V}_{p_1h_1h_2p_2} & \epsilon_{p_2} - \epsilon_{h_2} - \omega + \overline{V}_{p_2h_2h_2p_2} \end{pmatrix} \begin{pmatrix} X_{p_1h_1} \\ X_{p_2h_2} \end{pmatrix} = 0 \ ,$$

with $\omega = E - E_0$.

The system has solutions different from the trivial one only if the determinant of the known terms matrix is zero, therefore

$$(\epsilon_{p_1} - \epsilon_{h_1} - \omega + \overline{V}_{p_1h_1h_1p_1})(\epsilon_{p_2} - \epsilon_{h_2} - \omega + \overline{V}_{p_2h_2h_2p_2}) - \overline{V}_{p_2h_2h_1p_1}\overline{V}_{p_2h_2h_2p_2} = 0 \ .$$

The equation is quadratic in ω, therefore it is satisfied by two values.

In case $\overline{V} = 0$ the two values are $\omega = \epsilon_{p_1} - \epsilon_{h_1}$ and $\omega = \epsilon_{p_2} - \epsilon_{h_2}$, which are the energy of the possible excitations in a pure MF model. The presence of the residual interaction mixes the components of the two particle-hole pairs in each excited state, and also the energy eigenvalues are modified with respect to the MF solution. The stronger is the effect of $\overline{V}$ the larger is the mixing of the particle-hole pairs. These type of states with large mixing of particle-hole pairs are called collective states.

The TDA theory not only describes the energy spectrum of the system but also provides the many-body wave function for each excited state, expressed in terms of single-particle states. This formulation enables the calculation of the transition probability from the ground state to an excited state.

Let us assume that the action of the external field which excites the system is described by a one-body operator

$$\hat{F} = \sum_{\mu\mu'} \left\langle \mu | \hat{f} | \mu' \right\rangle \hat{a}^+_\mu \hat{a}_{\mu'} \equiv \sum_{\mu\mu'} f_{\mu\mu'} \hat{a}^+_\mu \hat{a}_{\mu'} \ . \tag{16.21}$$

The transition probability from the ground state to an excited state is

$$\begin{aligned}
\left\langle \Phi_\nu | \hat{F} | \Phi_0 \right\rangle &= \left\langle \Phi_0 | \hat{Q}_\nu \hat{F} | \Phi_0 \right\rangle \\
&= \left\langle \Phi_0 | \sum_{mi} X^{*\nu}_{mi} \hat{a}^+_i \hat{a}_m \sum_{\mu\mu'} f_{\mu\mu'} \hat{a}^+_\mu \hat{a}_{\mu'} | \Phi_0 \right\rangle \\
&= \sum_{mi} X^{*\nu}_{mi} \sum_{\mu\mu'} f_{\mu\mu'} \left\langle \Phi_0 | \hat{a}^+_i \hat{a}_m \hat{a}^+_\mu \hat{a}_{\mu'} | \Phi_0 \right\rangle \\
&= \sum_{mi} X^{*\nu}_{mi} \sum_{\mu\mu'} f_{\mu\mu'} \delta_{i\mu'} \delta_{m\mu} = \sum_{mi} X^{*\nu}_{mi} \left\langle m | \hat{f} | i \right\rangle \ .
\end{aligned} \tag{16.22}$$

16.3 Random Phase Approximation (RPA)

16.3.1 Limits of the TDA

The comparison between the TDA results and the experimental data is unsatisfactory, particularly in the context of nuclear physics. Consequently, since the latter half of the 1960s, the assumptions underlying TDA theory have been carefully analysed. These assumptions pertain to the choice of the expression (16.7) for the $\hat{Q}_\nu$ operator. From these studies, it has become evident that this choice is inconsistent with the equations of motion (16.5). This inconsistency can be highlighted as follows: the equations of motion (16.5) were derived without making any assumptions about the operator $\hat{\mathbb{O}}$. For an operator of the form $\hat{a}^+_m \hat{a}_i$, the equations of motion are:

$$\begin{aligned}
\left\langle \Psi_0 \left| \left[\hat{a}^+_m \hat{a}_i, [\hat{H}, \hat{Q}^+_\nu] \right] \right| \Psi_0 \right\rangle &= (E_\nu - E_0) \left\langle \Psi_0 | \hat{a}^+_m \hat{a}_i \hat{Q}^+_\nu | \Psi_0 \right\rangle \\
&= (E_\nu - E_0) \left\langle \Psi_0 \left| \left[\hat{a}^+_m \hat{a}_i, \hat{Q}^+_\nu \right] \right| \Psi_0 \right\rangle \ .
\end{aligned} \tag{16.23}$$

By inserting the expression of the TDA operator (16.7) for the right hand side of the above equation we obtain

$$\sum_{nj} X^{\nu}_{nj} \langle \Phi_0 | \left[\hat{a}^+_m \hat{a}_i, \hat{a}^+_n \hat{a}_j \right] | \Phi_0 \rangle$$

$$= \sum_{nj} X^{\nu}_{nj} \left\{ \langle \Phi_0 | \hat{a}^+_m \hat{a}_i \hat{a}^+_n \hat{a}_j | \Phi_0 \rangle - \langle \Phi_0 | \hat{a}^+_n \hat{a}_j \hat{a}^+_m \hat{a}_i | \Phi_0 \rangle \right\} = 0 \ . \qquad (16.24)$$

This result requires that also the left hand side of the Eq. (16.23) must be zero. In effects, the one-body term of the hamiltonian has a double commutator equal to zero

$$\sum_{\alpha\beta} h_{\alpha,\beta} \langle \Phi_0 | \hat{a}^+_m \hat{a}_i, (\hat{a}^+_\alpha \hat{a}_j \delta_{n\beta} - \hat{a}^+_n \hat{a}_\beta \delta_{j\alpha}) | \Phi_0 \rangle = 0 \ ,$$

but the double commutator of the interaction term, in general is not equal to zero,

$$\sum_{\alpha,\beta,\alpha',\beta'} \overline{V}_{\alpha,\beta,\alpha',\beta'} \left\langle \Phi_0 \middle| \left[\hat{a}^+_m \hat{a}_i, \left[\hat{\mathbb{N}}[\hat{a}^+_\alpha \hat{a}^+_\beta \hat{a}_{\beta'} \hat{a}_{\alpha'}], \hat{a}^+_n \hat{a}_j \right] \right] \middle| \Phi_0 \right\rangle \neq 0 \ .$$

In the evaluation of this double commutator there are terms of the type

$$\sum_{\alpha,\beta,\alpha',\beta'} \overline{V}_{\alpha,\beta,\alpha',\beta'} \left\langle \Phi_0 \middle| \hat{a}^+_m \hat{a}_i \hat{a}^+_\alpha \hat{a}^+_\beta \hat{a}_{\beta'} \hat{a}_{\alpha'} \hat{a}^+_n \hat{a}_j \middle| \Phi_0 \right\rangle \ ,$$

evidently different from zero.

16.3.2 The RPA Equations

The RPA excitation operator is defined as

$$\hat{Q}^+_\nu \equiv \sum_{ph} X^{\nu}_{ph} \hat{a}^+_p \hat{a}_h - \sum_{ph} Y^{\nu}_{ph} \hat{a}^+_h \hat{a}_p \ , \qquad (16.25)$$

where both X^{ν}_{ph} and Y^{ν}_{ph} are numbers, called *RPA amplitudes*.

The RPA ground state is defined by the equation $\hat{Q}_\nu |\nu_0\rangle = 0$. Clearly, $|\nu_0\rangle$ is not a MF ground state. In the latter case we would have

$$\hat{Q}_\nu |\Phi_0\rangle = \sum_{ph} X^{*\nu}_{ph} \hat{a}^+_h \hat{a}_p |\Phi_0\rangle - \sum_{ph} Y^{*\nu}_{ph} \hat{a}^+_p \hat{a}_h |\Phi_0\rangle \ .$$

The first term is certainly zero, while the second one is not zero. The RPA ground state $|\nu_0\rangle$ is more complex than the MF ground state, incorporating effects that extend beyond it, referred to as **correlations**. In this context, these correlations are described in terms of hole-particle excitations, which we will discuss in Sect. 16.3.6.

By using the definition (16.25) of the RPA amplitudes the equations of motion (16.5) becomes

$$\left\langle \nu_0 \left| \left[\delta \hat{Q}_\nu, \left[\hat{H}, \hat{Q}_\nu^+ \right] \right] \right| \nu_0 \right\rangle = (E_\nu - E_0) \left\langle \nu_0 \left| \left[\delta \hat{Q}_\nu, \hat{Q}_\nu^+ \right] \right| \nu_0 \right\rangle ,$$

therefore

$$\left\langle \nu_0 \left| \left[\left(\sum_{mi} \hat{a}_i^+ \hat{a}_m \delta X_{mi}^\nu - \sum_{mi} \hat{a}_m^+ \hat{a}_i \delta Y_{mi}^\nu \right), \left[\hat{H}, \hat{Q}_\nu^+ \right] \right] \right| \nu_0 \right\rangle$$
$$= (E_\nu - E_0) \left\langle \nu_0 \left| \left[\left(\sum_{mi} \hat{a}_i^+ \hat{a}_m \delta X_{mi}^\nu - \sum_{mi} \hat{a}_m^+ \hat{a}_i \delta Y_{mi}^\nu \right), \hat{Q}_\nu^+ \right] \right| \nu_0 \right\rangle ,$$

and making explicit the the variations on the RPA amplitudes we obtain

$$\sum_{mi} \delta X_{mi}^\nu \left\langle \nu_0 \left| \left[\hat{a}_i^+ \hat{a}_m, \left[\hat{H}, \hat{Q}_\nu^+ \right] \right] \right| \nu_0 \right\rangle - \sum_{mi} \delta Y_{mi}^\nu \left\langle \nu_0 \left| \left[\hat{a}_m^+ \hat{a}_i, \left[\hat{H}, \hat{Q}_\nu^+ \right] \right] \right| \nu_0 \right\rangle$$
$$= (E_\nu - E_0) \left\{ \sum_{mi} \delta X_{mi}^\nu \left\langle \nu_0 \left| \left[\hat{a}_i^+ \hat{a}_m, \hat{Q}_\nu^+ \right] \right| \nu_0 \right\rangle - \sum_{mi} \delta Y_{mi}^\nu \left\langle \nu_0 \left| \left[\hat{a}_m^+ \hat{a}_i, \hat{Q}_\nu^+ \right] \right| \nu_0 \right\rangle \right\} .$$

As in the TDA case, the above equation represent a sum of independent terms since each variation is independent of the other ones. By equating the terms related to the same variation we obtain the following relations

$$\left\langle \nu_0 \left| \left[\hat{a}_i^+ \hat{a}_m, \left[\hat{H}, \hat{Q}_\nu^+ \right] \right] \right| \nu_0 \right\rangle = (E_\nu - E_0) \left\langle \nu_0 \left| \left[\hat{a}_i^+ \hat{a}_m, \hat{Q}_\nu^+ \right] \right| \nu_0 \right\rangle \quad (16.26)$$
$$\left\langle \nu_0 \left| \left[\hat{a}_m^+ \hat{a}_i, \left[\hat{H}, \hat{Q}_\nu^+ \right] \right] \right| \nu_0 \right\rangle = (E_\nu - E_0) \left\langle \nu_0 \left| \left[\hat{a}_m^+ \hat{a}_i, \hat{Q}_\nu^+ \right] \right| \nu_0 \right\rangle . \quad (16.27)$$

Let us consider the left hand side of Eq. (16.26)

$$\left\langle \nu_0 \left| \left[\hat{a}_i^+ \hat{a}_m, \left[\hat{H}, \hat{Q}_\nu^+ \right] \right] \right| \nu_0 \right\rangle$$
$$= \sum_{nj} X_{nj}^\nu \left\langle \nu_0 \left| \left[\hat{a}_i^+ \hat{a}_m, \left[\hat{H}, \hat{a}_n^+ \hat{a}_j \right] \right] \right| \nu_0 \right\rangle - \sum_{nj} Y_{nj}^\nu \left\langle \nu_0 \left| \left[\hat{a}_i^+ \hat{a}_m, \left[\hat{H}, \hat{a}_j^+ \hat{a}_n \right] \right] \right| \nu_0 \right\rangle$$
$$\equiv \sum_{nj} X_{nj}^\nu A_{minj} + \sum_{nj} Y_{nj}^\nu B_{minj} . \quad (16.28)$$

These equations define the elements of the A and B matrices. Please observe the sign of the element of B.

To calculate the left hand side of Eq. (16.26), we use an approximation commonly referred to in the literature as *Quasi-Boson-Approximation* (QBA). This approach assumes that the expectation value of commutator between RPA states has the same value of the commutator between MF states $|\Phi_0\rangle$. In the specific case under study

we have that
$$\left\langle \nu_0 \left| \left[\hat{a}_i^+ \hat{a}_m, \hat{Q}_\nu^+ \right] \right| \nu_0 \right\rangle \simeq \left\langle \Phi_0 \left| \left[\hat{a}_i^+ \hat{a}_m, \hat{Q}_\nu^+ \right] \right| \Phi_0 \right\rangle \ . \tag{16.29}$$

It is worth to remark that the QBA can be applied only if the matrix element is expressed in terms of commutators. The idea is that the pairs of creation and destruction operators behave as
$$[\hat{a}_i^+ \hat{a}_m, \hat{a}_n^+ \hat{a}_j] = \delta_{mn} \delta_{ij} \ .$$

With this ansatz, the operators $\hat{\mathbb{O}}_{im} \equiv \hat{a}_i^+ \hat{a}_m$ and $\hat{\mathbb{O}}_{jn}^+ \equiv a_n^+ a_j$ would be a bosonic operators.

By using the QBA we obtain the following relations

$$\begin{aligned} & \left\langle \nu_0 \left| \left[\hat{a}_i^+ \hat{a}_m, \hat{Q}_\nu^+ \right] \right| \nu_0 \right\rangle \\ \simeq & \sum_{nj} X_{nj}^\nu \left\langle \Phi_0 | [\hat{a}_i^+ \hat{a}_m, \hat{a}_n^+ \hat{a}_j] | \Phi_0 \right\rangle - \sum_{nj} Y_{nj}^\nu \left\langle \Phi_0 | [\hat{a}_i^+ \hat{a}_m, \hat{a}_j^+ \hat{a}_n] | \Phi_0 \right\rangle \\ = & \sum_{nj} X_{nj}^\nu \left\{ \left\langle \Phi_0 | \hat{a}_i^+ \hat{a}_m \hat{a}_n^+ \hat{a}_j | \Phi_0 \right\rangle - \left\langle \Phi_0 | \hat{a}_n^+ \hat{a}_j \hat{a}_i^+ \hat{a}_m | \Phi_0 \right\rangle \right\} \\ - & \sum_{nj} Y_{nj}^\nu \left\{ \left\langle \Phi_0 | \hat{a}_i^+ \hat{a}_m \hat{a}_j^+ \hat{a}_n | \Phi_0 \right\rangle - \left\langle \Phi_0 | \hat{a}_j^+ \hat{a}_n \hat{a}_i^+ \hat{a}_m | \Phi_0 \right\rangle \right\} \\ = & \sum_{nj} X_{nj}^\nu \left\langle \Phi_0 | \hat{a}_i^+ \hat{a}_m, \hat{a}_n^+ \hat{a}_j | \Phi_0 \right\rangle = X_{mi}^\nu \delta_{mn} \delta_{ij} \ , \end{aligned} \tag{16.30}$$

where we have considered that the terms multiplying Y_{nj}^ν do not conserve the particle number and, furthermore, that $a_m \left| \Phi_0 \right\rangle = 0$. Equation (16.26) becomes

$$\sum_{nj} X_{nj}^\nu A_{minj} + \sum_{nj} Y_{nj}^\nu B_{minj} = (E_\nu - E_0) X_{mi}^\nu \ . \tag{16.31}$$

For the calculation of the left hand side of Eq. (16.27) we consider that:

$$[\hat{H}, \hat{a}_n^+ \hat{a}_j]^+ = (\hat{H} \hat{a}_n^+ \hat{a}_j - \hat{a}_n^+ \hat{a}_j \hat{H})^+ = \hat{a}_j^+ \hat{a}_n \hat{H} - \hat{H} \hat{a}_j^+ \hat{a}_n = -[\hat{H}, \hat{a}_j^+ \hat{a}_n] \ , \tag{16.32}$$

since $\hat{H} = \hat{H}^+$, and then

$$\left[\hat{a}_i^+ \hat{a}_m, [\hat{H}, \hat{a}_j^+ \hat{a}_n] \right]^+ = -\left[\hat{a}_m^+ \hat{a}_i, -[\hat{H}, \hat{a}_n^+ \hat{a}_j] \right] = \left[\hat{a}_m^+ \hat{a}_i, [\hat{H}, \hat{a}_n^+ \hat{a}_j] \right] \ . \tag{16.33}$$

The double commutator becomes

$$
\begin{aligned}
&\left\langle \nu_0 \right| \left[a_m^+ a_i, \left[H, Q_\nu^+ \right] \right] \left| \nu_0 \right\rangle \\
&= \sum X_{nj}^\nu \left\langle \nu_0 \right| \left[a_m^+ a_i, \left[H, a_n^+ a_j \right] \right] \left| \nu_0 \right\rangle - \sum Y_{nj}^\nu \left\langle \nu_0 \right| \left[a_m^+ a_i, \left[H, a_j^+ a_n \right] \right] \left| \nu_0 \right\rangle \\
&= \sum X_{nj}^\nu \left\langle \nu_0 \right| \left[a_i^+ a_m, \left[H, a_j^+ a_n \right]^+ \right] \left| \nu_0 \right\rangle - \sum Y_{nj}^\nu \left\langle \nu_0 \right| \left[a_i^+ a_m, \left[H, a_n^+ a_j \right]^+ \right] \left| \nu_0 \right\rangle \\
&= \sum_{nj} X_{nj}^\nu (-B_{minj}^*) + \sum_{nj} Y_{nj}^\nu (-A_{minj}^*) \quad , \qquad (16.34)
\end{aligned}
$$

where we used the definitions of the matrix elements A e B in Eq. (16.28).

For the calculation of the right hand side of Eq. (16.27) by using the QBA we have

$$
\begin{aligned}
\left\langle \nu_0 \right| \left[\hat{a}_m^+ \hat{a}_i, Q_\nu^+ \right] \left| \nu_0 \right\rangle \rightarrow \text{ (QBA) } \rightarrow &- \sum_{nj} Y_{nj}^\nu \left\langle \Phi_0 \right| \left[\hat{a}_m^+ \hat{a}_i, \hat{a}_j^+ \hat{a}_n \right] \left| \Phi_0 \right\rangle \\
&= Y_{mi}^\nu \delta_{ij} \delta_{mn} \quad , \qquad (16.35)
\end{aligned}
$$

therefore Eq. (16.27) becomes

$$
\sum_{nj} X_{nj}^\nu (-B_{minj}^*) + \sum_{nj} Y_{nj}^\nu (-A_{minj}^*) = (E_\nu - E_0) Y_{mi}^\nu \quad . \qquad (16.36)
$$

The Eqs. (16.31) and (16.36) represent a homogenous system of linear equations whose unknown are the RPA amplitudes X_{ph}^ν and Y_{ph}^ν. Usually, this system is presented as

$$
\begin{pmatrix} A & B \\ B^* & A^* \end{pmatrix} \begin{pmatrix} X^\nu \\ Y^\nu \end{pmatrix} = (E_\nu - E_0) \begin{pmatrix} I & 0 \\ 0 & -I \end{pmatrix} \begin{pmatrix} X^\nu \\ Y^\nu \end{pmatrix} = (E_\nu - E_0) \begin{pmatrix} X^\nu \\ -Y^\nu \end{pmatrix} \quad , \qquad (16.37)
$$

where A and B are square matrices whose dimensions are those of the number of the particle-hole pairs describing the excitation, and X e Y are column vectors of the same dimensions.

The expressions of the matrix elements of A and B in terms of effective interaction between two interacting particles are

$$
\begin{aligned}
A_{minj} \rightarrow \text{ (QBA) } \rightarrow &\left\langle \Phi_0 \right| \left[\hat{a}_i^+ \hat{a}_m, \left[\hat{H}, \hat{a}_n^+ \hat{a}_j \right] \right] \left| \Phi_0 \right\rangle \\
&= (\epsilon_m - \epsilon_i) \delta_{mn} \delta_{ij} + \overline{V}_{mjin} \quad , \qquad (16.38) \\
B_{minj} \rightarrow \text{ (QBA) } \rightarrow &- \left\langle \Phi_0 \right| \left[\hat{a}_i^+ \hat{a}_m, \left[\hat{H}, \hat{a}_j^+ \hat{a}_n \right] \right] \left| \Phi_0 \right\rangle = \overline{V}_{mnij} \quad . \qquad (16.39)
\end{aligned}
$$

The element A_{minj} is the same as that of the TDA, Eq. (16.18).

For the term B_{minj} we consider, again, the expression (16.11) of the hamiltonian. Also in this case, as in the case of the TDA, the scalar term $\overline{V}_{ijij}$ does not contribute to the double commutator. Also the contribution of the one-body term is zero. By considering the anti-commutation properties of the creation and destruction operators we obtain

$$[\hat{a}^+_\alpha \hat{a}_\beta, \hat{a}^+_j \hat{a}_n] = \delta_{\beta j} \hat{a}^+_\alpha \hat{a}_n - \delta_{n\alpha} \hat{a}^+_j \hat{a}_\beta \ ,$$

therefore

$$\begin{aligned} & \left\langle \Phi_0 \right| \left[\hat{a}^+_i \hat{a}_m, \left[\hat{a}^+_\alpha \hat{a}_\beta, \hat{a}^+_j a_n \right] \right] \left| \Phi_0 \right\rangle \\ = & \ \langle \Phi_0 | \hat{a}^+_i \hat{a}_m \hat{a}^+_\alpha \hat{a}_n | \Phi_0 \rangle \to 0 \\ - & \ \langle \Phi_0 | \hat{a}^+_\alpha \hat{a}_n \hat{a}^+_i \hat{a}_m | \Phi_0 \rangle \to 0 \\ - & \ \left\langle \Phi_0 | \hat{a}^+_i \hat{a}_m \hat{a}^+_j \hat{a}_\beta | \Phi_0 \right\rangle = \delta_{j\beta} \delta_{im} \to 0 \\ + & \ \left\langle \Phi_0 | \hat{a}^+_j \hat{a}_\beta \hat{a}^+_i \hat{a}_m | \Phi_0 \right\rangle \to 0 \ . \end{aligned}$$

For the two-body term we have to evaluate

$$\left\langle \Phi_0 \right| \left[a^+_i a_m, \left[N[a^+_\alpha a^+_\beta a_{\beta'} a_{\alpha'}, a^+_j a_n \right] \right] \left| \Phi_0 \right\rangle \ .$$

Three terms of the double commutators are zero since the contain $a_m \left| \Phi_0 \right\rangle = 0$. Only the term

$$- \left\langle \Phi_0 | \hat{a}^+_i \hat{a}_m \hat{a}^+_j \hat{a}_n \hat{a}^+_\alpha \hat{a}^+_\beta \hat{a}_{\beta'} \hat{a}_{\alpha'} | \Phi_0 \right\rangle \ ,$$

is different from zero, therefore

$$B_{minj} = \frac{1}{4} \sum_{\alpha\beta\alpha'\beta'} \overline{V}_{\alpha\beta\alpha'\beta'} \left\langle \Phi_0 | \hat{a}^+_i \hat{a}_m \hat{a}^+_j \hat{a}_n \hat{a}^+_\alpha \hat{a}^+_\beta \hat{a}_{\beta'} \hat{a}_{\alpha'} | \Phi_0 \right\rangle \ .$$

By considering the symmetry properties of $\overline{V}$ and all the possible contractions we obtain Eq. (16.39).

The RPA equations obtained by using the two-body Green's function in Sect. 10.6 seems to be a completely different with respect to those presented here. It is possible to show that the two formulations of the RPA generate the same set of equations. The details of the calculations are presented in Appendix C.

The key point which allows the identification of the two-formulations of the RPA is the following correspondences between the X and Y RPA amplitudes and the transition amplitudes of the two-body Green's function in Lehmann representation Eq. (9.76). The relations are:

$$X_{mi} = \left\langle \nu_0 | \hat{a}_m \hat{a}^+_i | \nu_n \right\rangle \quad ; \qquad X^*_{mi} = \left\langle \nu_n | \hat{a}_i \hat{a}^+_m | \nu_n \right\rangle ; \tag{16.40}$$

$$Y_{mi} = \left\langle \nu_0 | \hat{a}_i \hat{a}^+_m | \nu_n \right\rangle \quad ; \qquad Y^*_{mi} = \left\langle \nu_n | \hat{a}_m \hat{a}^+_i | \nu_0 \right\rangle . \tag{16.41}$$

where $|\Psi_n\rangle$ and $|\Psi_0\rangle$ are, respectively, RPA excited and ground states.

16.3.3 Properties of the RPA Equations

We consider the RPA equations in the form

$$\begin{pmatrix} A & B \\ B^* & A^* \end{pmatrix} \begin{pmatrix} X^\nu \\ Y^\nu \end{pmatrix} = \omega_\nu \begin{pmatrix} X^\nu \\ -Y^\nu \end{pmatrix} ,$$

where $\omega = E_\nu - E_0$ is the excitation energy.

- If $B = 0$ we obtain the TDA equations.
- The two RPA equations can be written as the system

$$\begin{aligned} A\, X^\nu + B\, Y^\nu &= \omega_\nu\, X^\nu \\ -B^*\, X^\nu - A^*\, Y^\nu &= \omega_\nu\, Y^\nu \ . \end{aligned}$$

  We take the complex conjugate of the above equations

$$\begin{aligned} A^*\, X^{*\nu} + B^*\, Y{*}^\nu &= \omega_\nu\, X^{*\nu} \\ B\, X^{*\nu} + A\, Y^{*\nu} &= -\omega_\nu\, Y^\nu \ , \end{aligned}$$

  which can be written as

$$\begin{pmatrix} A & B \\ B^* & A^* \end{pmatrix} \begin{pmatrix} Y^{*\nu} \\ X^{*\nu} \end{pmatrix} = -\omega_\nu \begin{pmatrix} Y^{*\nu} \\ -X^{*\nu} \end{pmatrix} .$$

  Therefore the RPA equations allow positive and negative eigenvalues with the same absolute value.
- The RPA matrix is not hermitian. A is hermitian but B is symmetric but not hermitian $B_{minj} = B_{njmi}$. It possible to show that the eigenvalues are real numbers. Normally real interactions are used, therefore the matrix element of the A and B matrices are real. Also the X and Y amplitudes are real.
- Eigenvectors corresponding to different eigenvalues are orthogonal.

$$\begin{pmatrix} A & B \\ B^* & A^* \end{pmatrix} \begin{pmatrix} X^\nu \\ Y^\nu \end{pmatrix} = \omega_\nu \begin{pmatrix} X^\nu \\ -Y^\nu \end{pmatrix} \ ; \ \begin{pmatrix} A & B \\ B^* & A^* \end{pmatrix} \begin{pmatrix} X^\mu \\ Y^\mu \end{pmatrix} = \omega_\mu \begin{pmatrix} X^\mu \\ -Y^\mu \end{pmatrix} .$$

  Let us calculate the hermitian conjugate of the second equation

$$(X^{\mu+}, Y^{\mu+}) \begin{pmatrix} A & B \\ B^* & A^* \end{pmatrix} = (X^{\mu+}, -Y^{\mu})\omega_\mu \ .$$

We multiply the first equation by $(X^{\mu+}, Y^{\mu+})$ on the left hand side, and the second equation on the right hand side by

$$\begin{pmatrix} X^{\nu} \\ -Y^{\nu} \end{pmatrix} ,$$

and we obtain

$$(X^{\mu+}, Y^{\mu+}) \begin{pmatrix} A & B \\ B^* & A^* \end{pmatrix} \begin{pmatrix} X^{\nu} \\ Y^{\nu} \end{pmatrix} = \omega_{\nu} (X^{\mu+}, Y^{\mu+}) \begin{pmatrix} X^{\nu} \\ -Y^{\nu} \end{pmatrix} ,$$

$$(X^{\mu+}, Y^{\mu+}) \begin{pmatrix} A & B \\ B^* & A^* \end{pmatrix} \begin{pmatrix} X^{\nu} \\ Y^{\nu} \end{pmatrix} = \omega_{\mu} (X^{\mu+}, -Y^{\mu}) \begin{pmatrix} X^{\nu} \\ Y^{\nu} \end{pmatrix} .$$

By subtracting the two equations we have

$$0 = (\omega_{\nu} - \omega_{\mu})(X^{\mu+} X^{\nu} - Y^{\mu+} Y^{\nu}) .$$

Since we have assumed $\omega_{\nu} \neq \omega_{\mu}$ we obtain

$$(X^{\mu+} X^{\nu} - Y^{\mu+} Y^{\nu}) = 0 .$$

- The normalization between two excited states requires

$$\begin{aligned} \delta_{\nu\nu'} &= \langle \nu | \nu' \rangle = \left\langle \nu_0 | \hat{Q}_{\nu} \hat{Q}^{+}_{\nu'} | \nu_0 \right\rangle = \left\langle \nu_0 | [\hat{Q}_{\nu}, \hat{Q}^{+}_{\nu'}] | \nu_0 \right\rangle \rightarrow \text{ (QBA)} \rightarrow \left\langle \Phi_0 | [\hat{Q}_{\nu}, \hat{Q}^{+}_{\nu'}] | \Phi_0 \right\rangle \\ &= \sum_{mi} \left(X^{\nu}_{mi} X^{\nu'}_{mi} - Y^{\nu}_{mi} Y^{\nu'}_{mi} \right) , \end{aligned}$$

where we used the fact that $\hat{Q}_{\nu} \, |\nu_0\rangle = 0$.

16.3.4 Transition Probabilities in RPA

In analogy to the TDA case, we assume that the action of the external field exciting the system is described by a one-body operator expressed as in Eq. (16.21). The transition probability between the RPA ground state and excited state is described by

$$\left\langle \nu | \hat{F} | \nu_0 \right\rangle = \left\langle \nu_0 | \hat{Q}_{\nu} \hat{F} | \nu_0 \right\rangle = \left\langle \nu_0 | [\hat{Q}_{\nu}, \hat{F}] | \nu_0 \right\rangle , \tag{16.42}$$

where we used the fact that $\hat{Q}_\nu |\nu_0\rangle = 0$. Since the equation is expressed in terms of commutator we can use the QBA

$$\left\langle \nu|\hat{F}|\nu_0\right\rangle \rightarrow \text{(QBA)} \rightarrow \left\langle \Phi_0|[\hat{Q}_\nu, \hat{F}]|\Phi_0\right\rangle$$
$$= \sum_{\mu\mu'} f_{\mu\mu'} \left\{ \sum_{mi} X^\nu_{mi} \langle \Phi_0|[\hat{a}^+_i \hat{a}_m, \hat{a}^+_\mu \hat{a}_{\mu'}]|\Phi_0\rangle \right.$$
$$\left. - \sum_{mi} Y^\nu_{mi} \langle \Phi_0|[\hat{a}^+_m \hat{a}_i, \hat{a}^+_\mu \hat{a}_{\mu'}]|\Phi_0\rangle \right\} .$$

The two matrix elements are

$$\langle \Phi_0|[\hat{a}^+_i \hat{a}_m, \hat{a}^+_\mu \hat{a}_{\mu'}]|\Phi_0\rangle = \langle \Phi_0|\hat{a}^+_i \hat{a}_m \hat{a}^+_\mu \hat{a}_{\mu'}|\Phi_0\rangle - \langle \Phi_0|\hat{a}^+_\mu \hat{a}_{\mu'} \hat{a}^+_i \hat{a}_m|\Phi_0\rangle$$
$$= \delta_{m\mu}\delta_{i\mu'} - 0 \ ,$$
$$\langle \Phi_0|[a^+_m a_i, a^+_\mu a_{\mu'}]|\Phi_0\rangle = \langle \Phi_0|a^+_m a_i a^+_\mu a_{\mu'}|\Phi_0\rangle - \langle \Phi_0|a^+_\mu a_{\mu'} a^+_m a_i|\Phi_0\rangle$$
$$= 0 - \delta_{m\mu'}\delta_{i\mu} \ .$$

Therefore

$$\left\langle \nu|\hat{F}|\nu_0\right\rangle \simeq \sum_{\mu\mu'} f_{\mu\mu'} \left(\sum_{mi} X^\nu_{mi} \delta_{m\mu}\delta_{i\mu'} + \sum_{mi} Y^\nu_{mi} \delta_{m\mu'}\delta_{i\mu} \right)$$
$$= \sum_{mi} \left(X^\nu_{mi} \langle m|f|i\rangle + Y^\nu_{mi} \langle i|f|m\rangle \right) . \qquad (16.43)$$

As in the TDA case, the transition amplitude of a many-body system is expressed as linear combination of single particle transitions.

16.3.5 Sum Rules

We consider the eigenstates $|\Psi_\nu\rangle$ of the hamiltonian $\hat{H}$

$$\hat{H} |\Psi_\nu\rangle = E_\nu |\Psi_\nu\rangle \ .$$

For an external operator $\hat{F}$ inducing a transition of the system from the ground state $|\Psi_0\rangle$ to the excited state $|\Psi_\nu\rangle$ we have that

$$2 \sum_\nu (E_\nu - E_0) \left| \left\langle \Psi_\nu|\hat{F}|\Psi_0\right\rangle \right|^2 = \left\langle \Psi_0\left| \left[\hat{F}, \left[\hat{H}, \hat{F} \right] \right] \right|\Psi_0\right\rangle \ . \qquad (16.44)$$

We derive here the above expression

$$\left\langle \Psi_0 \middle| \left[\hat{F}, (\hat{H}\hat{F} - \hat{F}\hat{H}) \right] \middle| \Psi_0 \right\rangle = \left\langle \Psi_0 \middle| \left[\hat{F}\hat{H}\hat{F} - \hat{F}\hat{F}\hat{H} - \hat{H}\hat{F}\hat{F} + \hat{F}\hat{H}\hat{F}) \right] \middle| \Psi_0 \right\rangle$$
$$= \left[2 \left\langle \Psi_0 \middle| \hat{F}\hat{H}\hat{F} \middle| \Psi_0 \right\rangle - \left\langle \Psi_0 \middle| \hat{F}\hat{F} \middle| \Psi_0 \right\rangle E_0 - E_0 \left\langle \Psi_0 \middle| \hat{F}\hat{F} \middle| \Psi_0 \right\rangle \right]$$
$$= 2 \left\langle \Psi_0 \middle| \hat{F}(\hat{H} - E_0) \middle| \Psi_0 \right\rangle \; .$$

We insert the completeness

$$2 \left\langle \Psi_0 \middle| \hat{F} \sum_\nu \middle| \Psi_\nu \right\rangle \left\langle \Psi_\nu \middle| (\hat{H} - E_0)\hat{F} \middle| \Psi_0 \right\rangle$$
$$= 2 \left\langle \Psi_0 \middle| \hat{F} \sum_\nu \middle| \Psi_\nu \right\rangle \left\langle \Psi_\nu \middle| (E_\nu - E_0)\hat{F} \middle| \Psi_0 \right\rangle = 2(E_\nu - E_0) \sum_\nu \left\langle \Psi_0 \middle| \hat{F} \middle| \Psi_\nu \right\rangle \left\langle \Psi_\nu \middle| \hat{F} \middle| \Psi_0 \right\rangle \; .$$

This expression establishes a quantitative limit on the total excitation strength of a many-body system in response to an external probe. This limit is determined solely by the properties of the ground state, meaning that knowledge of the excited states is not necessary. The validity of Eq. (16.44) is linked to the fact that the states $|\Psi_\nu\rangle$ are eigenstates of the Hamiltonian $\hat{H}$. In practical calculations, states derived from models or approximated solutions of the Schrödinger equation are utilized; therefore, the validity of Equation Eq. (16.44) is not always guaranteed. This must be verified on a case-by-case basis.

For example, for the RPA theory, it is possible to derive the following relation

$$2 \sum_\nu (E_\nu - E_0) \left| \left\langle \nu \middle| \hat{F} \middle| \nu_0 \right\rangle \right|^2 = \left\langle \Phi_0 \middle| \left[\hat{F}, \left[\hat{H}, \hat{F} \right] \right] \middle| \Phi_0 \right\rangle \; , \tag{16.45}$$

which, formally, is not a true sum rule since in the left hand side there are RPA states, both ground and excited states, while in the right hand side there is a MF ground state. The demonstration of Eq. (16.45) is presented in detail in Ref. [2].

When the residual interaction is neglected, one obtains MF excited states $\left|\Phi_{ph}\right\rangle$, i.e. single Slater determinants with particle-hole excitations. In this case, Eq. (16.44) is verified since all these MF states are eigenstates of the unperturbed hamiltonian $\hat{H}_0$

$$2 \sum_{ph} (\epsilon_p - \epsilon_h) \left| \left\langle \Phi_{ph} \middle| \hat{F} \middle| \Phi_0 \right\rangle \right|^2 = \left\langle \Phi_0 \middle| \left[\hat{F}, \left[\hat{H}_0, \hat{F} \right] \right] \middle| \Phi_0 \right\rangle , \tag{16.46}$$

where the excitation energies of the full system are given by the differences between the single-particle energies of the particle-hole excitation.

Since in the RPA the full hamiltonian $\hat{H} = \hat{H}_0 + \hat{V}_{\text{res}}$ is considered, by inserting this expression in Eq. (16.45) we obtain

$$2\sum_{\nu}(E_\nu - E_0)\left|\left\langle \nu|\hat{F}|\nu_0\right\rangle\right|^2$$
$$= \left\langle \Phi_0\left|\left[\hat{F},\left[\hat{H}_0,\hat{F}\right]\right]\right|\Phi_0\right\rangle + \left\langle \Phi_0\left|\left[\hat{F},\left[\hat{V}_{\rm res},\hat{F}\right]\right]\right|\Phi_0\right\rangle \ . \qquad (16.47)$$

For operators $\hat{F}$ which commute with $\hat{V}_{\rm res}$ the IPM and RPA sum rules coincide.

16.3.6 The RPA Ground State

We have already indicated that the RPA ground state is not a MF state but it contains effects beyond it, correlations, expressed in terms of hole-particle excitations. A more precise representation of the RPA ground state comes from a theorem demonstrated by D. J. Thouless [2] leading to an expression of the RPA ground state of the type [3]:

$$|\nu_0\rangle = \mathcal{N} e^{\hat{S}} \, |\Phi_0\rangle \ , \qquad (16.48)$$

where $\mathcal{N}$ is a normalisation constant and the operator $\hat{S}$ is defined as

$$\hat{S} \equiv \frac{1}{2}\sum_{\nu,minj} s_{\nu,minj}\hat{a}^+_m\hat{a}_i\hat{a}^+_j\hat{a}_n \ . \qquad (16.49)$$

The sum considers all the particle-hole $\hat{a}^+_m\hat{a}_i$ and hole-particle $\hat{a}^+_j\hat{a}_n$ pairs, and the index ν runs on all the possible angular momentum and parity combinations allowed by the particle-hole, and hole-particle, quantum numbers. We indicated with $s_{\nu,minj}$ an amplitude weighting the contribution of each pair.

Starting from the above expression it is possible to calculate the $s_{\nu,minj}$ from the knowledge of the RPA X^ν_{ph} e Y^ν_{ph} amplitudes [4]. What is relevant is that by using these expressions the expectation value of a one-body operator with respect to the RPA ground state can be expressed as

$$\left\langle \nu_0|\hat{F}|\nu_0\right\rangle = \left\langle \nu_0\left|\sum_{\mu\mu'}\langle\mu|f|\mu'\rangle\,\hat{a}^+_\mu\hat{a}_{\mu'}\right|\nu_0\right\rangle$$
$$= \sum_h \langle h|f|h\rangle\left[1 - \frac{1}{2}\sum_\nu\sum_p |Y^\nu_{ph}|^2\right]$$
$$+ \sum_p \langle p|f|p\rangle\left[\frac{1}{2}\sum_\nu\sum_h |Y^\nu_{ph}|^2\right] \ . \qquad (16.50)$$

This clearly shows that the Y^ν_{ph} amplitudes modify the value with respect to the MF result.

16.3.7 Application of the RPA

The RPA formulation involves a sum over all particle-hole pairs in the definition of the excitation operator (16.25). The hole states are well-defined, encompassing all states below the Fermi surface. In contrast, the number of particle states is infinite, as they include all states above the Fermi surface. Therefore, a truncation of the sum is necessary, which means using a limited set of particle states. This limitation defines what is known as the **configuration space**.

The selection of the configuration space is typically achieved by setting a maximum energy value for the single-particle states. The MF problem is addressed by obtaining single-particle wave functions and energies up to this chosen maximum. Once the configuration space is established, all compatible particle-hole pairs that align with the angular momentum and parity of the excited states under consideration are selected. At this stage, it becomes possible to calculate the A and B matrix elements defined in Eq. (16.28), and ultimately to perform the diagonalisation of the RPA matrix.

As previously noted in the TDA case, the outcome is a number of eigenstates and eigenvalues that matches the number of identified particle-hole pairs. In the absence of residual interactions, the eigenvalues correspond to the energy difference between the particle and hole states, given by $\omega_\nu = \epsilon_p - \epsilon_h$. However, the presence of residual interactions alters these energy values. In some cases, the effect of the residual interaction is significant, leading to RPA eigenstates that differ markedly from those predicted by the MF model. In such instances, the eigenstates exhibit a notable mixing of particle-hole excitations, as indicated by X^ν_{ph} values that are similar across various particle-hole pairs. These represent collective excitations of the system.

Conversely, there are excited states that are primarily single-particle excitations, characterised by a single $X^\nu_{ph} \simeq 1$, while the other amplitudes remain negligible. In these cases, the RPA eigenvalue is quite close to the energy difference of the dominant particle-hole pair.

Examples of this type of RPA results are presented in Tables 16.1 and 16.2, where we show a selection of results obtained for the $J^\pi = 3^-$ and $J^\pi = 14^-$ states of ^{208}Pb. In both tables, we only include particle-hole transitions with an X^ν_{ph} amplitude greater than 0.1. The single-particle states are identified using the conventional nuclear shell-model scheme [3].

For the calculation of the 3^- state, the selected configuration space predicted 1284 particle-hole pairs, which corresponds to the number of solutions obtained. The results displayed in Table 16.1 correspond to the solution with the lowest energy eigenvalue, $\omega = 3.69$ MeV. It is important to note that the minimum value of the particle-hole energy is 5.80 MeV, indicating that the residual interaction significantly modifies the energy values. There are 6 particle-hole pairs with $X^\nu_{ph} > 0.1$, suggesting a high degree of collectivity.

In contrast, the situation presented in Table 16.2 for the 14^- state is quite different. Here, we also show the results for the lowest energy eigenvalue. The chosen

Table 16.1 Extracted from the results of a RPA calculation for the $J^\pi = 3^-$ state of the ^{208}Pb nucleus. The RPA eigenvalue is 3.689 MeV, a number quite distant from the energies of the single particle excitations $\epsilon_p - \epsilon_h$. The calculation considered 1284 particle-hole pairs, and here only those with amplitude greater than 0.1 are shown

^{208}Pb	3^-	$\omega = 3.689$ MeV		
		Protons		
p	h	$\epsilon_p - \epsilon_h$ MeV	X	Y
$2f_{7/2}$	$3s_{1/2}$	5.8168	0.364778	−0.076573
$1h_{9/2}$	$2d_{3/2}$	5.8038	0.425067	−0.096266
$1i_{13/2}$	$1h_{11/2}$	8.8699	0.252087	−0.097650
		Neutrons		
p	h	$\epsilon_p - \epsilon_h$ MeV	X	Y
$1i_{11/2}$	$2f_{5/2}$	7.2150	0.326432	−0.101974
$2g_{9/2}$	$3p_{3/2}$	6.0838	0.410392	−0.096913
$1j_{15/2}$	$1i_{13/2}$	8.8621	0.277879	-0.110192

Table 16.2 The same as Table 16.1 for the state $J^\pi = 14^-$ of the ^{208}Pb nucleus. The RPA energy eigenvalue is 9.098 MeV rather close to the energy of the dominant particle-hole pair

^{208}Pb	14^-	$\omega = 9.098$ MeV		
		Neutrons		
p	h	$\epsilon_p - \epsilon_h$ MeV	X	Y
$1j_{15/2}$	$1i_{13/2}$	8.8699	−0.999661	0.012288

configuration space identified 306 particle-hole pairs, but only one has an $X^\nu_{ph} > 0.1$, contributing approximately 99% to the RPA eigenstate. The RPA energy eigenvalue is 9.10 MeV, compared to 8.87 MeV for the dominant particle-hole configuration. This scenario represents a typical single-particle excited state, where the contribution of the residual interaction is minimal. The RPA framework effectively describes both collective and single-particle excitations within the many-body system.

This fundamental approach of the Random Phase Approximation (RPA) has been extended to consider particles within a continuum, specifically in situations where a particle can be emitted from the nucleus. Another extension of the formalism, based on the Bardeen-Cooper-Schrieffer (BCS) theory of superconductivity, addresses the fact that single-particle states can be only partially occupied. This has led to the formalisation of the Quasi-Particle RPA, which is particularly useful for studying systems with open shells.

The primary limitation of RPA theory arises from the fact that the Q^+_ν operator only accounts for $1p - 1h$ and $1h - 1p$ types of excitations, as outlined in Eq. (16.25). In a many-body system, more complex excitation modes can occur, where n-particles and n-holes are created. The extension of the Q^+_ν operator to also include $2p - 2h$

(two-particle, two-hole) excitations is referred to as Second RPA. A presentation of these extensions of the RPA is given in [5].

Despite its relative simplicity, RPA remains a cornerstone of many-body theories.

References

1. D. Bohm, D. Pines, A collective description of electron interactions, III. Phys. Rev. **92**, 609 (1953)
2. D.J. Thouless, Vibrational states of nuclei in the random phase approximation. Nucl. Phys. **22**, 78 (1961)
3. P. Ring, P. Schuck, *The Nuclear Many-Body Problem* (Springer, Berlin, 1980)
4. J. Suhonen, *From Nucleons to Nucleus* (Springer, Berlin, 2007)
5. G. Co', Introducing the random phase approximation theory. Universe **9**, 141 (2023)

Chapter 17
The Fermi Liquid Theory

17.1 Introduction

One of the effective theories of significant success is that of Fermi liquids, formulated by L. V. Landau in the late 1950s and further developed by various authors, primarily from the Russian school [1, 2]. The fundamental idea is that a system of interacting particles can be described using a set of non-interacting quasi-particles. This concept is far from trivial. It is possible to add a particle to a collection of non-interacting quasi-particles, provided that the state of the new particle lies above the Fermi surface. In this case, the system remains stable.

When interactions are introduced, the added fermion can induce particle-hole $(p - h)$ excitations, which may lead the system to transition into a new ground state. Additionally, the energy of the added fermion will be altered, indicating that the state of the added fermion is unstable and subject to decay.

The notion that all states of non-interacting fermions can be transformed into quasi-particle states, essentially dressed by interactions, is incorrect. Conversely, if the energy of the added fermion is very close to the Fermi energy, the Pauli exclusion principle prevents decay, as the density of accessible final states is extremely low, if not zero. Therefore, it is reasonable to assume that those states near the Fermi surface can be effectively described in terms of quasi-particle states.

The theory of Fermi liquids was originally formulated by Landau to describe infinite fermionic systems, particularly the fermionic liquid helium, which is composed of fermionic atoms due to their nuclei being those of the isotope ^{3}He [3, 4]. The Landau theory serves as an excellent example of an effective theory, and its success has led to efforts to extend it for the treatment of finite systems and atomic nuclei [5].

© The Author(s), under exclusive license to Springer Nature Switzerland AG 2026

G. Co', *Concepts in Quantum Many-Body Physics*, UNITEXT for Physics,
https://doi.org/10.1007/978-3-032-08920-5_17

17.2 Adiabatic Continuity

The fundamental concept of Fermi liquid theory is the ability to transform a many-body system of non-interacting particles into a system of interacting particles by adiabatically turning on the interactions. We have already employed this technique in Sect. 6.4 as a theoretical and mathematical tool to connect the two systems. In the context of Fermi liquid theory, this assumption has a more phenomenological basis, warranting a more in-depth discussion to better define its validity and limitations.

To clarify the situation further, we will consider a one-dimensional toy model in which the particles interact through a potential that enables the factorisation of the time-dependent and space-dependent terms,

$$\hat{V}(x,t) = -\hat{V}_0(t)\hat{V}_x(x) \ . \tag{17.1}$$

The depth of the potential changes with the time between the initial value $V_0(1)$ and the final one $V_0(2)$.

The single-particle Schrödinger equation is

$$i\hbar\frac{\partial}{\partial t}\phi(x,t) = \hat{H}(t)\phi(x,t) = \left(-\frac{\hbar^2}{2m}\frac{\partial^2}{\partial x^2} + \hat{V}(x,t)\right)\phi(x,t) \ . \tag{17.2}$$

If we assume that $V_0(t)$ changes its value in a sensitive manner with respect to the initial value, and also with respect to the kinetic energy, much more slowly than the changes of the particle the wave function, we can search for an adiabatic solution which consider $V_0(t)$ as a constant in time

$$\phi_{\rm adia}(x,t) \simeq \phi_{V_0(t)}(x)e^{-\frac{i}{\hbar}E_{V_0(t)}t} \ , \tag{17.3}$$

where $E_{V_0(t)}$ is the eigenvalue of the time independent Schrödinger equation of which $\phi_{V_0(t)}(x)$ is eigenstate

$$\hat{H}(t)\,\phi_{V_0(t)}(x) = E_{V_0(t)}\,\phi_{V_0(t)}(x) \ . \tag{17.4}$$

By inserting the expression (17.3) in Eq. (17.2) we obtain

$$\begin{aligned} i\hbar\frac{\partial}{\partial t}\phi_{\rm adia}(x,t) &= \hat{H}(t)\phi_{\rm adia}(x,t) \\ &= E_{V_0(t)}\,\phi_{\rm adia}(x,t) + i\hbar\left(\frac{\partial\phi_{\rm adia}(x,t)}{\partial V_0(,t)}\right)\left(\frac{\partial V_0(t)}{\partial t}\right) \ . \end{aligned} \tag{17.5}$$

The assumption that the expression (17.3) represents an eigenstate of the Hamiltonian becomes increasingly valid as the second term of the expression diminishes. In other words, this validity increases when the time derivative of $V_0(t)$ becomes small. If the rate of change of $V(t)$ is sufficiently slow compared to the timescales involved in the study, it is possible to transition smoothly from the initial value $V_0(1)$ to the final value $V_0(2)$ by making gradual adjustments to the solution.

A crucial point is that this procedure does not permit discontinuities. For instance, if the initial state is a bound state, the final state must also be a bound state. It is not possible to arrive at a final state that includes particles in the continuum, such as an ionised atomic state. The same principle applies to phase transitions, which cannot be accurately described using this method.

17.3 The Concept of Quasi-particle

The properties of a fermionic system with translational invariance, known as a Fermi gas, have been discussed in Sect. 2.3. The single-particle wave functions are represented as plane waves characterized by a wave number $\mathbf{k}$, which is related to momentum by the equation $\mathbf{p} = \hbar\mathbf{k}$, see Eq. (2.38). The eigenstates of this system are constructed as Slater determinants formed from these plane waves.

To define the states of this system, it is sufficient to specify which single particle states, characterised by $\mathbf{k}$, are occupied. For this reason, it is convenient to use the distribution function $n(k)$, which indicates the probability density of finding a fermion with a wave vector magnitude between k and $k + dk$. The distribution function depends solely on the magnitude of the wave vector due to the isotropy of the medium. In the ground state, the distribution function is given by $n(k) = \Theta(k_F - k)$, where k_F is the Fermi wave number and Θ is the Heaviside step function. We show in Fig. 17.1 the distribution function $n(k)$ of the ground state of the Fermi gas.

Using the language of Chap. 5, we can state that the excited states of this system are generated by the creation of particle-hole $(p - h)$ pairs. The description of excited states with excitation energies that are small relative to the total energy of the system is achieved by considering small fluctuations, $\delta n(k)$, in the distribution $n(k)$ of the ground state. These fluctuations modify the ground state energy by the amount

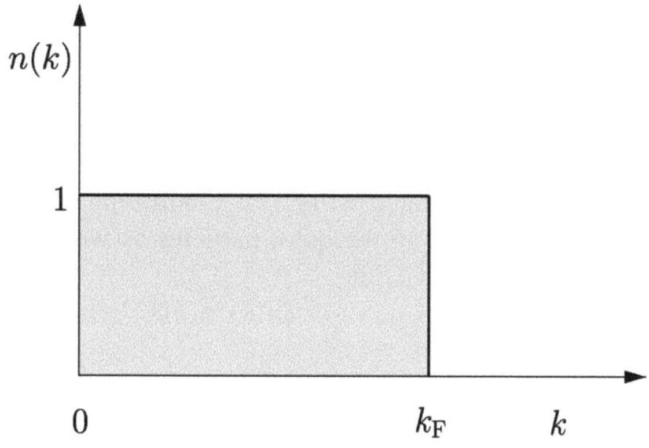

Fig. 17.1 Distribution function $n(k)$ for the Fermi gas ground state

$$\delta E = \sum_k \frac{\hbar^2 k^2}{2m} \delta n(k) \ . \tag{17.6}$$

The single particle energy is obtained as functional derivative of the energy $\delta E/\delta n(k) = \hbar^2 k^2/2m$. Since removing a particle, i.e. creating a hole state, reduces the value of the total energy, then $\delta n(k)$ must be negative. Obviously, the contrary is valid for the creation of a particle.

These considerations pertain to a gas of non-interacting fermions. The situation becomes more complex when interactions between fermions are introduced. Landau's idea is to transition from the description of a system of non-interacting fermions, which we will refer to as a Fermi gas, to that of an interacting fermion system, which we will call the real system, by adiabatically turning on the interaction. This process assumes that the evolution of the ground state of the Fermi gas leads to the ground state of the real system. This assumption enables for the formation of states in the real system, which then vanish when the interaction is turned off.

Let us introduce a particle with wave number $k > k_{\text{Fermi}}$ into the Fermi gas, and then activate the interaction. This process allows us to obtain a state representative of the real system. We can describe this as adding a quasi-particle with wave number k to the ground state of the real system. Similarly, we define a quasi-hole with wave number $k < k_{\text{Fermi}}$ as the scenario where a fermion is removed from the Fermi gas, followed by the activation of the interaction.

Some properties of the system are described by operators that commute with the Hamiltonian. These properties remain consistent between the Fermi gas and the real system. Examples of such properties include the number of particles, electric charge, current, and spin. Consequently, the quasi- particles can be characterised by quantum numbers that indicate their momentum $\mathbf{p}$, spin, electric charge, and so on.

The description of excitations in a real system through the creation and annihilation of quasi-particles is limited to a specific time window. The duration of these excitations must be shorter than the time required for the system to return to its ground state. This return is due to the elastic scattering of the quasi-particles, which means that the time needed to excite a pair of quasi-particles must be less than the typical time associated with their elastic scattering. Conversely, these durations must also be sufficiently long to ensure that the quasi-particles possess a well-defined energy, in accordance with the energy-time uncertainty relation, $\Delta E \Delta t \geq \hbar$.

Based on the points outlined above, the quasi-particles conserve all the quantum numbers associated with the particles, which means they also retain the characteristics of fermions. Consequently, the momentum distribution of the quasi-particles in the real system is illustrated in Fig. 17.1, confirming that the concept of the Fermi surface remains applicable to the quasi-particles as well. The excitation of the system is indicated by the deviation

$$\delta n(k) = n(k) - n^0(k), \tag{17.7}$$

where $n^0(k)$ indicates the quasi-particle distribution of the ground state, represented in Fig. 17.1.

The concept of quasi-particles is valid only when the values of $\delta n(k)$ are significantly different from zero for $k \approx k_F$. The assertion that a real system consists of quasi-particles filling all the states up to the Fermi surface is incorrect. Quasi-particles are elementary excitations of the real system that pertain to states close to the ground state. This implies that the excitation energies of these states are much smaller than the binding energy. Fermi liquid theory does not provide insights into the ground state of the system; rather, it focuses on describing small fluctuations around that state.

In this theory, the energy of the system E is assumed to be a functional of the momentum distribution function $n(k)$. In the case of a Fermi gas, this energy is the sum of the kinetic energies of individual particles, as described by Eq. (17.6). However, for a real system, the situation is more complex. If the momentum distribution $n^0(k)$ is perturbed by a quantity $\delta n(k)$, the first-order variation of the energy is given by.

$$\delta E = \sum_k \epsilon_k \, \delta n(k) \; , \tag{17.8}$$

this expression defines

$$\epsilon_k = \delta E / \delta n(k) \; . \tag{17.9}$$

For $k > k_F$, ϵ_k this is the variation of the energy when a particle is added, therefore it is the energy of the quasi-particle. The definition of ϵ_k is related to the fluctuation of the total energy of the system. In this theory, it is not possible to obtain information on the total energy of the system which is not the sum of the quasi-particle energies. This because, once a particle with k is added, the energy of the system changes, and the energy of a new particle $\epsilon_{k'}$ cannot be obtained by using (17.9) whose variation is done with respect to the old state which does not contain the particle k.

For $k = k_F$, ϵ_k is the energy acquired by adding one particle to the Fermi surface. The new state is the ground state of the system with $A + 1$ particles. We can write the equation

$$\epsilon_{k_F} = E_0(A+1) - E_0(A) \equiv \mu \; , \tag{17.10}$$

where $\mu = \partial E_0 / \partial A$ is, by definition, the chemical potential.

The expression of the energy variation (17.8) is valid up to the first order. This allows a description of situations dominated by excitations of a single quasi-particle. In general, this is not sufficient since the density of quasi-particle is so large that it is not possible to neglect their interaction. This indicates the need to go beyond the first order, and to consider also the second one

$$\delta E = \sum_k \epsilon_k^0 \, \delta n(k) + \frac{1}{2} \sum_k \sum_{k'} f(k, k') \delta n(k) \delta n(k') \; , \tag{17.11}$$

where the $f(k, k')$ term which describes the interaction between the quasi-particles is defined as the second functional derivative, with the property $f(k, k') = f(k', k)$. We called ϵ_k^0 the quasi-particle energy in absence of interaction.

Until now, we have not considered the presence of quasi-particle spin. The quasi-particles exhibit fermionic characteristics, and we focus on the spin-1/2 case, denoting its third component as s. In this scenario, the system is no longer homogeneous, as the spin orientation defines a specific direction in space. Consequently, all quantities that we previously defined as dependent solely on k must now be regarded as dependent on both $\mathbf{k}$ and s. This means that the sums over k will transform into sums over $\mathbf{k}$ and $s = \pm 1/2$. Specifically, we have $f(k, k') \to f(\mathbf{k}, s; \mathbf{k}', s')$, and the symmetry properties become

$$f(\mathbf{k}, s; \mathbf{k}', s') = f(-\mathbf{k}, -s; -\mathbf{k}', -s') = f(\mathbf{k}, -s; \mathbf{k}', -s') = f(\mathbf{k}', s'; \mathbf{k}, s) \ . \tag{17.12}$$

Once $\mathbf{k}$ and $\mathbf{k}'$ have been defined, the only differences between the f functions arise from whether the spins of the two interacting particles are parallel or anti-parallel. For this reason, it is useful to define the interaction in terms of symmetric and antisymmetric components

$$f(\mathbf{k}, s; \mathbf{k}', s) = f^s(\mathbf{k}, \mathbf{k}') + f^a(\mathbf{k}, \mathbf{k}') \ , \tag{17.13}$$

$$f(\mathbf{k}, s; \mathbf{k}', -s) = f^s(\mathbf{k}, \mathbf{k}') - f^a(\mathbf{k}, \mathbf{k}') \ . \tag{17.14}$$

The terms f^a and f^s depend solely on the vectors $\mathbf{k}$ and $\mathbf{k}'$, which means they are determined by the magnitudes of the two vectors and their relative angle. For this reason, they can be expanded in powers of Legendre polynomials P_l:

$$f^{s(a)}(\mathbf{k}, \mathbf{k}') = f^{s(a)}(k, k', cos\theta) = \sum_l f_l^{s(a)}(k, k') P_l(\cos\theta) \ . \tag{17.15}$$

Since the quasi-particles are defined on the Fermi surface, we have $k = k' = k_{\mathrm{F}}$. This allows us to eliminate the dependence on the magnitudes of the wave vectors in the coefficients f_l. The dimensions of these coefficients are those of energy. It is common practice to use dimensionless coefficients obtained by multiplying the f_l by the density of states at the Fermi surface

$$F_l \equiv \rho_\epsilon(\epsilon_{\mathrm{F}}) f_l = \frac{\mathcal{V}\mathcal{D}}{2\pi^2} \frac{k_{\mathrm{F}}}{\hbar^2} m^* f_l \ , \tag{17.16}$$

where we used the expression (2.63) of the density of states

$$\rho_\epsilon(\epsilon_{\mathrm{F}}) = \frac{\mathcal{V}\mathcal{D}}{4\pi^2} \left(\frac{2m^*}{\hbar^2} \right)^{3/2} \epsilon_{\mathrm{F}}^{1/2} = \frac{\mathcal{V}\mathcal{D}}{2\pi^2} \frac{m^*}{\hbar^2} k_{\mathrm{F}} \ . \tag{17.17}$$

The quantities $\mathcal{V}$ and $\mathcal{D}$, defined in Sect. 2.3, represent the volume of the system and the fermions degeneracy, respectively. In this chapter, we consider the degeneracy solely due to spin, which means that $\mathcal{D} = 2$. Since we are discussing quasi-particles, we denote their effective mass as m^*.

The usefulness of the theory lies in the fact that only a few F_l values are needed to achieve a good description of the fermionic system being studied. These F_l values are determined by fitting a limited set of empirical data. With just these few parameters, the theory can make predictions about other independent observables.

17.4 Equilibrium Properties

17.4.1 Effective Mass and Specific Heat

The velocity of the quasi-particle can be defined by drawing analogies with free particles. In the case of free particles, the single-particle energies correspond to their kinetic energies, given by the expression $mv^2/2$. Using this analogy, we can define one Cartesian component of the velocity, denoted as α, as follows

$$(v_k)_\alpha = \frac{1}{\hbar}\frac{\partial \epsilon_{\mathbf{k}}}{\partial k_\alpha} , \tag{17.18}$$

and, therefore, for an isotropic system, we have that

$$|\mathbf{v}_k| \equiv v_{\mathbf{k}} = \frac{\hbar k}{m^*}. \tag{17.19}$$

In principle, the effective mass m^* depends on the wave vector k. However, as we have previously discussed, quasi-particles are defined on the Fermi surface, which means that m^* has a single, well-defined value. The empirical data used to determine the value of m^* is the specific heat per unit volume, which is defined as the change in the internal energy of the system with respect to temperature, Eq. (2.71), divided by the volume.

$$c_v = \frac{1}{\mathcal{V}}C_v = \frac{1}{\mathcal{V}}\frac{\partial E}{\partial T} . \tag{17.20}$$

We consider temperature variations where $T/T_{\mathrm{F}} \ll 1$. The energy fluctuations of the energy can be expressed as

$$\delta E = \sum_{\mathbf{k},s} \epsilon_{k,s}\delta n(\mathbf{k}, s) . \tag{17.21}$$

An increase in temperature leads to a rise in the number of quasi-particles above the Fermi surface of the order of $\delta n(\mathbf{k}, s)$. In principle, there is also an increase in energy E due to interactions between quasi-particles; however, this increase is of the order of $(T/T_\mathrm{F})^3$ and, based on the assumptions made earlier, can be considered negligible.

According to Eq. (17.21), the specific heat of a real system can be calculated as if it were a system of non-interacting quasi-particles. We repeat the calculations from Sect. 2.3, keeping in mind that we are dealing with quasi-particles rather than traditional particles. Therefore, it is essential to consider the effective mass m^*. We use the expression (2.73) to obtain the result

$$c_v = \frac{1}{\mathcal{V}} C_v = \frac{1}{\mathcal{V}} \mathrm{k}_\mathrm{B}^2 \, T \rho_\epsilon(\epsilon_\mathrm{F}) \frac{\pi^2}{3} = \frac{1}{2} \mathcal{D} \frac{m^* k_\mathrm{F}}{3} \mathrm{k}_\mathrm{B}^2 \, T , \tag{17.22}$$

where we considered the expression (17.17) of the density of states. The value of the effective mass of the quasi-particles can be obtained by measuring the specific heat of a system of interacting fermions.

17.4.2 Sound Speed and Compressibility

In the Appendix E.3 we show that the relation between speed of sound v_s in a fluid and the compression modulus B is

$$v_s = \sqrt{\frac{B}{m\rho}} \; . \tag{17.23}$$

We calculate the compression modulus by using the definition of Eq. (2.59)

$$B = \frac{1}{K} = -\mathcal{V} \frac{\partial P}{\partial \mathcal{V}} \; . \tag{17.24}$$

Since the pressure is related to the variation of the energy with respect to the variation of the volume, it is necessary to calculate the energy variations. We assume that the ground state energy E depends on a function $\mathcal{F}$ of the particle density as

$$E = \mathcal{V} \mathcal{F}(\rho) = \mathcal{V} \mathcal{F}\left(\frac{A}{\mathcal{V}}\right) \; . \tag{17.25}$$

Since

$$\frac{\partial \rho}{\partial \mathcal{V}} = A \left(\frac{-1}{\mathcal{V}^2}\right) = \frac{-1}{\mathcal{V}} \rho \; , \tag{17.26}$$

using the traditional definition of pressure we obtain

$$P = -\frac{\partial E}{\partial \mathcal{V}} = -\left[\mathcal{F} - \rho \frac{\partial \mathcal{F}}{\partial \rho}\right] , \tag{17.27}$$

and, therefore, the variation of the pressure becomes

$$\frac{\partial P}{\partial \mathcal{V}} = -\left[\frac{\partial \mathcal{F}}{\partial \rho}\frac{\partial \rho}{\partial \mathcal{V}} - \frac{\partial \rho}{\partial \mathcal{V}}\frac{\partial \mathcal{F}}{\partial \rho} - \rho\frac{\partial}{\partial \rho}\left(\frac{\partial \mathcal{F}}{\partial \rho}\right)\frac{\partial \rho}{\partial \mathcal{V}}\right]$$
$$= -\frac{\rho^2}{\mathcal{V}}\frac{\partial^2 \mathcal{F}}{\partial \rho^2}\,, \tag{17.28}$$

and the compression modulus is

$$B = \rho^2\frac{\partial^2 \mathcal{F}}{\partial \rho^2}\,. \tag{17.29}$$

We insert the expression (17.25) in the definition of chemical potential

$$\mu = \frac{\partial E}{\partial A} = \frac{\partial}{\partial A}[\mathcal{V}\mathcal{F}] = \frac{\partial \mathcal{F}}{\partial \rho}\,. \tag{17.30}$$

The variation of the chemical potential is

$$\frac{\partial \mu}{\partial A} = \frac{\partial}{\partial \rho}\left(\frac{\partial \mathcal{F}}{\partial \rho}\right)\frac{\partial \rho}{\partial A} = \frac{\partial^2 \mathcal{F}}{\partial \rho^2}\frac{1}{\mathcal{V}}\,. \tag{17.31}$$

From the above equation we extract the expression of the second derivative of $\mathcal{F}$ and we substitute it in the Eq. (17.29) and we obtain

$$B = A\rho\frac{\partial \mu}{\partial A}\,, \tag{17.32}$$

and for the sound speed

$$v_s^2 = \frac{B}{m\rho} = \frac{A}{m}\frac{\partial \mu}{\partial A}\,. \tag{17.33}$$

The Fermi liquid theory provides the expression of $\partial A/\partial \mu$. In the box here below we show that A is related to the variation of $n(k)$ by the expression

$$dA = \sum_{\mathbf{k},s} \delta n(\mathbf{k}, s) = \sum_{\mathbf{k},s} \delta(\epsilon_k - \mu)\,\hbar v_k\,dk\,. \tag{17.34}$$

Let us derive Eq. (17.34).

$$\delta n(\mathbf{k}, s) = -\frac{dn^0(\mathbf{k}, s)}{d\epsilon_k}\frac{\partial \epsilon_k}{\partial k}dk\,. \tag{17.35}$$

At the end the calculation of the variation we shall make the limit $k \to k_F$.

The variation of the single-particle energy can be written as

$$d\epsilon_k = d\left(\frac{\hbar^2 k^2}{2m}\right) = \hbar\frac{\hbar k}{m}dk = \hbar\frac{m v_k}{m}dk = \hbar v_k dk \ . \tag{17.36}$$

Let us consider now the variation of the momentum distribution $n^T(k, s)$ for an excited state, and for $T > 0$. Since the system is composed by fermions, the expression of $n^T(\mathbf{k}, s)$ is that of a Fermi-Dirac distribution

$$n^T(\mathbf{k}, s) = \left[\exp\left(\frac{\epsilon - \mu}{k_B T}\right) + 1\right]^{-1} , \tag{17.37}$$

therefore

$$\frac{d\, n^T(\mathbf{k}, s)}{d\epsilon} = \frac{-\frac{1}{k_B T}\exp\left[(\epsilon - \mu)/(k_B T)\right]}{\{\exp\left[(\epsilon - \mu)/(k_B T)\right] + 1\}^2} \ . \tag{17.38}$$

In the limit $T \to 0$, therefore for the ground state, the Fermi-Dirac distribution becomes a step function Θ. For the property of the Dirac distribution δ, $d\Theta/dx = \delta(x)$, we can write

$$\lim_{T\to 0}\frac{d\, n^T(\mathbf{k}, s)}{d\epsilon} = -\delta(\epsilon - \mu) \ , \tag{17.39}$$

therefore, on the Fermi surface the expression is

$$\delta n(\mathbf{k}_F, s) = \delta(\epsilon_F - \mu)\hbar v_{k_F} dk_F \ . \tag{17.40}$$

A quasi-particle added to the system on top to the new Fermi energy must have an energy $\epsilon(\mu + d\mu)$ satisfying

$$d\mu = \epsilon(\mu + d\mu) - \epsilon(\mu) = \hbar v_{k_F} dk_F + \sum_{\mathbf{k}', s'} f(\mathbf{k}, s; \mathbf{k}', s')\delta n(\mathbf{k}', s'). \tag{17.41}$$

By using the expression (17.40) we have that

$$\frac{\partial\mu}{\partial\mu} = 1 = \hbar v_{k_F}\frac{\partial k_F}{\partial\mu} + \sum_{\mathbf{k}', s} f(\mathbf{k}, s; \mathbf{k}', s)\delta(\epsilon_{k'} - \mu)\hbar v_{k'}\frac{\partial k'}{\partial\mu} \ . \tag{17.42}$$

The calculation of the sum on $\mathbf{k}'$ and s' is presented in the box here below.

The sum on $\mathbf{k}'$ is transformed by using the conventions presented in Sect. 2.3.

$$\sum_{\mathbf{k}', s'} f(\mathbf{k}, s; \mathbf{k}', s')\delta(\epsilon_{k'} - \mu)\hbar v_{k'}\frac{\partial k'}{\partial\mu}$$

$$= \sum_{s'}\frac{\mathcal{V}}{(2\pi)^3}\int d^3k' \ f(\mathbf{k}, s; \mathbf{k}', s')\delta(\epsilon_{k'} - \mu)\hbar v_{k'}\frac{\partial k'}{\partial\mu}$$

$$= \frac{\mathcal{V}}{(2\pi)^3}\sum_{s'}\int dk' k'^2\int_{-1}^{1} d(\cos\theta)\int_0^{2\pi} d\phi f(\mathbf{k}, s; \mathbf{k}', s')\delta(\epsilon_{k'} - \mu)\hbar v_{k'}\frac{\partial k'}{\partial\mu} \ .$$

Since the variation between ϵ_{k_F} and μ is very small, we can write,

$$\epsilon_k - \mu|_{k=k_\mathrm{F}} \simeq d\epsilon_k|_{k=k_\mathrm{F}} = d\left(\frac{\hbar^2 k}{2m^*}\right)_{k=k_\mathrm{F}} = \frac{\hbar^2 k_\mathrm{F}}{m^*} dk_\mathrm{F} \simeq \frac{\hbar^2 k_\mathrm{F}}{m^*}(k - k_\mathrm{F}) ,$$

where we have considered that we are describing a system of quasi-particles of effective mass m^*. For the properties of Dirac's δ distribution

$$\delta(ax) = \frac{1}{|a|}\delta(x) ,$$

we can substitute in the integral

$$\delta(\epsilon_k' - \mu) = \frac{m^*}{\hbar^2 k_\mathrm{F}} \delta(k' - k_\mathrm{F}) .$$

The expression of the integral is now

$$\begin{aligned}
&\sum_{\mathbf{k}',s'} f(\mathbf{k}, s; \mathbf{k}', s')\delta(\epsilon_{k'} - \mu)\hbar v_{k'} \frac{\partial k'}{\partial \mu} \\
&= \frac{V 2\pi}{(2\pi)^3} \sum_{s'} \int dk' k'^2 \int_{-1}^{1} d(\cos\theta) f(\mathbf{k}, s; \mathbf{k}', s') \frac{m^*}{\hbar^2 k_\mathrm{F}} \delta(k' - k_\mathrm{F}) \hbar v_{k'} \frac{\partial k'}{\partial \mu} \\
&= \frac{V}{4\pi^2} \frac{m^* {k_\mathrm{F}}^2}{\hbar^2 k_\mathrm{F}} \hbar v_{k_\mathrm{F}} \frac{\partial k_\mathrm{F}}{\partial \mu} \int_{-1}^{1} d(\cos\theta)\left[f(\mathbf{k}_\mathrm{F}, s; \mathbf{k}_\mathrm{F}, s) + f(\mathbf{k}, s; \mathbf{k}_\mathrm{F}, -s)\right] .
\end{aligned}$$

Let us utilize the expressions (17.13) and (17.14) for the function f. We note that, on the Fermi surface, the interaction depends solely on the angle between $\mathbf{k}$ and $\mathbf{k}'$, both of which have magnitudes equal to k_F. Therefore, we can apply the expansion in Legendre polynomials as shown in (17.15). We obtain the equation

$$\begin{aligned}
&\sum_{\mathbf{k}',s'} f(\mathbf{k}, s; \mathbf{k}', s')\delta(\epsilon_{k'} - \mu)\hbar v_{k'} \frac{\partial k'}{\partial \mu} \\
&= \frac{V}{4\pi^2} \frac{m^*}{\hbar^2} k_\mathrm{F} \hbar v_{k_\mathrm{F}} \frac{\partial k_\mathrm{F}}{\partial \mu} \int_{-1}^{1} d(\cos\theta)\, 2 \sum_l f_l^s(k_\mathrm{F}, k_\mathrm{F}) P_l(\cos\theta) .
\end{aligned}$$

Because of the orthogonality of the Legendre polynomials

$$\int_{-1}^{1} dx\, P_l(x) = 2\delta_{l,0} ,$$

we obtain

$$\sum_{\mathbf{k}',s'} f(\mathbf{k}, s; \mathbf{k}', s')\delta(\epsilon_{k'} - \mu)\hbar v_{k'} \frac{\partial k'}{\partial \mu} = \frac{V}{\pi^2} \frac{m^* k_\mathrm{F}}{\hbar^2} \hbar v_{k_\mathrm{F}} \frac{\partial k_\mathrm{F}}{\partial \mu} f_0^s(k_\mathrm{F}, k_\mathrm{F}) . \tag{17.43}$$

We insert the result (17.43) in Eq. (17.42) and use the expression (17.17) of the density of states, and the definition (17.16) of the dimensionless Landau coefficients. We obtain

$$
\begin{aligned}
1 &= \hbar v_{k_F} \frac{\partial k_{\rm F}}{\partial \mu} + \frac{V}{\pi^2} \frac{m^* k_{\rm F}}{\hbar^2} \hbar v_{k_{\rm F}} \frac{\partial k_{\rm F}}{\partial \mu} f_0^s(k_{\rm F}, k_{\rm F}) \\
&= \hbar v_{k_F} \frac{\partial k_{\rm F}}{\partial \mu} + \rho_\epsilon(\epsilon_{\rm F}) f_0^s(k_{\rm F}, k_{\rm F}) \hbar v_{k_F} \frac{\partial k_{\rm F}}{\partial \mu} = \hbar v_{k_F} \frac{\partial k_{\rm F}}{\partial \mu} \left[1 + F_0^s\right] . \qquad (17.44)
\end{aligned}
$$

From Eq. (17.34) we have that

$$
\frac{\partial A}{\partial \mu} = \sum_{\mathbf{k},s} \delta(\epsilon_k - \mu)\, \hbar v_k \frac{\partial k}{\partial \mu} . \qquad (17.45)
$$

We use the definition (17.33) of speed of sound and consider that the sum on $\delta(\epsilon_k - \mu)$ is the density of states ρ_ϵ, therefore we have that

$$
\frac{A}{m v_s^2} = \frac{\rho_\epsilon(\epsilon_{\rm F})}{1 + F_0^s} . \qquad (17.46)
$$

From this expression we obtain

$$
\begin{aligned}
v_s^2 &= \frac{A}{m} \frac{\pi^2 \hbar^2}{V m^* k_{\rm F}} (1 + F_0^s) = \frac{\pi^2 \hbar^2}{m m^* k_{\rm F}}\, \rho\, (1 + F_0^s) \\
&= \frac{\pi^2 \hbar^2}{m m^* k_{\rm F}} \left(\frac{{k_{\rm F}}^3}{3\pi^2}\right) (1 + F_0^s) = \frac{\hbar^2 {k_{\rm F}}^2}{3\, m m^*} (1 + F_0^s) . \qquad (17.47)
\end{aligned}
$$

It is remarkable to observe the simultaneous presence of two masses: the mass of the particle m and the mass of the quasi-particle m^*. The former mass is included due to the definition of mass density in the fluid, as is standard in traditional statistical mechanics. The latter, the effective mass of the quasi-particle, accounts for the fact that interactions between particles modify the density of states.

Clearly, particle interactions also influence the expression for the speed of sound, particularly through the term F_0^s. When all other variables are held constant, a repulsive interaction (i.e., $F_0^s > 0$) leads to an increase in the speed of sound. Conversely, values of $F_0^s < -1$ result in imaginary sound speeds, which indicate that density fluctuations accumulate and create instability within the system.

17.4.3 Magnetic Susceptibility

Another observable used to determine the values of the free parameters in Fermi liquid theory is the magnetic susceptibility, denoted as χ_M. In the presence of a magnetic field $\mathbf{H}$, a particle alters its energy by an amount of $-g\mu_{\rm B} s|\mathbf{H}|$, where g is the Landé factor, which we assume to be equal to 2, and $\mu_{\rm B} = e\hbar/mc$ is the Bohr magneton. We define $s = 1/2$ for the case where the fermion spin is aligned with the direction of the magnetic field $\mathbf{H}$, and $s = -1/2$ for the opposite case.

In a real gas, a change in energy also leads to a change in the momentum distribution $n(\mathbf{k}, s)$. The system contains particles with both values of s. When the system is in equilibrium, the chemical potential μ must be the same for all particles, regardless of the value of s. In other words, the energy required to add a particle should be uniform across the system.

For particles with $s = -1/2$, the presence of the magnetic field $\mathbf{H}$ causes them to gain energy, resulting in a decrease in the Fermi energy by a factor of δk_{F} compared to the scenario without a magnetic field. Conversely, for particles with $s = 1/2$, the situation is reversed, leading to an increase in δk_{F} at the Fermi surface.

The isotropy of the system implies that $n(\mathbf{k}, s) = n(k, s)$, therefore, the variation of the momentum distribution is given by

$$\delta n(k, s) = -\left(\frac{\partial n(k, s)}{\partial k}\right)_{k=k_{\mathrm{F}}} \delta k_{\mathrm{F}} = -\left[-\delta(k - k_{\mathrm{F}})\right](2s)\delta k_{\mathrm{F}} \ . \tag{17.48}$$

The Dirac δ distribution arises from the observation that, in the ground state, the function $n(k, s)$ behaves like a step function of k. Consequently, its derivative corresponds to the Dirac delta function (refer to the box in Sect. 17.4.2). The term $2s = \pm 1$ accounts for the increase or decrease of the Fermi surface due to the influence of the magnetic field (Fig. 17.2).

The variation of the quasi-particle energy is given by

$$\delta\epsilon_{k_{\mathrm{F}}} = -g\mu_{\mathrm{B}} s|\mathbf{H}| + (2s)\sum_{\mathbf{k}', s'} f(\mathbf{k}, s; \mathbf{k}', s')\delta n(\mathbf{k}', s') \ , \tag{17.49}$$

the term $2s = \pm 1$ is inserted since $s = 1/2$ and $s = -1/2$ give different contributions. Let us search for a solution of the type

$$\delta\epsilon_{k_{\mathrm{F}}} = -\eta s|\mathbf{H}| \ , \tag{17.50}$$

where η is a constant to be defined. We consider that

$$\frac{d\epsilon_k}{d|\mathbf{k}|} = \frac{\hbar^2|\mathbf{k}|}{m^*} \ , \tag{17.51}$$

therefore

$$\delta k_{\mathrm{F}} = \left|\frac{d\epsilon_k}{d|\mathbf{k}|}\right|^{-1}_{k=k_{\mathrm{F}}} |\delta\epsilon_{k_{\mathrm{F}}}| = \frac{m^*}{\hbar^2 k_{\mathrm{F}}}\eta\frac{1}{2}|\mathbf{H}| \ , \tag{17.52}$$

for $s = 1/2$.

We sum over $\mathbf{k}'$ and s', taking into account that the momentum distributions $n(k, s)$ (17.48) differ for fermions with spins aligned parallel or antiparallel to the direction of the magnetic field. The calculation follows the same strategy used for evaluating the speed of sound. In this case, we utilise the definition of the antisymmetric interaction from Eqs. (17.13) and (17.14), along with the expansion in Legendre

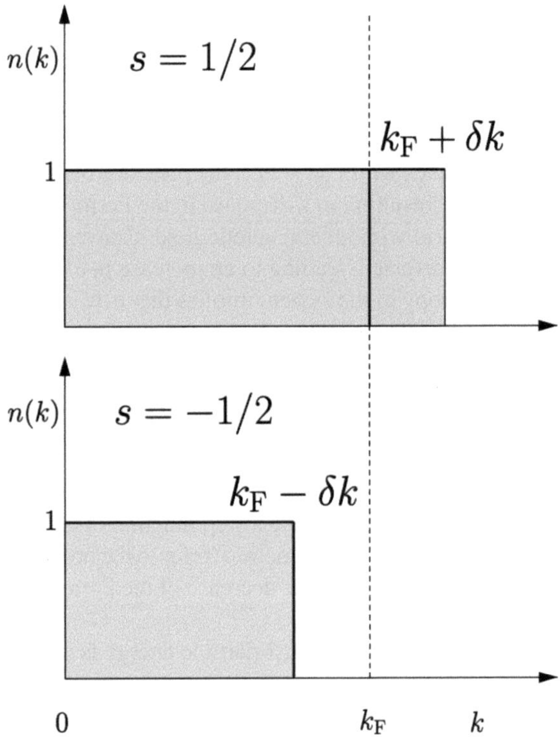

Fig. 17.2 Variation of the Fermi surface due to the presence of the magnetic field **H**

polynomials as described in (17.15). This approach allows us to derive the following expressions.

$$
\begin{aligned}
&(2s)\sum_{\mathbf{k}',s'} f(\mathbf{k}, s; \mathbf{k}', s')\delta n(\mathbf{k}', s') \\
&= (2s)\sum_{s'=\pm 1/2} \frac{\mathcal{V}}{(2\pi^3)}\int d^3k' f(\mathbf{k}, s, \mathbf{k}', s')\delta(k'-k_{\mathrm{F}})(2s')\left[\frac{m^*}{\hbar^2 k_{\mathrm{F}}}\eta\frac{1}{2}|\mathbf{H}|\right] \\
&= \frac{\mathcal{V}}{(2\pi^3)}(2\pi)(2s)\int dk' k'^2\delta(k'-k_{\mathrm{F}}) \\
&\qquad \int_{-1}^{1} d(\cos\theta) 2 f^a(k, k', \cos\theta)\left[\frac{m^*}{\hbar^2 k_{\mathrm{F}}}\eta\frac{1}{2}|\mathbf{H}|\right] \\
&= \frac{\mathcal{V}}{2\pi^2}s\left[\frac{m^*}{\hbar^2 k_{\mathrm{F}}}\eta|\mathbf{H}|\right]k_{\mathrm{F}}{}^2\int_{-1}^{1} d(\cos\theta)\sum_l f_l^a(k, k')P_l(\cos\theta) \\
&= \frac{\mathcal{V}}{2\pi^2}s\frac{m^*}{\hbar^2 k_{\mathrm{F}}}\eta|\mathbf{H}|k_{\mathrm{F}}{}^2 f_0^a 2 = F_0^a s\eta|\mathbf{H}| \ .
\end{aligned}
$$

We insert this result in Eq. (17.49) and obtain

$$\delta\epsilon_{k_F} = -\eta s|\mathbf{H}| = -g\mu_B s|\mathbf{H}| + F_0^a s\eta|\mathbf{H}| \ , \tag{17.53}$$

from which we have

$$\eta\left[1 + F_0^a\right] = g\mu_B \ ; \ \ \eta = \frac{g\mu_B}{1 + F_0^a} \ . \tag{17.54}$$

The magnetisation energy per unit volume is expressed as the product of the magnetic susceptibility χ_M and the magnitude of the magnetic field $|\mathbf{H}|$. This relationship corresponds to the sum of the variations in the quasi-particle energies, which are influenced by the magnetic field $gs\mu_B$, due to changes in the momentum distribution $n(\mathbf{k}, s)$.

$$\begin{aligned} M &= \frac{1}{V}\chi_M|\mathbf{H}| = \frac{1}{V}\sum_{\mathbf{k},s} gs\mu_B\delta n(\mathbf{k}, s) \\ &= \frac{1}{V} g\mu_B \sum_s s\frac{V}{(2\pi^3)}\int d^3k\delta(k - k_F)\left[\frac{m^*}{\hbar^2 k_F}(2s)\eta\frac{1}{2}|\mathbf{H}|\right] \\ &= g\mu_B\frac{1}{4\pi^2}\frac{m^* k_F}{\hbar^2}\eta|\mathbf{H}| \ . \end{aligned}$$

The magnetic susceptibility per unit of volume is

$$\frac{1}{V}\chi_M = \frac{M}{|\mathbf{H}|} = \frac{m^* k_F}{4\pi^2\hbar^2}\frac{(g\mu_B)^2}{1 + F_0^a}. \tag{17.55}$$

The values of the fundamental parameters in Landau's theory, such as effective mass and the scalar and spin-dependent components of the interaction f are determined by their relationships with empirical quantities: specific heat, speed of sound, and magnetic susceptibility. It is possible to connect these parameters to more fundamental properties of the Hamiltonian [6–9]. Ultimately, the primary goal of the theory is to make predictions about phenomena that are not directly related to the observable quantities used to establish the effective mass and the interaction.

17.5 Excitations

17.5.1 Transport Equation

In this section, we shall describe how a fermionic fluid responds to compression waves, using the framework of Fermi liquid theory.

Let us consider a system that contains some non-homogeneous regions, which we treat as small units in local equilibrium. These units are small compared to the overall size of the system, but large enough to contain a sufficient number of fermions to define a momentum distribution function, $n_{\mathbf{k},s}(\mathbf{r}, t)$. Here, $\mathbf{r}$ represents the position of the sub-unit.

The presence of a quantity defined precisely by the momentum $\hbar\mathbf{k}$ and the position $\mathbf{r}$ violates the Heisenberg uncertainty principle. Therefore, this description is valid only when the phenomena we study are macroscopic. In other words, the energy and momentum of the excitations must be much smaller than those of the individual particles, and the distances involved should be much larger than the typical spacing between particles in the system. For example, in the case of liquid helium, these assumptions hold because the typical helium binding energies are much smaller than the excitation energies of helium atoms.

Now, let us assume that the external perturbation, which locally changes the density of the system, behaves harmonically with a frequency $\hbar\omega$ and imparts a momentum $\mathbf{q}$ to the system. The momentum distribution of the fermion system subject to this perturbation can be expressed as

$$n_{\mathbf{k},s}(\mathbf{r}, t) = n^0_{\mathbf{k},s} + \delta n_{\mathbf{k},s}(\mathbf{r}, t) = n^0_{\mathbf{k},s} + \delta n_{\mathbf{k},s}(\mathbf{q}, \omega) e^{i(\mathbf{q}\cdot\mathbf{r}-\hbar\omega t)} \ , \tag{17.56}$$

where $n^0_{\mathbf{k},s}$ is the momentum distribution of the system in absence of perturbation. The approach is valid for energy much smaller than the chemical potential $\omega << \mu$. We can express the energy fluctuations at the time t as

$$E(t) = E_0 + \sum_{\mathbf{k},s} \int d^3r \, \epsilon_{k,s} \, \delta n_{\mathbf{k},s}(\mathbf{r}, t)$$
$$+ \frac{1}{2} \sum_{\mathbf{k},s,\mathbf{k}',s} \int \int d^3r d^3r' f(\mathbf{r}, \mathbf{k}, s; \mathbf{r}', \mathbf{k}', s') \delta n_{\mathbf{k},s}(\mathbf{r}, t) \delta n_{\mathbf{k}',s}(\mathbf{r}', t) \ . \tag{17.57}$$

We make some assumptions regarding the properties of the interactions between particles. The first assumption, which was already used in the previous expression, is that the interaction does not depend on time. The second assumption relates to Galilean invariance: we suppose that the interaction depends only on the distance $\mathbf{r} - \mathbf{r}'$ between the particles. Finally, we assume that the interaction is short-ranged, as explained in Chap. 3. It's important to note that this last assumption does not hold for the Coulomb interaction. The charged Fermi liquid must be treated differently, see Ref. [6].

Let us consider an extremely short-ranged interaction and assume

$$f(\mathbf{r}, \mathbf{k}, s; \mathbf{r}', \mathbf{k}', s') = f(\mathbf{k}, s; \mathbf{k}', s') \delta(\mathbf{r} - \mathbf{r}') \ , \tag{17.58}$$

therefore, the second term of Eq. (17.57) can be written as

$$\int\int d^3r d^3r' f(\mathbf{r}, \mathbf{k}, s; \mathbf{r}', \mathbf{k}', s')\delta n_{\mathbf{k},s}(\mathbf{r}, t)\delta n_{\mathbf{k}',s'}(\mathbf{r}', t)$$
$$= \int d^3r f(\mathbf{k}, s; \mathbf{k}', s')\delta n_{\mathbf{k},s}(\mathbf{r}, t)\delta n_{\mathbf{k}',s'}(\mathbf{r}, t) \ , \tag{17.59}$$

and

$$E(t) = E_0 + \int d^3r \delta E(\mathbf{r}, t) \ , \tag{17.60}$$

with

$$\delta E(\mathbf{r}, t) = \sum_{\mathbf{k},s} \epsilon_{k,s}\delta n_{\mathbf{k},s}(\mathbf{r}, t) + \frac{1}{2}\sum_{\mathbf{k},s,\mathbf{k}',s} f(\mathbf{k}, s; \mathbf{k}', s')\delta n_{\mathbf{k},s}(\mathbf{r}, t)\delta n_{\mathbf{k}',s'}(\mathbf{r}, t) \ . \tag{17.61}$$

The local excitation energy of the quasi-particle is defined as

$$\tilde{\epsilon}_{\mathbf{k},s}(\mathbf{r}, t) \equiv \frac{\partial E(\mathbf{r}, t)}{\partial n_{\mathbf{k},s}(\mathbf{r}, t)} = \epsilon_{k,s} + \sum_{\mathbf{k}',s'} f(\mathbf{k}, s; \mathbf{k}', s')\delta n_{\mathbf{k}',s'}(\mathbf{r}, t) \ . \tag{17.62}$$

In the approach of Landau the gas of quasi-particle gas is considered as a classic gas following the Maxwell-Boltzmann statistics. Each particle, whose energy is $\tilde{\epsilon}_{\mathbf{k},s}$, is described by a classical hamiltonian. Let us consider a volume $d\mathbf{r}\ d\mathbf{k}$ in the six-dimensional phase space and apply to the $n_{\mathbf{k},s}(\mathbf{r}, t)$ distribution the transport equation (F.24)

$$\mathcal{I}(n_{\mathbf{k},s}(\mathbf{r}, t)) = \frac{\partial n_{\mathbf{k},s}(\mathbf{r}, t)}{\partial t} + \nabla_{\mathbf{r}} n_{\mathbf{k},s}(\mathbf{r}, t)\cdot\nabla_{\mathbf{p}}\tilde{\epsilon}_{\mathbf{k},s}(\mathbf{r}, t) - \nabla_{\mathbf{p}} n_{\mathbf{k},s}(\mathbf{r}, t)\cdot\nabla_{r}\tilde{\epsilon}_{\mathbf{k},s}(\mathbf{r}, t) \ . \tag{17.63}$$

This expression can be immediately recognised as corresponding to Eq. (F.24) when considering that

$$\nabla_{\mathbf{p}}\tilde{\epsilon}_{\mathbf{k},s} = \mathbf{v}_{\mathbf{k},s} \quad \text{and} \quad -\nabla_{\mathbf{r}}\tilde{\epsilon}_{\mathbf{k},s}(\mathbf{r}, t) = \mathbf{F} \ .$$

Let us explicitly write Eq. (17.63) by using Eqs. (17.56) and (17.62)

$$\mathcal{I}(n_{\mathbf{k},s}(\mathbf{r}, t)) = \frac{\partial n_{\mathbf{k},s}(\mathbf{r}, t)}{\partial t} + \nabla_{\mathbf{r}}\delta n_{\mathbf{k},s}(\mathbf{r}, t)\cdot\nabla_{\mathbf{p}}\tilde{\epsilon}_{\mathbf{k},s}(\mathbf{r}, t)$$
$$-\left\{\nabla_{\mathbf{p}}[n^0_{\mathbf{k},s} + \delta n_{\mathbf{k},s}(\mathbf{r}, t)]\cdot\nabla_{\mathbf{r}}[\epsilon_{k,s} + \sum_{\mathbf{k}',s'} f(\mathbf{k}, s; \mathbf{k}', s')\delta n_{\mathbf{k}',s'}(\mathbf{r}, t)]\right\}$$

$$
\begin{aligned}
&= \frac{\partial n_{\mathbf{k},s}(\mathbf{r},t)}{\partial t} + \nabla_{\mathbf{r}} \delta n_{\mathbf{k},s}(\mathbf{r},t) \cdot \mathbf{v}_{\mathbf{k},s} \\
&\quad - \left[\nabla_{\mathbf{p}} n^0_{\mathbf{k},s} + \nabla_{\mathbf{p}} \delta n_{\mathbf{k},s}(\mathbf{r},t) \right] \cdot \sum_{\mathbf{k}',s'} f(\mathbf{k},s;\mathbf{k}',s') \nabla_{\mathbf{r}} \delta n_{\mathbf{k}',s'}(\mathbf{r},t) \\
&\simeq \frac{\partial n_{\mathbf{k},s}(\mathbf{r},t)}{\partial t} + \nabla_{\mathbf{r}} \delta n_{\mathbf{k},s}(\mathbf{r},t) \cdot \mathbf{v}_{\mathbf{k},s} \\
&\quad - \nabla_{\mathbf{p}} n^0_{\mathbf{k},s} \cdot \sum_{\mathbf{k}',s'} f(\mathbf{k},s;\mathbf{k}',s') \nabla_{\mathbf{r}} \delta n_{\mathbf{k}',s'}(\mathbf{r},t) \ , \qquad (17.64)
\end{aligned}
$$

where we considered that $\nabla_{\mathbf{r}} \epsilon_{k,s} = 0$ since $\epsilon_{k,s}$ is constant and, in the last step, we neglected the second order terms in $\delta n_{\mathbf{k}}$. The first two terms describe a flux of quasi-particles moving independently of each other. The last term, which includes the interaction, can be thought as the flux of ground state particles drifting due to their interaction with the non-homogeneous parts of the system. Without these non-homogeneous components, δn would remain constant, and therefore, this term would be zero.

17.5.2 Continuity Equation

Let us consider the case when the collision integral $\mathcal{I}$ of the Boltzmann transport equation (F.24) is negligible. This implies that the number of quasi-particle collisions which that take them out of the counting of $n_{\mathbf{k},s}$ is very small. At zero temperature we have that

$$
\nabla_{\mathbf{p}} n^0_{\mathbf{k},s} = -\mathbf{v}_{k,s} \delta(\epsilon_{k,s} - \mu) \ , \qquad (17.65)
$$

and we remember that the values of $\mathbf{p} = \hbar \mathbf{k}$ which have to be considered are closed to the Fermi surface. By inserting this expression in Eq. (17.64) we obtain

$$
\begin{aligned}
&\frac{\partial n_{\mathbf{k},s}(\mathbf{r},t)}{\partial t} \\
&+ \ \mathbf{v}_k \delta(\epsilon_{k,s} - \mu) \cdot \nabla_{\mathbf{r}} \left[\delta n_{\mathbf{k},s}(\mathbf{r},t) + \sum_{\mathbf{k}',s'} f(\mathbf{k},s;\mathbf{k}',s') \delta n_{\mathbf{k}',s'}(\mathbf{r},t) \right] = 0 \ .
\end{aligned}
$$

An equation describing the motion of all the particles of the system is obtained by summing on all the values of $\mathbf{k}$ and s.

$$
\begin{aligned}
&\sum_{\mathbf{k},s} \frac{\partial n_{\mathbf{k},s}(\mathbf{r},t)}{\partial t} + \\
&\sum_{\mathbf{k},s} \mathbf{v}_{k,s} \cdot \nabla_{\mathbf{r}} \left[\delta n_{\mathbf{k},s}(\mathbf{r},t) \delta(\epsilon_{k,s} - \mu) \sum_{\mathbf{k}',s'} f(\mathbf{k},s;\mathbf{k}',s') \delta n_{\mathbf{k}',s'}(\mathbf{r},t) \right]
\end{aligned}
$$

$$
\begin{aligned}
&= \frac{\partial}{\partial t}\sum_{\mathbf{k}} n_{\mathbf{k},s}(\mathbf{r},t) + \sum_{\mathbf{k},s} \nabla_{\mathbf{r}}\delta n_{\mathbf{k},s}(\mathbf{r},t)\cdot \mathbf{v}_{k,s} \\
&+ \sum_{\mathbf{k},s,\mathbf{k}',s'} \delta(\epsilon_{k,s}-\mu) f(\mathbf{k},s;\mathbf{k}',s')\nabla_{\mathbf{r}}\delta n_{\mathbf{k}',s'}(\mathbf{r},t)\cdot \mathbf{v}_{k,s} \\
&= \frac{\partial}{\partial t}\sum_{\mathbf{k},s} n_{\mathbf{k},s}(\mathbf{r},t) + \sum_{\mathbf{k},s} \nabla_{\mathbf{r}}\delta n_{\mathbf{k},s}(\mathbf{r},t)\cdot \mathbf{v}_{k,s} \\
&+ \sum_{\mathbf{k},s,\mathbf{k}',s'} \delta(\epsilon_{k',s'}-\mu) f(\mathbf{k},s;\mathbf{k}',s')\nabla_{\mathbf{r}}\delta n_{\mathbf{k},s}(\mathbf{r},t)\cdot \mathbf{v}_{k',s'} \\
&= \frac{\partial}{\partial t}\sum_{\mathbf{k},s} n_{\mathbf{k},s}(\mathbf{r},t) \\
&+ \nabla_{\mathbf{r}}\cdot \sum_{\mathbf{k},s} \delta n_{\mathbf{k},s}(\mathbf{r},t)\left[\mathbf{v}_{k,s} + \sum_{\mathbf{k}',s'} \delta(\epsilon_{k',s'}-\mu) f(\mathbf{k},s;\mathbf{k}',s')\mathbf{v}_{k',s'}\right] \\
&= \frac{\partial \rho(\mathbf{r},t)}{\partial t} + \nabla\cdot \mathbf{J}(\mathbf{r},t) = 0 \ , \qquad (17.66)
\end{aligned}
$$

where we have defined

$$
\rho(\mathbf{r},t) = \sum_{\mathbf{k},s} n_{\mathbf{k},s}(\mathbf{r},t) \ , \qquad (17.67)
$$

and

$$
\begin{aligned}
\mathbf{J}(\mathbf{r},t) &= \sum_{\mathbf{k},s} \delta n_{\mathbf{k},s}(\mathbf{r},t)\left[\mathbf{v}_{k,s} + \sum_{\mathbf{k}',s'} \delta(\epsilon_{k',s'}-\mu) f(\mathbf{k},s;\mathbf{k}',s')\mathbf{v}_{\mathbf{k}',s'}\right] \\
&\equiv \sum_{\mathbf{k},s} \delta n_{\mathbf{k},s}(\mathbf{r},t)\mathbf{j}_{\mathbf{k},s} \ . \qquad (17.68)
\end{aligned}
$$

In the above calculation the properties

$$
f(\mathbf{k},s;\mathbf{k}',s') = f(\mathbf{k}',s';\mathbf{k},s) \quad \text{and} \quad |\delta n_{\mathbf{k}',s'}(\mathbf{r},t)| = |\delta n_{\mathbf{k},s}(\mathbf{r},t)|,
$$

have been taken into account. We also need to consider that the variations are always evaluated on the Fermi surface, therefore, $|\mathbf{k}| = |\mathbf{k}'| = k_{\mathrm{F}}$.

Equation (17.66) is evidently a continuity equation, and describes the conservation of the number of particles of the system. The current generated by the quasi-particle motion is

$$
\mathbf{j}_{\mathbf{k},s} = \mathbf{v}_{k,s} + \sum_{\mathbf{k}',s'} f(\mathbf{k},s;\mathbf{k}',s')\delta(\epsilon_{k',s'}-\mu)\mathbf{v}_{k',s'} \ . \qquad (17.69)
$$

In the absence of interactions, the current would correspond to $\mathbf{v}_{k,s}$, as it would in a classical treatment. The second term appears because we are describing an interacting system. When adding a particle to the system, besides the trivial contribution of the current generated by $\mathbf{v}_{k,s}$, there is also a contribution due to the interaction of the quasi-particle with all the other particles. This second term is sometimes called the **drag current**, as it describes the fact that the medium is dragged along by the quasi-particle added.

In an infinite system, with translational invariance, we can use the relation (17.19) in Eq. (17.69) to relate the effective mass of the quasi-particle to the bare mass

$$\frac{\hbar k}{m} = \frac{\hbar k}{m^*} + \sum_{\mathbf{k}',s'} f(\mathbf{k}, s; \mathbf{k}', s')\delta(\epsilon_{k,s} - \mu)\mathbf{v}_{k',s'}. \tag{17.70}$$

The evaluation of the sum is presented in the box.

In this calculation the sum is transformed into an integral by using the conventions presented in Sect. 2.3. We consider the z-axis parallel to $\mathbf{k}$ without loss of generality. We indicate with θ the angle between $\mathbf{k}$ and $\mathbf{k}'$ and $v_{k,s} = |\mathbf{v}_{k,s}|$.

$$\begin{aligned}
&\sum_{\mathbf{k}',s'} f(\mathbf{k}, s; \mathbf{k}', s')\delta(\epsilon_{k'} - \mu)v_{k',s'}\cos\theta \\
&= \sum_{s'} \frac{\mathcal{V}}{(2\pi)^3} \int d^3k'\, f(\mathbf{k}, s; \mathbf{k}', s)\frac{m^*}{\hbar^2 k}\delta(k' - k_{\mathrm{F}})v_{k',s'}\cos\theta \\
&= \frac{\mathcal{V}}{(2\pi)^3} \sum_{s'} \int dk' k'^2 \int_{-1}^{1} d(\cos\theta) \\
&\int_0^{2\pi} d\phi\, f(\mathbf{k}, s; \mathbf{k}', s)\frac{m^*}{\hbar^2 k}\delta(k' - k_{\mathrm{F}})v_{k',s'} P_1(\cos\theta) \\
&= \frac{\mathcal{V}}{(2\pi)^3}(2)(2\pi)\frac{m^*}{\hbar^2 k_{\mathrm{F}}} {k_{\mathrm{F}}}^2 v_{k_{\mathrm{F}}} \int_{-1}^{1} d(\cos\theta) \sum_l f_l^s P_l(\cos\theta)P_1(\cos\theta) \\
&= \frac{\mathcal{V} m^* k_{\mathrm{F}}}{2\pi^2} \sum_l f_l^s \int_{-1}^{1} dx\, P_l(x)P_1(x) = F_1^s \frac{1}{2}\frac{2}{3}\delta_{l,1} v_k = \frac{1}{3} v_k F_1^s,
\end{aligned}$$

where we have considered that $v_{k_{\mathrm{F}},1/2} = v_{k_{\mathrm{F}},-1/2} = v_{k_{\mathrm{F}}}$.

$$\frac{\hbar k}{m} = \frac{\hbar k}{m^*} + \frac{1}{3}F_1^s v_k = \frac{\hbar k}{m^*}\left(1 + \frac{F_1^s}{3}\right), \tag{17.71}$$

therefore

$$\frac{m^*}{m} = 1 + \frac{F_1^s}{3}. \tag{17.72}$$

Since the value of m^* is related to the specific heat, see Sect. 17.4.1, this relation gives new information on the interaction $f(\mathbf{k}, s; \mathbf{k}', s')$. We note that when $F_1^s \leq -3$, the system becomes unstable because it would result having negative masses.

17.5.3 Collective Excitations

Let us suppose that the external perturbation of the system is harmonic ad has momentum $\mathbf{q}$ and energy ω. The variation (17.56) of the distribution can be expressed as

$$n_{\mathbf{k},s}(\mathbf{r},t) = n^0_{\mathbf{k},s} + \delta n_{\mathbf{k},s}(\mathbf{q},\omega)e^{i(\mathbf{q}\cdot\mathbf{r}-\hbar\omega t)}. \tag{17.73}$$

By inserting this expression in the transport equation (17.64) we obtain

$$\begin{aligned} &\frac{\partial}{\partial t}\left[\delta n_{\mathbf{k},s}(\mathbf{q},\omega)e^{i(\mathbf{q}\cdot\mathbf{r}-\omega t)}\right] + \nabla_{\mathbf{r}}\left[\delta n_{\mathbf{k},s}(\mathbf{q},\omega)e^{i(\mathbf{q}\cdot\mathbf{r}-\omega t)}\right]\cdot\mathbf{v}_{k,s} \\ &- \nabla_{\mathbf{p}} n^0_{\mathbf{k},s}\cdot\sum_{\mathbf{k}',s'} f(\mathbf{k},s;\mathbf{k}',s')\nabla_{\mathbf{r}}\left[\delta n_{\mathbf{k}',s'}(\mathbf{q},\omega)e^{i(\mathbf{q}\cdot\mathbf{r}-\omega t)}\right] = 0 \ , \end{aligned} \tag{17.74}$$

therefore

$$(\mathbf{q}\cdot\mathbf{v}_{k,s}-\omega)\delta n_{\mathbf{k},s}(\mathbf{q},\omega) - \nabla_{\mathbf{p}} n^0_{\mathbf{k},s}\cdot\mathbf{q}\sum_{\mathbf{k}',s'} f(\mathbf{k},s;\mathbf{k}',s')\delta n_{\mathbf{k}',s'}(\mathbf{q},\omega) = 0 \ , \tag{17.75}$$

and also

$$(\mathbf{q}\cdot\mathbf{v}_{k,s}-\omega)\delta n_{\mathbf{k},s}(\mathbf{q},\omega) + \mathbf{v}_{k,s}\cdot\mathbf{q}\sum_{\mathbf{k}',s'} f(\mathbf{k},s;\mathbf{k}',s')\delta n_{\mathbf{k}',s'}(\mathbf{q},\omega) = 0 \ . \tag{17.76}$$

Since for Eq. (17.65)

$$\delta(\epsilon_p - \mu) = \delta(k-k_{\mathrm{F}})m^*/\hbar^2 k_{\mathrm{F}},$$

we can rewrite the variation $\delta n_{\mathbf{k},s}$ as

$$\delta n_{\mathbf{k},s}(\mathbf{q},\omega) = \delta(k-k_{\mathrm{F}})v_{k_{\mathrm{F}}}u_{\mathbf{k},s} \ , \tag{17.77}$$

and, then, we obtain the equation

$$(\mathbf{q}\cdot\mathbf{v}_{k,s}-\omega)u_{\mathbf{k},s}(\mathbf{q},\omega) + \mathbf{v}_{k,s}\cdot\mathbf{q}\sum_{\mathbf{k}',s'} f(\mathbf{k},s;\mathbf{k}',s')u_{\mathbf{k}',s'}(\mathbf{q},\omega) = 0 \ . \tag{17.78}$$

Without loss of generality, we defined the z in the direction of $\mathbf{q}$. We indicate with θ the angle between $\mathbf{q}$ and $\mathbf{k}$ and with θ' the angle between $\mathbf{k}'$ and $\mathbf{q}$. We use the symbols q, k, k' to indicate the magnitudes of the relative vectors. Since everything happens at the Fermi surface then $k = k' = k_{\mathrm{F}}$. The interaction depends only on the angle between the two vectors $\mathbf{k}$ and $\mathbf{k}'$, therefore

$$f(\mathbf{k},s;\mathbf{k}',s') \equiv f[\cos(\theta-\theta');s,s'] = \frac{\pi^2\hbar^2}{\mathcal{V}m^*}F[\cos(\theta-\theta');s,s'] \ , \tag{17.79}$$

where we used the definition (17.65) of F. We transform the sums in Eq. (17.78) as integrals following the conventions of Sect. 2.3, and obtain

$$
\begin{aligned}
&(q v_{k_{\mathrm{F}},s} \cos\theta - \omega) u(\theta,\phi,\sigma) \\
&+ q v_{k_{\mathrm{F}},s} \frac{\cos\theta}{8\pi} \sum_{s'} \int_0^{2\pi} d\phi' \\
&\int_{-1}^{1} d(\cos\theta') F[\cos(\theta-\theta'); s, s'] u(\theta',\phi',\sigma') = 0 \ . \qquad (17.80)
\end{aligned}
$$

In analogy to what has been done for the interaction in the Eqs. (17.13) and (17.14) we separate u in a symmetric and antisymmetric parts of the spin

$$
u(\theta,\phi,\pm 1/2) = u^s(\theta,\phi) \pm u^a(\theta,\phi) \ . \qquad (17.81)
$$

We insert these definitions in Eq. (17.80), we divide by $q v_{k_{\mathrm{F}},s}$ and by using a new variable $\xi \equiv \omega / q v_{k_{\mathrm{F}},s}$ we obtain two independent equations for the unknowns u^s and u^a.

$$
\begin{aligned}
&(\cos\theta - \xi) u^s(\theta,\phi) \\
&+ \frac{\cos\theta}{4\pi} \int_0^{2\pi} d\phi' \int_{-1}^{1} d(\cos\theta') F^s\big(\cos(\theta-\theta')\big) u^s(\theta',\phi') = 0 \ , \qquad (17.82) \\
&(\cos\theta - \xi) u^a(\theta,\phi) \\
&+ \frac{\cos\theta}{4\pi} \int_0^{2\pi} d\phi' \int_{-1}^{1} d(\cos\theta') F^a\big(\cos(\theta-\theta')\big) u^a(\theta',\phi') = 0 \ . \qquad (17.83)
\end{aligned}
$$

A formal solution to these equations can be achieved by expanding $u^{s,a}(\theta,\phi)$ in spherical harmonics $Y_{l,\mu}(\theta,\phi)$. Since the known terms are independent of the angle ϕ, this leads to a set of independent equations, each characterised by the quantum number μ. This allows us to classify the solutions based on the value of μ. Specifically, solutions with $\mu = 0$ are called longitudinal solutions, those with $\mu = 1$ are transverse solutions, $\mu = 2$ corresponds to quadrupole solutions, and so on.

The most interesting excitation mode is the longitudinal one for the symmetric solution. This mode describes a density compression wave that propagates in the $\mathbf{q}$ direction and involves all particles regardless of their spin orientation. It is very similar to the traditional sound wave in classical systems. Because of this similarity, it is called **zero sound**, in contrast to first sound, which is the conventional mode described within the hydrodynamical regime.

We present here a simple model to describe the zero sound. Let us assume that the interaction is independent of θ and θ', therefore, in expansion in terms of Lagrange polynomials $F^s\big(\cos(\theta-\theta')\big)$, only the term F_0^s remains. By inserting this interaction in Eq. (17.82), and working out the integrals we obtain

$$(\cos\theta - \xi)u^s(\theta,\phi) + \frac{\cos\theta}{2} F_0^s \int_{-1}^{1} d(\cos\theta')u^s(\theta',\phi') = 0 \ . \tag{17.84}$$

We search for a solution of the type

$$u^s(\theta,\phi) = C\frac{\cos\theta}{\xi - \cos\theta} \ , \tag{17.85}$$

where C is a constant. By inserting this solution in Eq. (17.84) we obtain

$$\frac{\xi}{2}\log\frac{\xi+1}{\xi-1} = \frac{1}{F_0^s} \ . \tag{17.86}$$

The integral to be calculated is

$$\int_{-1}^{1} d(\cos\theta)\frac{\cos\theta}{\xi - \cos\theta} \ .$$

Since

$$\int dx \, \frac{x}{\xi - x} = -\left[x + \xi \log(x - \xi)\right] + \text{constant} \ ,$$

Equation (17.84) becomes

$$(\cos\theta - \xi)C\frac{\cos\theta}{\xi - \cos\theta} + \frac{\cos\theta}{2}F_0^s C\left\{\xi[\log(\xi+1) - \log(\xi-1)]\right\} = 0 \ ,$$

from which we obtain Eq. (17.86).

In the case of a repulsive interaction where $F_0^s > 0$, there is always a real solution with $\xi > 1$. This corresponds to a wave that does not experience damping. The phase velocity in this scenario exceeds the Fermi velocity v_{k_F}. When there is a weak attraction between the quasi-particles, with $-1 < F_0^s < 0$, Eq. (17.86) admits a complex solution, which indicates a damped oscillation. If $F_0^s < -1$, the zero sound oscillation becomes unstable, meaning it does not form.

It is important to analyse the differences between zero sound and first sound to better understand their physical nature. There are two key aspects to consider: the interaction between particles and their elastic collisions. Both effects are present in a realistic description of the system. Zero sound occurs when particle interactions dominate, whereas first sound appears when elastic collisions play the leading role.

The first sound wave propagates following a local change in density caused by a perturbation. Its oscillation frequency, given by ω, is much lower than the frequency of elastic collisions between quasi-particles. Since each oscillation involves many collisions, it becomes possible for a small, localised region of the system, however large compared to the interparticle distance, to reach a state of local equilibrium. This perturbation then moves through the system as a result of particle collisions. These collisions tend to dampen the amplitude of the wave because they transfer energy from the localised region to the entire system. Additionally, as the temperature increases, the speed of propagation also increases because the number of elastic collisions rises. Consequently, the damping effect becomes more pronounced at higher temperatures. Mathematically, this situation is described by the transport equation (17.64), where the collision integral plays a crucial role in capturing the effects of these particle interactions.

The opposite physical conditions allow the formation and propagation of zero sound and are characterised by an oscillation frequency, ω, that is much higher than the frequency of elastic collisions between particles. In this scenario, the interaction between quasi-particles, active at times much shorter than ω^{-1}, and the typical average times between elastic collisions, dominates the behaviour. Zero sound propagates because particles are correlated through their interactions. Since the number of elastic collisions per oscillation is relatively small, the perturbation is not damped. In the limit of zero temperature, elastic collisions effectively disappear, allowing zero sound to propagate indefinitely. To analyse zero sound propagation, we used the transport equation (17.64) and neglected the collision integral, since collisions are negligible in this regime.

The transition between these two collective modes of sound propagation occurs when ω is comparable to the frequency of elastic collisions. In this situation, zero sound cannot exist because the collisions are frequent enough to immediately damp the oscillations. Conversely, when collisions are too rare to thermalise the system and produce collective motion, first sound cannot propagate either.

The existence of zero sound was predicted by Landau at the end of the 1950s and was experimentally observed in 1966 in fermionic liquid helium [10]. The experiment was carried out by analysing the absorption of sound waves at different frequencies while keeping the temperature fixed. This analysis allowed researchers to identify the three physical regimes described above: first sound, the transition region, and zero sound.

References

1. L.D. Landau, On the theory of a Fermi liquid. Sov. Phys. JETP **3**, 920 (1956)
2. A.A. Abrikosov, I.M. Khalatnikov, On a model for a non-ideal Fermi gas. Sov. Phys. JETP **6**, 888 (1958)
3. L.D. Landau, The theory of superfluidity of helium II. Eksp. Teor. Fiz. **11**, 592 (1941)
4. L.D. Landau, On the theory of superfluidity of helium II. J. Phys. USSR **11**, 91 (1947)
5. A.B. Migdal, *Theory of Finite Fermi Systems and Applications to Atomic Nuclei* (Wiley Interscience, New York, 1967)
6. D. Pines, P. Nozières, *The Theory of Quantum Liquids* (Benjamin, New York, 1966)
7. A.A. Abrikosov, L.P. Gorkov, I.E. Dzyaloshinski, *Methods of Quantum Field Theory in Statistical Physics* (Dover, New York, 1975)
8. E.K.U. Gross, E. Runge, O. Heinonen, *Many-Particle Theory* (Adam Hilger, Bristol, 1991)
9. H. Bruus, K. Flensbeg, *Many-Body Quantum Theory in Condensed Matter Physics* (Oxford University Press, Oxford, 2004)
10. W.R. Abel, A.C. Anderson, J.C. Wheatley, Propagation of zero sound in liquid He^3 at low temperatures. Phys. Rev. Lett. **17**, 74 (1966)

Appendix A
Variational Principle

Let us address the problem of finding the ground state of a system described by the Schrödinger equation

$$\hat{H}|\Psi\rangle = E|\Psi\rangle \ . \tag{A.1}$$

We shall show that, by considering the energy of the system as functional of $|\Psi\rangle$, the search for its minimum, i.e. the solution of the equation

$$\delta E[\Psi] = \delta \frac{\langle\Psi|\hat{H}|\Psi\rangle}{\langle\Psi|\Psi\rangle} = 0 \ , \tag{A.2}$$

corresponds to the solution of the Schrödinger equation (A.1).

From Eq. (A.2) we can write

$$E[\Psi]\,\langle\Psi|\Psi\rangle = \langle\Psi|\hat{H}|\Psi\rangle \ , \tag{A.3}$$

By doing the variation we obtain

$$\begin{aligned} \delta\left(E[\Psi]\right)\langle\Psi|\Psi\rangle + E[\Psi]\delta\left(\langle\Psi|\Psi\rangle\right) &= \delta\left(\langle\Psi|\hat{H}|\Psi\rangle\right) \\ \delta\left(E[\Psi]\right)\langle\Psi|\Psi\rangle &= \delta\left(\langle\Psi|\hat{H}|\Psi\rangle\right) - E[\Psi]\delta\left(\langle\Psi|\Psi\rangle\right) \ , \end{aligned} \tag{A.4}$$

therefore

$$\delta E[\Psi] = \frac{1}{\langle\Psi|\Psi\rangle}\left[\delta\left(\langle\Psi|\hat{H}|\Psi\rangle\right) - E[\Psi]\delta\left(\langle\Psi|\Psi\rangle\right)\right] = 0 \ . \tag{A.5}$$

This term is zero if the part between squared brackets is zero

$$\delta\left(\langle\Psi|\hat{H}|\Psi\rangle\right) - E[\Psi]\delta\left(\langle\Psi|\Psi\rangle\right) = 0 \ , \tag{A.6}$$

© The Editor(s) (if applicable) and The Author(s), under exclusive license to Springer Nature Switzerland AG 2026

G. Co', *Concepts in Quantum Many-Body Physics*, UNITEXT for Physics,
https://doi.org/10.1007/978-3-032-08920-5

and, because E is a number, we obtain

$$\langle\delta\Psi|\hat{H}-E|\Psi\rangle+\langle\Psi|\hat{H}-E|\delta\Psi\rangle=0\ . \tag{A.7}$$

Since $|\Psi\rangle$ is a complex function, the variation of $\langle\Psi|$ is independent of that of $|\Psi\rangle$. This because the real and the imaginary parts of $|\Psi\rangle$ vary independently of each other. This can be seen by substituting in (A.7) $|\delta\Psi\rangle$ with $|i\delta\Psi\rangle$.

$$\begin{aligned}&-i\langle\delta\Psi|\hat{H}-E|\Psi\rangle+i\langle\Psi|\hat{H}-E|\Psi\delta\rangle\\&=i\Big[-\langle\delta\Psi|\hat{H}-E|\Psi\rangle+\langle\Psi|\hat{H}-E|\Psi\delta\rangle\Big]=0\ .\end{aligned} \tag{A.8}$$

Equations (A.7) and (A.8) must be satisfied simultaneously, then we have

$$\langle\delta\Psi|\hat{H}-E|\Psi\rangle=0\quad\text{and}\quad\langle\Psi|\hat{H}-E|\delta\Psi\rangle=0\ . \tag{A.9}$$

The variation $|\delta\Psi\rangle$ is arbitrary, therefore the above equations are always satisfied if

$$[H-E]\,|\Psi\rangle=0\ , \tag{A.10}$$

which is the Schrödinger equation (A.1).

Let us consider now a trial wave function $|\Phi\rangle$. We can express it as linear combination of eigenstates $|\Psi_n\rangle$ of $\hat{H}$

$$\hat{H}|\Psi_n\rangle=E_n|\Psi_n\rangle\ , \tag{A.11}$$

which form a complete basis:

$$|\Phi\rangle=\sum_{n=0}^{\infty}D_n|\Psi_n\rangle\ , \tag{A.12}$$

where the D_n are numbers. We write the energy functional as

$$\begin{aligned}E[\Phi]=\frac{\langle\Phi|\hat{H}|\Phi\rangle}{\langle\Phi|\Phi\rangle}&=\frac{\sum\limits_{n,n'}\langle\Psi_{n'}|D^*_{n'}\hat{H}D_n|\Psi_n\rangle}{\sum\limits_{n,n'}\langle\Psi_{n'}|D^*_{n'}D_n|\Psi_n\rangle}\\&=\frac{\sum\limits_{n,n'}D^*_{n'}D_n\langle\Psi_{n'}|\hat{H}|\Psi_n\rangle}{\sum\limits_{n,n'}D^*_{n'}D_n\langle\Psi_{n'}|\Psi_n\rangle}=\frac{\sum\limits_{nn'}D^*_{n'}D_nE_n\delta_{n,n'}}{\sum\limits_{n}|D_n|^2}\\&\geq\frac{\sum\limits_{n}|D_n|^2E_0}{\sum\limits_{n}|D_n|^2}=E_0\ .\end{aligned} \tag{A.13}$$

This inequality demonstrates that the energy obtained by minimising the energy functional within a restricted subspace of the full Hilbert space, spanned by the trial wave functions, results in an energy value that is greater than the true ground state eigenvalue of the Hamiltonian $\hat{H}$.

Appendix B
Creation and Destruction Operators in Angular Momentum Coupling

In this appendix, we explore the definition of fermion creation and annihilation operators, especially highlighting when it is beneficial to describe systems with spherical symmetry.

For a one-body Hamiltonian exhibiting spherical symmetry and spin-orbit interaction, the quantum numbers that characterise the single-particle wave function include: the principal quantum number n, the quantum number l associated with orbital angular momentum, j related to total angular momentum, and m, which represents the projection of j onto the quantization axis z. In the following discussion, we will omit the explicit dependence on n and l.

The action of the creation operators is:

$$\hat{a}^+_{j,m}|0\rangle = |jm\rangle \quad , \tag{B.1}$$

which indicates that the $\hat{a}^+_{j,m}$ adds a particle on the $|jm\rangle$ state. All states with the same j have the same energy, resulting in a degeneracy of $2j + 1$. For this reason it is necessary to specify also m.

The $2j + 1$ components of $\hat{a}^+_{j,m}$ behave such as to form an irreducible spherical tensor satisfying the equations

$$\left[\hat{J}_z, \hat{a}^+_{j,m}\right] = m\hat{a}^+_{j,m} \quad , \tag{B.2}$$

$$\left[\hat{J}_\pm, \hat{a}^+_{j,m}\right] = [j(j+1) - m(m \pm 1)]^{1/2} \, \hat{a}^+_{j,m\pm1} \quad , \tag{B.3}$$

where we indicated with $J_\pm$ and J_z the spherical components of the generalized angular momentum of the system.

One of the properties characterizing the irreducible spherical tensor of rank k is that its $2k + 1$ components transform in those of its hermitian conjugate as:

$$\left(\hat{T}^k_q\right)^+ = \left(\hat{T}^k_{-q}\right)(-)^q \quad . \tag{B.4}$$

© The Editor(s) (if applicable) and The Author(s), under exclusive license to Springer Nature Switzerland AG 2026

G. Co', *Concepts in Quantum Many-Body Physics*, UNITEXT for Physics,
https://doi.org/10.1007/978-3-032-08920-5

This means that the conjugate of $\hat{a}^{+}_{j,m}$, i.e. $\hat{a}_{j,m}$, is not a component of an irreducible spherical tensor, while the operator

$$\tilde{\hat{a}}_{j,m} = (-)^{j+m} a_{j,-m} \quad , \tag{B.5}$$

satisfies this requirement.

It is useful to use operators which are irreducible spherical tensor in the description of system with rotational invariance. For this reason, the use of $\tilde{\hat{a}}_{j,m}$ is preferred with respect to $\hat{a}_{j,m}$. The anticommutation properties are:

$$\left\{\hat{a}^{+}_{j,m}, \hat{a}^{+}_{j,m'}\right\} = 0 \quad , \quad \left\{\tilde{\hat{a}}_{j,m}, \tilde{\hat{a}}_{j',m'}\right\} = 0 \quad , \tag{B.6}$$

$$\left\{\tilde{\hat{a}}_{j,m}, \hat{a}^{+}_{j',m'}\right\} = (-)^{j+m} \delta_{j,j'} \delta_{-m,m'} \quad . \tag{B.7}$$

It is possible to create a state composed by many particles by iteratively apply $a^{+}_{j,m}$ to the vacuum state and by using the rules of the angular momentum coupling. For example, the creation operator of two particles becomes:

$$\hat{A}^{+}(j_1 j_2; JM) = \frac{1}{\left(1+\delta_{j_1 j_2}\right)^{1/2}} \left[\hat{a}^{+}_{j_1} \otimes \hat{a}^{+}_{j_2}\right]^{J}_{M} \quad , \tag{B.8}$$

where we have defined:

$$\left[\hat{a}^{+}_{j_1} \otimes \hat{a}^{+}_{j_2}\right]^{J}_{M} = \sum_{m_1 m_2} \langle j_1 m_1 j_2 m_2 | JM \rangle \hat{a}^{+}_{j_1,m_1} \hat{a}^{+}_{j_2,m_2} \quad , \tag{B.9}$$

where $\langle j_1 m_1 j_2 m_2 | JM \rangle$ is a Clebsch-Gordan coefficient. The destruction operator for a particle pair becomes:

$$\begin{aligned} \tilde{\hat{A}}(j_1 j_2; JM) &= (-)^{J+M} \left[\hat{A}^{+}(j_1 j_2; J-M)\right]^{+} \\ &= -\frac{1}{\left(1+\delta_{j_1 j_2}\right)^{1/2}} \left[\tilde{\hat{a}}_{j_1} \otimes \tilde{\hat{a}}_{j_2}\right]^{J}_{M} \quad , \end{aligned} \tag{B.10}$$

and the operator creating a particle-hole pair is:

$$\hat{U}(j_1 j_2; JM) = [\hat{a}^{+}_{j_1} \otimes \tilde{\hat{a}}_{j_2}]^{J}_{M} \quad . \tag{B.11}$$

When $j_1 = j_2$, the two particles are identified by $m_1 \neq m_2$ therefore, since the above equations include a sum on the m_1 and m_2 it is necessary to divide by $\sqrt{2}$ since there is a double counting of identical configurations. For example, the configuration where the particle 1 has m_1 and that of particle 2 m_2, is identical to that where the particle 1 has m_2 and particle 2 m_1.

Appendix C
RPA with Equation of Motions and with the Green's Function

We use the general expression of the two-body Green's function in mixed representation, Eq. (9.76), to express the non interacting two-body Green's function

$$\begin{aligned}
&\tilde{G}^0(\nu_1, \nu_2, \nu_3, \nu_4, E) \\
&= i\hbar \sum_n \left[\frac{\langle \Phi_0 | \hat{a}_{\nu_1} \hat{a}^+_{\nu_3} | \Phi_n \rangle \langle \Phi_n | \hat{a}_{\nu_2} \hat{a}^+_{\nu_4} | \Phi_0 \rangle}{E - (E_n - E_0) + i\eta} \right. \\
&\quad \left. - \frac{\langle \Phi_0 | \hat{a}_{\nu_2} \hat{a}^+_{\nu_4} | \Phi_n \rangle \langle \Phi_n | \hat{a}_{\nu_1} \hat{a}^+_{\nu_3} | \Phi_0 \rangle}{E + (E_n - E_0) - i\eta} \right] . \qquad \text{(C.1)}
\end{aligned}$$

In mixed representation the RPA equations are

$$\begin{aligned}
&\tilde{G}^{\text{RPA}}(\nu_1, \nu_2, \nu_3, \nu_4, E) = \tilde{G}^0(\nu_1, \nu_2, \nu_3, \nu_4, E) \\
&+ \sum_{\mu_1,\mu_2,\mu_3,\mu_4} \tilde{G}^0(\nu_1, \nu_2, \mu_1, \mu_2, E) \langle \mu_1\mu_2 | V | \mu_3\mu_4 \rangle \tilde{G}^{\text{RPA}}(\mu_3, \mu_4, \nu_3, \nu_4, E) \\
&- \sum_{\mu_1,\mu_2,\mu_3,\mu_4} \tilde{G}^0(\nu_1, \nu_2, \mu_1, \mu_2, E) \langle \mu_1\mu_2 | V | \mu_4\mu_3 \rangle \tilde{G}^{\text{RPA}}(\mu_3, \mu_4, \nu_3, \nu_4, E) \\
&= \sum_{\mu_1,\mu_2,\mu_3,\mu_4} \tilde{G}^0(\nu_1, \nu_2, \mu_1, \mu_2, E) \Big\{ \delta_{\mu_1,\nu_3}\delta_{\mu_2,\nu_4} \\
&\quad + \langle \mu_1\mu_2 | V | \mu_4\mu_3 \rangle \tilde{G}^{\text{RPA}}(\mu_3, \mu_4, \nu_3, \nu_4, E) \\
&\quad - \langle \mu_1\mu_2 | V | \mu_4\mu_3 \rangle \tilde{G}^{\text{RPA}}(\mu_3, \mu_4, \nu_3, \nu_4, E) \Big\} . \qquad \text{(C.2)}
\end{aligned}$$

Equation (C.1) establishes precise conditions on the creation and destruction operators. The non-interacting Green's function $\tilde{G}^0$ is different from zero only if ν_1 and ν_4 are particle states and ν_1 and ν_2 are hole states. There are four possible configurations that yield non-zero Green's functions. By following the usual convention of indicating with i, j, k, l the hole states and with m, n, p, q the particle states, we can express them as

© The Editor(s) (if applicable) and The Author(s), under exclusive license to Springer Nature Switzerland AG 2026

G. Co', *Concepts in Quantum Many-Body Physics*, UNITEXT for Physics,
https://doi.org/10.1007/978-3-032-08920-5

$$\tilde{G}^0(m,i,n,j,E) = \tilde{G}^0(i,m,n,j,E) = \tilde{G}^0(i,m,j,n,E) = \tilde{G}^0(m,i,j,n,E) \\ = \frac{\delta_{m,n}\delta_{i,j}}{\epsilon_m - \epsilon_i E}. \tag{C.3}$$

By inserting these expressions in Eq. (C.2) we obtain four equations:

$$\sum_{q,l} \Big\{ \big[(\epsilon_m - \epsilon_i - E)\delta_{m,q}\delta_{i,l} - \langle mi|V|lq\rangle\big]\, \tilde{G}^{\rm RPA}(q,l,n,j,E) \\ - \langle mi|V|ql\rangle\, \tilde{G}^{\rm RPA}(l,qn,j,E)\Big\} = \delta_{m,n}\delta_{i,j},$$

$$\sum_{q,l} \Big\{ \big[(\epsilon_m - \epsilon_i + E)\delta_{m,q}\delta_{i,l} - \langle im|V|ql\rangle\big]\, \tilde{G}^{\rm RPA}(l,q,n,j,E) \\ - \langle im|V|lq\rangle\, \tilde{G}^{\rm RPA}(q,l,n,j,E)\Big\} = \delta_{m,n}\delta_{i,j},$$

$$\sum_{q,l} \Big\{ \big[(\epsilon_m - \epsilon_i - E)\delta_{m,q}\delta_{i,l} - \langle mi|V|lq\rangle\big]\, \tilde{G}^{\rm RPA}(q,l,j,n,E) \\ - \langle mi|V|ql\rangle\, \tilde{G}^{\rm RPA}(l,q,j,n,E)\Big\} = \delta_{m,n}\delta_{i,j},$$

$$\sum_{q,l} \Big\{ \big[(\epsilon_m - \epsilon_i + E)\delta_{m,q}\delta_{i,l} - \langle im|V|ql\rangle\big]\, \tilde{G}^{\rm RPA}(l,q,j,n,E) \\ - \langle im|V|lq\rangle\, \tilde{G}^{\rm RPA}(q,l,j,n,E)\Big\} = \delta_{m,n}\delta_{i,j},$$

We define the matrices

$$A_{miql} = (\epsilon_m - \epsilon_i)\delta_{m,q}\delta_{i,l} - \langle mi|V|lq\rangle\,, \tag{C.4}$$

$$B_{miql} = -\langle mi|V|ql\rangle \quad , \tag{C.5}$$

therefore we write the above equations as

$$\sum_{q,l} \Big\{ \big[A_{miql} - E\,\delta_{m,q}\,\delta_{i,l}\big] \quad \tilde{G}^{\rm RPA}(q,l,n,j,E) \\ + B_{miql} \quad \tilde{G}^{\rm RPA}(l,q,n,j,E)\Big\} = \delta_{m,n}\delta_{i,j} \quad ,$$

$$\sum_{q,l} \Big\{ \big[A^*_{miql} + E\,\delta_{m,q}\delta_{i,l}\big] \quad \tilde{G}^{\rm RPA}(l,q,n,j,E) \\ + B^*_{miql} \quad \tilde{G}^{\rm RPA}(q,l,n,j,E)\Big\} = \delta_{m,n}\delta_{i,j} \quad ,$$

$$\sum_{q,l} \left\{ \left[A_{miql} - E)\delta_{m,q}\delta_{i,l} \right] \quad \tilde{G}^{\mathrm{RPA}}(q, l, j, n, E) \right.$$
$$\left. + B_{miql} \quad \tilde{G}^{\mathrm{RPA}}(l, q, j, n, E) \right\} = \delta_{m,n}\delta_{i,j} \quad ,$$

$$\sum_{q,l} \left\{ \left[A^*_{miql} + E\, \delta_{m,q}\delta_{i,l} \right] \quad \tilde{G}^{\mathrm{RPA}}(l, q, j, n, E) \right.$$
$$\left. + B^*_{miql} \quad \tilde{G}^{\mathrm{RPA}}(q, l, j, n) \right\} = \delta_{m,n}\delta_{i,j} \quad .$$

These equations can be written in matrix form. By defining

$$\begin{aligned} G_1(E) &\equiv \tilde{G}^{\mathrm{RPA}}(m, i, j, n, E) \ ; \quad G_2(E) \equiv \tilde{G}^{\mathrm{RPA}}(m, i, n, j, E) \ ; \\ G_3(E) &\equiv \tilde{G}^{\mathrm{RPA}}(i, m, j, n, E) \ ; \quad G_4(E) \equiv \tilde{G}^{\mathrm{RPA}}(i, m, n, j, E) \ , \end{aligned} \tag{C.6}$$

we obtain

$$\begin{pmatrix} A - E\,\mathbb{I} & B \\ B^* & A^* + E\,\mathbb{I} \end{pmatrix} \begin{pmatrix} G_1(E)\; G_2(E) \\ G_3(E)\; G_4(E) \end{pmatrix} = \begin{pmatrix} \mathbb{I}\; 0 \\ 0\; \mathbb{I} \end{pmatrix} \ , \tag{C.7}$$

The two-body Green's functions depend on the energy E. The poles $\omega_n = E_n - E_0$ of these Green's functions correspond to the excitation energies of the RPA excited states $|\Psi_n\rangle$. When the energy E matches the value of a pole, the Green's function diverges. Equation (C.7) remains valid if the matrix of the coefficients goes to zero. Consequently, the excitation energies are defined by the non trivial solution of the homogeneous system of equations

$$\begin{pmatrix} A - \omega_n\,\mathbb{I} & B \\ B^* & A^* + \omega_n\,\mathbb{I} \end{pmatrix} \begin{pmatrix} X_n \\ Y_n \end{pmatrix} = 0 \ ; \quad \begin{pmatrix} A & B \\ B^* & A^* \end{pmatrix} \begin{pmatrix} X_n \\ Y_n \end{pmatrix} = \omega_n \begin{pmatrix} X_n \\ -Y_n \end{pmatrix} \ , \tag{C.8}$$

which is the expression (16.37) of the RPA equations.

In Sect. 16.3.3 we have shown that the RPA equations, for each positive eigenvalue ω_n admit also a negative eigenvalue $-\omega$. The set of the vectors of the X and Y amplitudes is complete and orthogonal

$$(X^*_m, -Y^*_m) \begin{pmatrix} X_n \\ Y_n \end{pmatrix} = \delta_{m,n} \ ; \quad \sum_n^{\pm} \begin{pmatrix} X_n \\ -Y_n \end{pmatrix} (X^*_n, -Y^*_n) = \mathbb{I} \ ,$$

where we have indicated with $\pm$ the sum on both eigenstates of positive and negative eigenvalues. By using the above expressions we can write Eq. (C.7) as:

$$\begin{pmatrix} A - E\mathbb{I} & B \\ B^* & A^* + E\mathbb{I} \end{pmatrix} \begin{pmatrix} G_1(E) & G_2(E) \\ G_3(E) & G_4(E) \end{pmatrix} = \sum_n \begin{pmatrix} X_n \\ Y_n \end{pmatrix} (X_m^*, Y_m^*) - \sum_n \begin{pmatrix} Y_n \\ X_n \end{pmatrix} (Y_n^*, X_n^*),$$

where we have explicitly written the sum on positive and negative eigenvalues. Solution of the above equation is

$$\begin{pmatrix} G_1(E) & G_2(E) \\ G_3(E) & G_4(E) \end{pmatrix} = \sum_n \frac{1}{\omega_n - E} \begin{pmatrix} X_n \\ Y_n \end{pmatrix} (X_m^*, Y_m^*) - \sum_n \frac{1}{\omega_n - E} \begin{pmatrix} Y_n^* \\ X_n \end{pmatrix} (Y_n^*, X_n^*)$$

$$= \begin{pmatrix} \sum_n \left(\frac{X_n X_n^*}{\omega_n - E} + \frac{Y_n Y_n^*}{\omega_n + E} \right) & \sum_n \left(\frac{X_n Y_n^*}{\omega_n - E} + \frac{X_n Y_n^*}{\omega_n + E} \right) \\ \sum_n \left(\frac{Y_n X_n^*}{\omega_n - E} + \frac{X_n^* Y_n}{\omega_n + E} \right) & \sum_n \left(\frac{Y_n Y_n^*}{\omega_n - E} + \frac{X_n X_n^*}{\omega_n + E} \right) \end{pmatrix}. \quad \text{(C.9)}$$

The comparison with the expression of the two-body Green's function in Lehmann representation, Eq. (9.76) allow the identification of the X and Y amplitudes as indicated by Eq. (16.41).

Appendix D
Time-Dependent Hartree-Fock and RPA

Another method for obtaining RPA equations consists in utilizing time-dependent Hartree-Fock equations in conjunction with the variational principle. We consider the time-dependent Schrödinger equation of the form

$$\hat{H} \, |\Psi(t)\rangle = i\hbar \frac{\partial}{\partial t} \, |\Psi(t)\rangle \quad . \tag{D.1}$$

We apply the variational principle to this equation

$$\delta \left\langle \Psi(t) | \hat{H} - i\hbar \frac{\partial}{\partial t} | \Psi(t) \right\rangle = 0 \ . \tag{D.2}$$

We search for the minimum of the above equation in the Hilbert subspace spanned by many-body wave functions of the form

$$|\Psi(t)\rangle = e^{\sum_{mi} C_{mi}(t) \hat{a}^+_m \hat{a}_i} \, |\Phi_0(t)\rangle \quad , \tag{D.3}$$

with

$$|\Phi_0(t)\rangle = e^{-\frac{i}{\hbar}\mathcal{E}_0 t} \, |\Phi_0\rangle \quad , \tag{D.4}$$

where $|\Phi_0\rangle$ is the stationary MF ground state and $\mathcal{E}_0$ is the MF energy (15.16).

Equation (D.2) is equivalent to

$$\left\langle \delta\Psi(t) | \hat{H} - i\hbar \frac{\partial}{\partial t} | \Psi(t) \right\rangle + \left\langle \Psi(t) | \hat{H} - i\hbar \frac{\partial}{\partial t} | \delta\Psi(t) \right\rangle = 0 \ . \tag{D.5}$$

Since $|\Psi(t)\rangle$ is a complex function it is possible to vary the real and the imaginary part independently. This implies that the variations of the two components of the

© The Editor(s) (if applicable) and The Author(s), under exclusive license to Springer Nature Switzerland AG 2026

G. Co', *Concepts in Quantum Many-Body Physics*, UNITEXT for Physics,
https://doi.org/10.1007/978-3-032-08920-5

above equation are independent, meaning that each of the two components must be set to zero separately. The variation should be applied to the $C_{mi}(t)$ and $C^*_{mi}(t)$ terms, as these are the only ones that depend on time. Consequently, we obtain a system consisting of two equations

$$\frac{\delta}{\delta C^*_{mi}(t)}\left\langle \Psi(t)|\hat{H} - i\hbar\frac{\partial}{\partial t}|\Psi(t)\right\rangle = 0 \tag{D.6}$$

$$\frac{\delta}{\delta C_{mi}(t)}\left\langle \Psi(t)|\hat{H} - i\hbar\frac{\partial}{\partial t}|\Psi(t)\right\rangle = 0\ . \tag{D.7}$$

These two equations are the complex conjugate of each other, therefore we consider only the first one. By considering the expansion of the exponential

$$\exp\left(\sum_{mi} C_{mi}(t)\hat{a}^+_m\hat{a}_i\right) = \\ \hat{\mathbb{I}} + \sum_{mi} C_{mi}(t)\hat{a}^+_m\hat{a}_i + \frac{1}{2}\sum_{minj} C_{mi}(t)\hat{a}^+_m\hat{a}_i C_{nj}(t)\hat{a}^+_n\hat{a}_j + \dots\ , \tag{D.8}$$

we obtain for the hamiltonian expectation value

$$\begin{aligned}\left\langle\Psi(t)|\hat{H}|\Psi(t)\right\rangle &= \left\langle\Phi_0(t)|\hat{H}|\Phi_0(t)\right\rangle \\ &+ \sum_{mi} C^*_{mi}(t)\left\langle\Phi_0(t)|\hat{a}^+_i\hat{a}_m\hat{H}|\Phi_0(t)\right\rangle \\ &+ \sum_{mi} C_{mi}(t)\left\langle\Phi_0(t)|\hat{H}\hat{a}^+_m\hat{a}_i|\Phi_0(t)\right\rangle \\ &+ \frac{1}{2}\sum_{minj} C^*_{mi}(t)C^*_{nj}(t)\left\langle\Phi_0(t)|\hat{a}^+_j\hat{a}_n\hat{a}^+_i\hat{a}_m\hat{H}|\Phi_0(t)\right\rangle \\ &+ \frac{1}{2}\sum_{minj} C_{mi}(t)C_{nj}(t)\left\langle\Phi_0(t)|\hat{H}\hat{a}^+_m\hat{a}_i\hat{a}^+_n\hat{a}_j|\Phi_0(t)\right\rangle \\ &+ \sum_{minj} C^*_{mi}(t)C_{nj}(t)\left\langle\Phi_0(t)|\hat{a}^+_i\hat{a}_m\hat{H}\hat{a}^+_n\hat{a}_j\hat{H}|\Phi_0(t)\right\rangle \\ &+ \dots\ .\end{aligned} \tag{D.9}$$

The first term of the above equation is $\mathcal{E}_0$ (15.16). The linear terms in $C_{mi}(t)$ are all zero since they are overlap of orthogonal Slater determinants.

Let us calculate the matrix element of the 5th term by using the expression of the hamiltonian given in Eq. (15.15)

$$\left\langle \Phi_0(t)|\hat{H}|\Phi_0(t)\right\rangle = \sum_{\nu} \epsilon_\nu \left\langle \Phi_0(t)|\hat{a}^+_\nu \hat{a}_\nu \hat{a}^+_m \hat{a}_i \hat{a}^+_n \hat{a}_j|\Phi_0(t)\right\rangle$$
$$- \frac{1}{2}\sum_{kl} \overline{V}_{klkl} \left\langle \Phi_0(t)|\hat{a}^+_m \hat{a}_i \hat{a}^+_n \hat{a}_j|\Phi_0(t)\right\rangle$$
$$+ \frac{1}{4}\sum_{\mu\mu'\nu\nu'} \overline{V}_{\nu\mu\nu'\mu'} \left\langle \Phi_0(t)|\hat{\mathbb{N}}[\hat{a}^+_\nu \hat{a}^+_\mu \, \hat{a}_{\mu'} \, \hat{a}_{\nu'}]\hat{a}^+_m \hat{a}_i \hat{a}^+_n \hat{a}_j|\Phi_0(t)\right\rangle \; . \tag{D.10}$$

The first and second terms are zero because of the orthogonality of the Slater determinants. With a calculation analogous to that leading to Eq. (16.18) we obtain

$$\frac{1}{4}\sum_{\mu\mu'\nu\nu'} \overline{V}_{\nu\mu\nu'\mu'} \left\langle \Phi_0(t)|\hat{\mathbb{N}}[\hat{a}^+_\nu \hat{a}^+_\mu \, \hat{a}_{\mu'} \, \hat{a}_{\nu'}]\hat{a}^+_m \hat{a}_i \hat{a}^+_n \hat{a}_j|\Phi_0(t)\right\rangle = \overline{V}_{ijmn} \; , \tag{D.11}$$

therefore we have

$$\frac{1}{2}\sum_{minj} C_{mi}(t) C_{nj}(t) \left\langle \Phi_0(t)|\hat{H}\hat{a}^+_m \hat{a}_i \hat{a}^+_n \hat{a}_j|\Phi_0(t)\right\rangle$$
$$= \frac{1}{2}\sum_{minj} C_{mi}(t) C_{nj}(t) \overline{V}_{ijmn} \; . \tag{D.12}$$

By working in analogous manner we obtain

$$\frac{1}{2}\sum_{minj} C^*_{mi}(t) C^*_{nj}(t) \left\langle \Phi_0(t)|\hat{a}^+_j \hat{a}_n \hat{a}^+_i \hat{a}_m \hat{H}|\Phi_0(t)\right\rangle$$
$$= \frac{1}{2}\sum_{minj} C^*_{mi}(t) C^*_{nj}(t) \overline{V}_{mnij} \; . \tag{D.13}$$

The expression of the last term of Eq. (D.9) is

$$\sum_{minj} C^*_{mi}(t) C_{nj}(t) \left\langle \Phi_0(t)|\hat{a}^+_i \hat{a}_m \hat{H}\hat{a}^+_n \hat{a}_j \hat{H}|\Phi_0(t)\right\rangle$$
$$= \sum_{mi} |C_{mi}|^2 \sum_k \epsilon_k + \sum_{mi} |C_{mi}|^2 (\epsilon_m - \epsilon_i)$$
$$-\frac{1}{2}\sum_{mi} |C_{mi}|^2 \sum_{kl} \overline{V}_{klkl} + \frac{1}{2}\sum_{minj} C_{mi} C^*_{nj} \overline{V}_{mjin}$$
$$\equiv \sum_{mi} |C_{mi}|^2 \mathcal{E}_0 + \sum_{mi} |C_{mi}|^2 (\epsilon_m - \epsilon_i) + \tag{D.14}$$
$$\sum_{minj} C_{mi} C^*_{nj} \overline{V}_{mjin} \; .$$

The final expression of Eq. (D.9) is

$$\left\langle \Psi(t)|\hat{H}|\Psi(t)\right\rangle = \mathcal{E}_0\left(1+\sum_{mi}|C_{mi}|^2\right)$$
$$+\sum_{mi}|C_{mi}|^2(\epsilon_m-\epsilon_i)+\sum_{minj}C_{mi}C^*_{nj}\overline{V}_{mjin}$$
$$+\frac{1}{2}\sum_{minj}C_{mi}(t)C_{nj}(t)\overline{V}_{ijmn}+\frac{1}{2}\sum_{minj}C^*_{mi}(t)C^*_{nj}(t)\overline{V}_{mnij}\ . \quad \text{(D.15)}$$

Let us calculate the second term of Eq. (D.6). By considering the expression (D.3) we have

$$i\hbar\left\langle \Psi(t)|\frac{\partial}{\partial t}|\Psi(t)\right\rangle = \mathcal{E}_0\left\langle \Psi(t)|\Psi(t)\right\rangle$$
$$+\,i\hbar\sum_{mi}\frac{d}{dt}C_{mi}(t)\left\langle \Psi(t)|\hat{a}^+_m\hat{a}_i|\Psi(t)\right\rangle\ . \quad \text{(D.16)}$$

We make a power expansion of the exponential function in Eq. (D.3) and consider terms up to the second order in C

$$\langle \Psi(t)|\Psi(t)\rangle = \langle \Phi_0(t)|\Phi_0(t)\rangle$$
$$+\sum_{minj}C^*_{mi}(t)C_{nj}(t)\left\langle \Phi_0(t)|\hat{a}^+_i\hat{a}_m\hat{a}^+_n\hat{a}_j|\Phi_0(t)\right\rangle+\dots\ , \quad \text{(D.17)}$$

and, after the application of the Wick's theorem

$$\langle \Psi(t)|\Psi(t)\rangle = 1+\sum_{mi}|C_{mi}(t)|^2+\dots\ . \quad \text{(D.18)}$$

By using the power expansion of the exponential to calculate the second term of Eq. (D.16) we have

$$\left\langle \Psi(t)|\hat{a}^+_m\hat{a}_i|\Psi(t)\right\rangle = \sum_{nj}C^*_{nj}\left\langle \Phi_0(t)|\hat{a}^+_j\hat{a}_n\hat{a}^+_m\hat{a}_i|\Phi_0(t)\right\rangle+\dots$$
$$= C^*_{mi}+\dots\ . \quad \text{(D.19)}$$

The term related to the time derivative becomes

$$i\hbar\left\langle \Psi(t)|\frac{\partial}{\partial t}|\Psi(t)\right\rangle = \mathcal{E}_0\left(1+\sum_{mi}|C_{mi}(t)|^2\right)$$
$$+\sum_{mi}C^*_{mi}(t)\frac{d}{dt}C_{mi}(t)+\dots\ . \quad \text{(D.20)}$$

Combining the results of Eqs. (D.15) and (D.20) we obtain

$$\left\langle \Psi(t)|\hat{H} - i\hbar\frac{\partial}{\partial t}|\Psi(t)\right\rangle = \sum_{mi} |C_{mi}(t)|^2(\epsilon_m - \epsilon_i) + \sum_{minj} C_{mi}C^*_{nj}\overline{V}_{mjin}$$
$$+ \frac{1}{2}\sum_{minj} C_{mi}(t)C_{nj}(t)\overline{V}_{ijmn}$$
$$+ \frac{1}{2}\sum_{minj} C^*_{mi}(t)C^*_{nj}(t)\overline{V}_{mnij}$$
$$- i\hbar\sum_{mi} C^*_{mi}(t)\frac{d}{dt}C_{mi}(t) + \dots \quad \text{(D.21)}$$

We have to impose the variational condition

$$\frac{\delta}{\delta C^*_{mi}(t)}\left\langle \Psi(t)|\hat{H} - i\hbar\frac{\partial}{\partial t}|\Psi(t)\right\rangle = \frac{\partial}{\partial C^*_{mi}(t)}\left\langle \Psi(t)|\hat{H} - i\hbar\frac{\partial}{\partial t}|\Psi(t)\right\rangle = 0 \ , \quad \text{(D.22)}$$

where the variational derivative have been changed in partial derivatives since the C's are the only terms depending on time. By working out the derivative we obtain the expression

$$C_{mi}(t)(\epsilon_m - \epsilon_i) + \sum_{nj} C^*_{nj}\overline{V}_{mnij} + \sum_{nj} C_{nj}\overline{V}_{mjin} = i\hbar\sum_{mi}\frac{d}{dt}C_{mi}(t) \ . \quad \text{(D.23)}$$

We consider small vibrations around the ground state, and we simulate them by considering harmonic oscillations

$$C_{mi}(t) = X_{mi}e^{-i\omega t} + Y_{mi}e^{i\omega t} \ . \quad \text{(D.24)}$$

After inserting this expression in Eq. (D.23) and separating the positive and negative frequencies we obtain the two equations

$$X_{mi}(\epsilon_m - \epsilon_i) + \sum_{nj}\overline{V}_{mjin}X_{nj} + \sum_{nj}\overline{V}_{mnij}Y_{nj} = \hbar\omega X_{mi} \quad \text{(D.25)}$$
$$Y^*_{mi}(\epsilon_m - \epsilon_i) + \sum_{nj}\overline{V}_{mjin}Y^*_{nj} + \sum_{nj}\overline{V}_{mnij}X^*_{nj} = -\hbar\omega^* X_{mi} \ . \quad \text{(D.26)}$$

We calculate the complex conjugate of the second equation above, and, by considering the properties of the antisymmetrized matrix element of the interaction,

$$\overline{V}^*_{mnij} = \overline{V}_{ijmn} \ ; \ \overline{V}^*_{mjin} = \overline{V}_{imnj} \ , \quad \text{(D.27)}$$

we obtain the equations

$$\sum_{nj}\left[(\epsilon_m - \epsilon_i)\delta_{nm}\delta_{ij} + \overline{V}_{mjin}\right]X_{nj} + \sum_{nj}\overline{V}_{mnij}Y_{nj} = \hbar\omega X_{mi} \qquad \text{(D.28)}$$

$$\sum_{nj}\left[(\epsilon_m - \epsilon_i)\delta_{nm}\delta_{ij} + \overline{V}_{imnj}\right]Y_{nj} + \sum_{nj}\overline{V}_{ijmn}X_{nj} = -\hbar\omega Y_{mi} \ . \qquad \text{(D.29)}$$

This system of equation is identical to that obtained by Eq. (16.37) where the A and B matrices have been defined by the Eqs. (16.38) and (16.39).

Appendix E
Speed of Sound in Fluids

In this appendix, we derive the expression for the speed of sound in a classical fluid. We focus on classical fluids, excluding quantum fluids, and we combine the continuity equation with the Euler equations.

Let $\mathbf{u}(\mathbf{r}, t)$ represent the local velocity of an infinitesimal mass of fluid, and let $\rho_{\rm m}$ denote the mass density. This mass density is related to the number density of a many-body system of identical particles by the equation $\rho_{\rm m} = m\rho$, where m is the mass of a particle, and ρ represents the probability of finding a particle within a unit volume.

E.1 Continuity Equation

The mass of a fluid that exits a closed surface S in unit of time is given by:

$$\int_S \rho_{\rm m} \, \mathbf{u} \cdot d\mathbf{S} \ . \tag{E.1}$$

This is a surface integral and the direction of $d\mathbf{S}$ is given by a vector orthogonal to the plane touching the infinitesimal surface element. The loss of flux indicates a decrease in the fluid density within the volume enclosed by the surface S

$$\int_S \rho_{\rm m} \, \mathbf{u} \cdot d\mathbf{S} = -\frac{\partial}{\partial t} \int_V \rho_{\rm m} \, dV, \tag{E.2}$$

where the integral in the right hand side represents a volume integral.

This is the continuity equation expressed in the integral form. We obtain a differential expression of this equation by applying the divergence theorem to the term of left hand side

© The Editor(s) (if applicable) and The Author(s), under exclusive license to Springer Nature Switzerland AG 2026

G. Co', *Concepts in Quantum Many-Body Physics*, UNITEXT for Physics, https://doi.org/10.1007/978-3-032-08920-5

$$\int_S \rho_\mathrm{m}\, \mathbf{u} \cdot d\mathbf{S} = \int_V \rho_\mathrm{m}\, \nabla \cdot \mathbf{u} dV = -\int_V \frac{\partial \rho_\mathrm{m}}{\partial t} dV \ , \tag{E.3}$$

where we have inverted the order of the integration and differentiation operations in the last integral. By equating the two integrands we obtain

$$\nabla \cdot (\rho_\mathrm{m} \mathbf{u}) = -\frac{\partial \rho_\mathrm{m}}{\partial t} \ , \tag{E.4}$$

where $\rho_\mathrm{m}\mathbf{u}$ is commonly called current density. The one-dimensional expression is

$$\frac{\partial(\rho_\mathrm{m} u)}{\partial x} = -\frac{\partial \rho_\mathrm{m}}{\partial t} \ . \tag{E.5}$$

E.2 Euler Equation

In a fluid, the rate of change of a vector property $\mathbf{X}(\mathbf{r}, t)$ depending on the position and on the time t is described by a **convective derivative** which is a total derivative with respect to the time.

$$\frac{D\mathbf{X}}{Dt} = \frac{\partial \mathbf{X}}{\partial t} + \sum_{i=1}^{3} \frac{\partial \mathbf{X}}{\partial x_i} \frac{\partial x_i}{\partial t} = \frac{\partial \mathbf{X}}{\partial t} + (\mathbf{u} \cdot \nabla)\mathbf{X} \ . \tag{E.6}$$

We consider an infinitesimal element of the fluid. The external force acting on this element is given by the convective derivative of the velocity $\mathbf{u}$. This force induce a gradient on the pressure P generated by the considered element on the other parts of the fluid. The pressure gradient per mass units is given by the equation

$$\frac{D\mathbf{u}}{Dt} = -\frac{1}{\rho_\mathrm{m}} \nabla P \ , \tag{E.7}$$

where the minus sign is due the fact that the force is external to the volume considered and a positive force induces a negative pressure gradient.

By writing the explicit expression of the convective derivative we obtain

$$-\frac{1}{\rho_\mathrm{m}} \nabla P = \frac{\partial \mathbf{u}}{\partial t} + (\mathbf{u} \cdot \nabla)\mathbf{u} \ , \tag{E.8}$$

and in one dimension, by indicating with u the x component of the velocity $\mathbf{u}$, we have

$$-\frac{1}{\rho_\mathrm{m}} \frac{\partial P}{\partial x} = \frac{\partial u}{\partial t} + u \frac{\partial u}{\partial x} \ . \tag{E.9}$$

E.3 Velocity of the Sound

Let us consider a situation where the sound propagates in a specific direction. We use the one-dimensional expressions of the continuity and Euler equations. The latter one can be written as

$$u\frac{\partial \rho_{\rm m}}{\partial x} + \rho_{\rm m}\frac{\partial u}{\partial x} = -\frac{\partial \rho_{\rm m}}{\partial t} \ . \tag{E.10}$$

We divide by $\rho_{\rm m}$ and use the variable $s = (\delta\rho_{\rm m})/\rho_{\rm m}$

$$u\frac{\partial s}{\partial x} + \frac{\partial u}{\partial x} = -\frac{\partial s}{\partial t} \ . \tag{E.11}$$

We consider that changes in density occur over a small distance scale compared to the overall size of the fluid, and also with respect to the distances covered by the propagation of the sound. The term where the fluid velocity u multiplies the density variation $\delta s/\delta x$ can be considered negligible, therefore the previous equation becomes

$$\frac{\partial u}{\partial x} = -\frac{\partial s}{\partial t}, \tag{E.12}$$

and the Euler equation

$$-\frac{1}{\rho_{\rm m}}\frac{\partial P}{\partial x} = \frac{\partial u}{\partial t}. \tag{E.13}$$

By using the definition of compression modulus (2.59) we have

$$B = -\mathcal{V}\frac{\partial P}{\partial \mathcal{V}} = -\frac{A}{\rho_{\rm m}}\frac{\partial P}{\left(-\frac{\mathcal{V}}{\rho_{\rm m}}\partial\rho_{\rm m}\right)} = \rho_{\rm m}\frac{\partial P}{\partial \rho_{\rm m}} \ , \tag{E.14}$$

from which $\delta P = B\delta s$, We can express the equation (E.13) as

$$\frac{\partial u}{\partial t} = -\frac{B}{\rho_{\rm m}}\frac{\partial s}{\partial x} \ . \tag{E.15}$$

We insert the expression of δu obtained by this equation in (E.12), and obtain

$$\frac{B}{\rho_{\rm m}}\frac{\partial^2 s}{\partial x^2} = \frac{\partial^2 s}{\partial t^2} \ . \tag{E.16}$$

Possible solutions of this equation are

$$s \propto e^{i(kx-\omega t)} \ , \tag{E.17}$$

which inserted in (E.16) gives the relation

$$\frac{B}{\rho_{\mathrm{m}}}k^2 = \omega^2 \; . \tag{E.18}$$

Therefore, the speed of propagation of the sound wave is

$$v_s = \frac{\omega}{k} = \sqrt{\frac{B}{\rho_{\mathrm{m}}}} \; . \tag{E.19}$$

Appendix F
Boltzmann Transport Equation

In this appendix, we present a derivation of the classical Boltzmann transport equation. In this derivation we describe the motion of a single particle using the three components of its position, denoted as $\mathbf{r}$, and the three components of its velocity, denoted as $\mathbf{v}$. Our objective is to determine the variations in the distribution function $f(\mathbf{r}, \mathbf{v}, t)$. At time t, the number of particles located within the infinitesimal volume $d\mathcal{V}$ of the six-dimensional phase space, characterised by $\mathbf{r}$ and $\mathbf{v}$, is given by $f(\mathbf{r}, \mathbf{v}, t)$. In other words, if A represents the total number of particles, the probability of finding a particle within the volume $d\mathcal{V}$ at the time t is given by $(1/A)f(\mathbf{r}, \mathbf{v}, t)$.

The distribution function completely defines the state of the system. For instance, the local density of particles, $\rho(\mathbf{r}, t)$, regardless of their velocities, can be obtained by integrating over the velocity space

$$\rho(\mathbf{r}, t) = \int d^3v\, f(\mathbf{r}, \mathbf{v}, t) \ . \tag{F.1}$$

Let us consider the scenario in which there are no collisions among the particles. External forces, represented by the vector sum $\mathbf{F}(\mathbf{r}, t)$ at time t, are acting on the system. The time variation of position and velocity of the individual particle is given by

$$\mathbf{r}' = \mathbf{r}(t + dt) = \mathbf{r}(t) + \mathbf{v}(t)dt \ , \tag{F.2}$$

$$\mathbf{v}' = \mathbf{v}(t + dt) = \mathbf{v}(t) + \frac{\mathbf{F}(\mathbf{r}, t)}{m}dt \ . \tag{F.3}$$

The distribution function that describes the system within a volume $d\mathcal{V}$ at time t evolves at time $t + dt$ not only due to its explicit dependence on time but also because of the variations of $\mathbf{r}$ and $\mathbf{v}$. We can write the new distribution function as

$$f(\mathbf{r}', \mathbf{v}', t + dt) = f(\mathbf{r}', \mathbf{v}', t)dt + \frac{\partial f}{\partial t}dt + \sum_{i=1}^{3} \frac{\partial f}{\partial x_i}\dot{x}_i dt + \sum_{i=1}^{3} \frac{\partial f}{\partial v_i}\dot{v}_i dt \ , \tag{F.4}$$

© The Editor(s) (if applicable) and The Author(s), under exclusive license to Springer Nature Switzerland AG 2026

G. Co', *Concepts in Quantum Many-Body Physics*, UNITEXT for Physics,
https://doi.org/10.1007/978-3-032-08920-5

where the dot on the letter indicates the partial derivative on the time. We can define the total derivative as

$$\frac{Df}{Dt} = \frac{\partial f}{\partial t} + \sum_{i=1}^{3} \frac{\partial f}{\partial x_i}\dot{x}_i + \sum_{i=1}^{3} \frac{\partial f}{\partial v_i}\dot{v}_i \ , \tag{F.5}$$

or in vector form

$$\frac{Df}{Dt} = \frac{\partial f}{\partial t} + \mathbf{v} \cdot (\nabla_{\mathbf{r}} f) + \frac{1}{m}\mathbf{F} \cdot (\nabla_{\mathbf{v}} f) \ , \tag{F.6}$$

where the subscripts indicate that the gradients are calculated with respect to the coordinates or to the velocities.

In a collisionless situation and for conservative external forces, i.e. independent of the velocities, there is a conservation of the particle flux in the volume $d\mathcal{V}$. This statement can be expressed in terms of current conservation

$$\frac{\partial f}{\partial t} + \nabla \cdot (f\mathbf{v}) = 0 \ , \tag{F.7}$$

which in the six-dimensional **r**, **v** space becomes

$$\frac{\partial f}{\partial t} + \sum_{i=1}^{3} \frac{\partial (f\dot{x}_i)}{\partial x_i} + \sum_{i=1}^{3} \frac{\partial (f\dot{v}_i)}{\partial v_i} = 0 \ . \tag{F.8}$$

Therefore

$$\frac{\partial f}{\partial t} + \sum_{i=1}^{3} \left[\frac{\partial f}{\partial x_i}\dot{x}_i + f\frac{\partial \dot{x}_i}{\partial x_i}\right] + \sum_{i=1}^{3} \left[\frac{\partial f}{\partial v_i}\dot{v}_i + f\frac{\partial \dot{v}_i}{\partial v_i}\right] = 0 \ , \tag{F.9}$$

and

$$\frac{Df}{Dt} + f\sum_{i=1}^{3} \left(\frac{\partial \dot{x}_i}{\partial x_i} + \frac{\partial \dot{v}_i}{\partial v_i}\right) = 0 \ . \tag{F.10}$$

Since the forces are conservative, also the hamiltonian is conservative, therefore the Hamilton equations are valid

$$\dot{x}_i = \frac{\partial H}{\partial p_i} = \frac{\partial H}{m \partial v_i} \ , \tag{F.11}$$

$$-\dot{p}_i = -m\dot{v}_i = \frac{\partial H}{\partial x_i} \ , \tag{F.12}$$

$$\frac{\partial \dot{x}}{\partial x_i} = \frac{\partial}{\partial x_i}\left[\frac{\partial H}{m\partial v_i}\right] = \frac{1}{-m}\left[\frac{\partial^2 H}{\partial v_i \partial x_i}\right] = \frac{\partial}{\partial v_i}\left[\frac{1}{-m}\frac{\partial H}{\partial x_i}\right] = -\frac{\partial \dot{v}_i}{\partial v_i} \ . \tag{F.13}$$

In this case we have that

$$\frac{Df}{Dt} = 0 \ . \tag{F.14}$$

The presence of collision between the particle modifies Eq. (F.14) as

$$\frac{Df(\mathbf{r}, \mathbf{v}, t)}{Dt} = \Delta^{+} - \Delta^{-} \ , \tag{F.15}$$

We denote by Δ^{-} the decrease in the value of f due to particles within the physical volume $dV = d\mathbf{r}$ that, after the collision, have a velocity different from $\mathbf{v}$. Conversely, we denote by Δ^{+} the increase in the value of f resulting from particles that initially have a velocity different from $\mathbf{v}$ but, after the collision, have a velocity that falls between $\mathbf{v}$ and $\mathbf{v} + d\mathbf{v}$. All of these changes occur within a unit of time.

Let us make some assumptions to evaluate these two terms.

1. The mean free path of each particle is much larger than its dimensions.
2. The interactions between the particles are short-ranged.
3. We consider only binary collisions. Collisions between three, or even more particles are negligible.
4. The particle collisions are always elastic.

We evaluate first Δ^{-}. Let us consider a particle with velocity $\mathbf{v}$ having a collision with another particle of initial velocity $\mathbf{w}$. We call $\mathbf{v}'$ and $\mathbf{w}'$ the velocities after the collision. Since we assumed that the collision is elastic, because of the momentum and kinetic energy conservation we obtain,

$$\mathbf{v} + \mathbf{w} = \mathbf{v}' + \mathbf{w}' \tag{F.16}$$

$$\frac{m}{2}\mathbf{v}^2 + \frac{m}{2}\mathbf{w}^2 = \frac{m}{2}(\mathbf{v}')^2 + \frac{m}{2}(\mathbf{w}')^2 \ . \tag{F.17}$$

The magnitudes of the relative velocities $\mathbf{V} = \mathbf{v} - \mathbf{w}$ e $\mathbf{V}' = \mathbf{v}' - \mathbf{w}'$ do not change

$$|\mathbf{V}| = |\mathbf{V}'| \ . \tag{F.18}$$

We call $\sigma_{\mathbf{v},\mathbf{w}\to\mathbf{v}',\mathbf{w}'}$ the elastic cross section of this collision which is invariant under time reversal, $t \to -t$, therefore,

$$\sigma_{\mathbf{v},\mathbf{w}\to\mathbf{v}',\mathbf{w}'} = \sigma_{\mathbf{v}',\mathbf{w}'\to\mathbf{v},\mathbf{w}} \ , \tag{F.19}$$

this means that, once the relative velocity value $|\mathbf{V}|$ is selected, the cross section depends only on the incoming and outgoing directions, in other words on the solid angle Ω where the particles are scattered. In the specific case of unpolarised particles, such as those without spin, the cross section depends only on the angle θ between incoming and outgoing directions. Therefore, we have

$$\int d^3v' d^3w' \sigma_{\mathbf{v},\mathbf{w}\to\mathbf{v}',\mathbf{w}'} = \int d\Omega \sigma(\Omega) \ . \tag{F.20}$$

The number of particles that, because of the collisions, in the unit of time, make a transition from $(\mathbf{v}, \mathbf{w})$ to $(\mathbf{v}', \mathbf{w}')$ is proportional to the cross section and the flux. This relationship also depend on the relative speed $\mathbf{v} - \mathbf{w}$, and the joint probability of finding a particle with velocity $\mathbf{v}$ and another one with velocity $\mathbf{w}$. Now, we introduce a new assumption: we will consider the joint probability as the product of the two individual probabilities. This implies that we are neglecting any correlations between the particles, treating the motion of each particle as independent of the presence of the others. This probability is, therefore, the product of $f(\mathbf{r}, \mathbf{v}, t)$ times $f(\mathbf{r}, \mathbf{w}, t)$ which we write as $f_\mathbf{v} f_\mathbf{w}$. By combining all the assumptions we have gathered we obtain

$$\Delta^- = \int d^3w\, d^3v'\, d^3w'\, |\mathbf{v} - \mathbf{w}| \sigma_{\mathbf{v},\mathbf{w}\to\mathbf{v}',\mathbf{w}'} f_\mathbf{v} f_\mathbf{w} \ . \tag{F.21}$$

The path to obtain Δ^+ is similar. One has to substitute $\mathbf{v}$ and $\mathbf{w}$ with $\mathbf{v}'$ and $\mathbf{w}'$ respectively

$$\Delta^+ = \int d^3w\, d^3v'\, d^3w'\, |\mathbf{v}' - \mathbf{w}'| \sigma_{\mathbf{v}',\mathbf{w}'\to\mathbf{v},\mathbf{w}} f_{\mathbf{v}'} f_{\mathbf{w}'} \ . \tag{F.22}$$

By considering the invariance of the relative velocity magnitudes (F.18) the time reversal invariance (F.19) and the relation (F.20), we can write the transport equation (F.15) as

$$\begin{aligned} \frac{Df}{Dt} &= \int d^3w\, d^3dv'\, d^3w'\, |\mathbf{v} - \mathbf{w}| \sigma_{\mathbf{v}',\mathbf{w}'\to\mathbf{v},\mathbf{w}} (f_{\mathbf{v}'} f_{\mathbf{w}'} - f_\mathbf{v} f_\mathbf{w}) \\ &= \int d^3w\, d\Omega\, |\mathbf{v} - \mathbf{w}| \sigma(\Omega) (f_{\mathbf{v}'} f_{\mathbf{w}'} - f_\mathbf{v} f_\mathbf{w}) \ , \end{aligned} \tag{F.23}$$

and in the more extended expression

$$\frac{\partial f}{\partial t} + \mathbf{v} \cdot (\nabla_\mathbf{r} f) + \frac{1}{-} m \mathbf{F} \cdot \nabla_\mathbf{v} f = \int d^3w\, d\Omega\, |\mathbf{v} - \mathbf{w}| \sigma(\Omega) (f_{\mathbf{v}'} f_{\mathbf{w}'} - f_\mathbf{v} f_\mathbf{w}) \ . \tag{F.24}$$

The right hand side of this equation is usually called collision integral, $\mathcal{I}(f)$.

Index

© The Editor(s) (if applicable) and The Author(s), under exclusive license
to Springer Nature Switzerland AG 2026
G. Co', *Concepts in Quantum Many-Body Physics*, UNITEXT for Physics,
https://doi.org/10.1007/978-3-032-08920-5

MIX
Papier aus verantwortungsvollen Quellen
Paper from responsible sources
FSC® C105338

If you have any concerns about our products,
you can contact us on
ProductSafety@springernature.com

In case Publisher is established outside the EU,
the EU authorized representative is:
Springer Nature Customer Service Center GmbH
Europaplatz 3, 69115 Heidelberg, Germany

Printed by Libri Plureos GmbH
in Hamburg, Germany